AF872871

ADVANCES IN BIOFUEL PRODUCTION

Algae and Aquatic Plants

ADVANCES IN BIOFUEL PRODUCTION

Algae and Aquatic Plants

Edited by

Barnabas Gikonyo, PhD

Apple Academic Press

TORONTO NEW JERSEY

Apple Academic Press Inc.
3333 Mistwell Crescent
Oakville, ON L6L 0A2
Canada

Apple Academic Press Inc.
9 Spinnaker Way
Waretown, NJ 08758
USA

©2014 by Apple Academic Press, Inc.
Exclusive worldwide distribution by CRC Press, a member of Taylor & Francis Group

No claim to original U.S. Government works

International Standard Book Number-13: 978-1-926895-95-6 (Hardcover)

This book contains information obtained from authentic and highly regarded sources. Reprinted material is quoted with permission and sources are indicated. Copyright for individual articles remains with the authors as indicated. A wide variety of references are listed. Reasonable efforts have been made to publish reliable data and information, but the authors, editors, and the publisher cannot assume responsibility for the validity of all materials or the consequences of their use. The authors, editors, and the publisher have attempted to trace the copyright holders of all material reproduced in this publication and apologize to copyright holders if permission to publish in this form has not been obtained. If any copyright material has not been acknowledged, please write and let us know so we may rectify in any future reprint.

Trademark Notice: Registered trademark of products or corporate names are used only for explanation and identification without intent to infringe.

Library of Congress Control Number: 2013949391

Library and Archives Canada Cataloguing in Publication

Advances in biofuel production: algae and aquatic plants/edited by Barnabas Gikonyo, PhD.

Chapters in this book were previously published in various formats and in various places.
Includes bibliographical references and index.
ISBN 978-1-926895-95-6
1. Algae--Biotechnology. 2. Microalgae--Biotechnology. 3. Aquatic plants--Biotechnology.
4. Biomass energy. I. Gikonyo, Barnabas, editor of compilation

TP248.27.A46A38 2013 579.8 C2013-906173-8

Apple Academic Press also publishes its books in a variety of electronic formats. Some content that appears in print may not be available in electronic format. For information about Apple Academic Press products, visit our website at **www.appleacademicpress.com** and the CRC Press website at **www.crcpress.com**

ABOUT THE EDITOR

BARNABAS GIKONYO, PhD

Barnabas Gikonyo graduated from Southern Illinois University Carbondale, Illinois (2007) with a PhD in organic and materials chemistry. He teaches organic and general chemistry classes plus corresponding laboratories and oversees the running of general chemistry labs. His research interests range from application of various biocompatible polymeric materials as "biomaterial bridging surfaces" for the repair of spinal cord injuries to the use of osteoconductive cements for the repair of critical sized bone defects/fractures to (currently) the development of alternative non-food biofuels.

CONTENTS

ACKNOWLEDGMENT AND HOW TO CITE

The chapters in this book were previously published in various places and in various formats. By bringing them together here in one place, we offer the reader a comprehensive perspective on recent investigations of biofuels from algae and aquatic plants. Each chapter is added to and enriched by being placed within the context of the larger investigative landscape.

We wish to thank the authors who made their research available for this book, whether by granting their permission individually or by releasing their research as open source articles. When citing information contained within this book, please do the authors the courtesy of attributing them by name, referring back to their original articles, using the credits provided at the end of each chapter.

LIST OF CONTRIBUTORS

Douglas Aitken
Institute for Infrastructure and Environment, School of Engineering, University of Edinburgh, Edinburgh EH9 3JL, UK

Firoz Alam
School of Aerospace, Mechanical and Manufacturing Engineering, RMIT University, Plenty Road, Bundoora, Melbourne, VIC 3083, Australia

Blanca Antizar-Ladislao
Institute for Infrastructure and Environment, School of Engineering, University of Edinburgh, Edinburgh EH9 3JL, UK

Alina Mariana Balu
Departamento de Química Orgánica, Universidad de Córdoba, Edificio Marie Curie (C-3), Ctra Nnal IV-A, Km 396, E-14014 Córdoba, Spain

Abdul Baqui
Department of Mechanical Engineering Technology, Yanbu Industrial College, Yanbu Al-Sinaiyah, Saudi Arabia

Colin M. Beal
Department of Mechanical Engineering, Cockrell School of Engineering, The University of Texas at Austin, 1 University Station, C2200, Austin, TX 78712, USA and The Center for Electromechanics, The University of Texas at Austin, 1 University Station, R7000, Austin, TX 78712, USA

Ashish Bhatnagar
Department of Biological and Agricultural Engineering, The University of Georgia, Athens, GA 30602, USA

Juan Manuel Campelo
Departamento de Química Orgánica, Universidad de Córdoba, Edificio Marie Curie (C-3), Ctra Nnal IV-A, Km 396, E-14014 Córdoba, Spain

David Casero
Department of Molecular, Cell, and Developmental Biology, University of California, Los Angeles, CA, USA

Senthil Chinnasamy
Department of Biological and Agricultural Engineering, The University of Georgia, Athens, GA 30602, USA

Shawn J Cokus
Department of Molecular, Cell, and Developmental Biology, University of California, Los Angeles, CA, USA

Stijn Cornelissen
ZinInZin, van Koetsveldstraat 112, 3532 ET, Utrecht, The Netherlands

Keshav C. Das
Department of Biological and Agricultural Engineering, The University of Georgia, Athens, GA 30602, USA

Abhijit Date
School of Aerospace, Mechanical and Manufacturing Engineering, RMIT University, Plenty Road, Bundoora, Melbourne, VIC 3083, Australia

Yvonne Y. Deng
Ecofys UK, 1 Alie St, London E1 8DE, United Kingdom.

Zhenyi Du
Center for Biorefining and Department of Bioproducts and Biosystems Engineering, University of Minnesota, St. Paul, MN 55108, USA

Van Thang Duong
Algae Biotechnology Laboratory, School of Agriculture and Food Sciences, The University of Queensland, Brisbane 4072, Australia

Berat Z Haznedaroglu
Department of Chemical and Environmental Engineering, Yale University, 9 Hillhouse Ave, New Haven, CT 06520, USA

Robert E. Hebner
Department of Mechanical Engineering, Cockrell School of Engineering, The University of Texas at Austin, 1 University Station, C2200, Austin, TX 78712, USA and The Center for Electromechanics, The University of Texas at Austin, 1 University Station, R7000, Austin, TX 78712, USA

Carol Hsin
Department of Chemical and Environmental Engineering, Yale University, 9 Hillhouse Ave, New Haven, CT 06520, USA

Ryan W. Hunt
Department of Biological and Agricultural Engineering, The University of Georgia, Athens, GA 30602, USA

Jessica Jones
Department of Biomedical Engineering, University of Texas at Austin, Austin, TX 78712, USA

Carey W. King
Center for International Energy and Environmental Policy, Jackson School of Geosciences, The University of Texas at Austin, 1 University Station, C9000, Austin, TX 78712, USA

Michèle Koper
Ecofys bv, P.O. Box 8408, 3503 RK Utrecht, The Netherlands

Maggie Law
NGS Sequencing Department, Beijing Genomics Institute (BGI), 4th Floor, Building 11, Beishan Industrial Zone, Yantian District, Guangdong, Shenzhen 518083, China

Cheng-Han Lee
Section of Molecular, Cell and Developmental Biology, University of Texas at Austin, Austin, TX 78712, USA

Siliang Li
NGS Sequencing Department, Beijing Genomics Institute (BGI), 4th Floor, Building 11, Beishan Industrial Zone, Yantian District, Guangdong, Shenzhen 518083, China

Yan Li
Algae Biotechnology Laboratory, School of Agriculture and Food Sciences, The University of Queensland, Brisbane 4072, Australia

Yinhu Li
NGS Sequencing Department, Beijing Genomics Institute (BGI), 4th Floor, Building 11, Beishan Industrial Zone, Yantian District, Guangdong, Shenzhen 518083, China

Lin Liu
NGS Sequencing Department, Beijing Genomics Institute (BGI), 4th Floor, Building 11, Beishan Industrial Zone, Yantian District, Guangdong, Shenzhen 518083, China

Yuhuan Liu
The State Key Laboratory of Food Science and Technology, Nanchang University, Nanchang 330047, China and The Engineering Research Center for Biomass Conversion, MOE, Nanchang University, Nanchang 330047, China

David Lopez
Department of Molecular, Cell, and Developmental Biology, University of California, Los Angeles, CA, USA

Rafael Luque
Departamento de Química Orgánica, Universidad de Córdoba, Edificio Marie Curie (C-3), Ctra Nnal IV-A, Km 396, E-14014 Córdoba, Spain

Sabeeha S. Merchant
Department of Chemistry and Biochemistry, University of California, Los Angeles, CA, USA and Institute of Genomics and Proteomics, University of California, Los Angeles, CA, USA

Saleh Mobin
Department of Higher Education Primary Industries, Northern Melbourne Institute of TAFE (NMIT), Epping, Melbourne, VIC 3076, Australia

Hazim Moria
School of Aerospace, Mechanical and Manufacturing Engineering, RMIT University, Plenty Road, Bundoora, Melbourne, VIC 3083, Australia

Ekaterina Nowak
Algae Biotechnology Laboratory, School of Agriculture and Food Sciences, The University of Queensland, Brisbane 4072, Australia

Jordan Peccia
Department of Chemical and Environmental Engineering, Yale University, 9 Hillhouse Ave, New Haven, CT 06520, USA

Matteo Pellegrini
Department of Molecular, Cell, and Developmental Biology, University of California, Los Angeles, CA, USA and Institute of Genomics and Proteomics, University of California, Los Angeles, CA, USA

Antonio Pineda
Departamento de Química Orgánica, Universidad de Córdoba, Edificio Marie Curie (C-3), Ctra Nnal IV-A, Km 396, E-14014 Córdoba, Spain

Martin Poenie
Section of Molecular, Cell and Developmental Biology, University of Texas at Austin, Austin, TX 78712, USA

Jose Manuel Ramos-Fernández
Dpto. Química Inorgánica, Universidad de Alicante, Apto 99, E-03690 Alicante, Spain

Roesfiansjah Rasjidin
School of Aerospace, Mechanical and Manufacturing Engineering, RMIT University, Plenty Road, Bundoora, Melbourne, VIC 3083, Australia

Hamid Rismani-Yazdi
Department of Chemical and Environmental Engineering, Yale University, 9 Hillhouse Ave, New Haven, CT 06520, USA and Department of Chemical Engineering, Massachusetts Institute of Technology, Cambridge, MA, 02139, USA

Antonio Angel Romero
Departamento de Química Orgánica, Universidad de Córdoba, Edificio Marie Curie (C-3), Ctra Nnal IV-A, Km 396, E-14014 Córdoba, Spain

Rongsheng Ruan
The State Key Laboratory of Food Science and Technology, Nanchang University, Nanchang 330047, China, The Engineering Research Center for Biomass Conversion, MOE, Nanchang University, Nanchang 330047, China, and Center for Biorefining and Department of Bioproducts and Biosystems Engineering, University of Minnesota, St. Paul, MN 55108, USA

Rodney S. Ruoff
Department of Mechanical Engineering, Cockrell School of Engineering, The University of Texas at Austin, 1 University Station, C2200, Austin, TX 78712, USA and Texas Materials Institute, The University of Texas at Austin, 1 University Station, C2201, Austin, TX 78712, USA

Peer M. Schenk
Algae Biotechnology Laboratory, School of Agriculture and Food Sciences, The University of Queensland, Brisbane 4072, Australia

A. Frank Seibert
Center for Energy and Environmental Resources, Cockrell School of Engineering, The University of Texas at Austin, 1 University Station, R7100, Austin, TX 78712, USA

Juan Carlos Serrano-Ruiz
Departamento de Química Orgánica, Universidad de Córdoba, Edificio Marie Curie (C-3), Ctra Nnal IV-A, Km 396, E-14014 Córdoba, Spain

James Wang
Section of Molecular, Cell and Developmental Biology, University of Texas at Austin, Austin, TX 78712, USA

Wenqin Wang
Waksman Institute of Microbiology, Rutgers University, 190 Frelinghuysen Road, Piscataway, NJ 08854, USA and Department of Plant Biology and Pathology, Rutgers University, 59 Dudley Road, New Brunswick, NJ 08901, USA

Michael E. Webber
Department of Mechanical Engineering, Cockrell School of Engineering, The University of Texas at Austin, 1 University Station, C2200, Austin, TX 78712, USA and Center for International Energy and Environmental Policy, Jackson School of Geosciences, The University of Texas at Austin, 1 University Station, C9000, Austin, TX 78712, USA

Xiaodan Wu
The State Key Laboratory of Food Science and Technology, Nanchang University, Nanchang 330047, China and The Engineering Research Center for Biomass Conversion, MOE, Nanchang University, Nanchang 330047, China

Andrey Zavalin
Mass Spectrometry Research Center, Vanderbilt University Medical Center, Nashville, TN 37232, USA

INTRODUCTION

A few short decades ago, if someone told you that used cooking oil, inedible plant materials, trash, algae, etc. would one day be crucial in job creation, improving global economies, keeping the air clean and reducing pollution, solving global energy needs, and in matters of national security, you could have easily classified such an individual as being out of touch with reality. Today, however, this is the new reality. This book describes how production and use of biofuels (defined as fuels produced from previously living organisms) is helping meet this new reality. In particular, we look at biofuels from algae and aquatic plants.

The reader will explore how biomass, specifically sugars, nonedible plant materials, and algae (which are designated first, second and third generation biofuels respectively), are used in production of fuel. A description of the feasibility of such projects, current methodologies, and how to optimize biofuel production is presented.

Ever since the oil crisis of the 1970s, tremendous efforts have been devoted into seeking alternative fuels for the modern industrial, transport, and agricultural systems as they are heavily dependent on oil. The world population continues to increase rapidly while emergent economies such as India and China coupled with the fast rate of urbanization have put a severe strain on the current sources of fuel. This has also led to a concomitant rise in pollution with far-reaching environmental impacts. It is this realization that has made it necessary to publish this book.

This book starts with a clear and succinct description of biofuel production from microalgae (also referred to as phytoplankton), the progress made in this field, limitations of current methodologies, and sustainability issues. The book then delves into the role of bioenergy in a fully sustainable global energy system. In particular, it examines the supply potential and use of biomass with the aim of achieving a transition to a fully renewable global energy system by 2050. Important factors such as land use, food security, residues, and waste are also addressed.

The text not only discusses common types of biofuels and relatively simple technologies involved, it goes into detail about advanced biofuel technologies in some very unique ways. It describes plausible ways of optimizing biofuel production and ends with a detailed and captivating look at future research involving gene discovery in biofuel production. This features technological advances that make it possible to economically cultivate microalgae that have a high lipid or starch content. Other efforts devoted into optimizing specific microalgae strains and environments in order to increase the per cell enrichment of lipids or starch are also discussed in vivid detail.

Chapter 1, by Wu and colleagues, explores the role of bioenergy in the global energy system. They argue that microalgae represent a sustainable energy source because of their high biomass productivity and ability to remove air and water born pollutants. This paper reviews the current status of production and conversion of microalgae, including the advantages of microalgae biodiesel, high density cultivation of microalgae, high-lipid content microalgae selection and metabolic control, and innovative harvesting and processing technologies. The key barriers to commercial production of microalgae biodiesel and future perspective of the technologies are also discussed.

In chapter 2, Aitken and Antizar-Ladislao investigate the potential of producing biofuels from algae, which has been enjoying a recent revival due to heightened oil prices, uncertain fossil fuel sources, and legislative targets aimed at reducing our contribution to climate change. If the concept is to become a reality, however, many obstacles need to be overcome. It is necessary to minimize energetic inputs to the system and maximize energy recovery. The cultivation process can be one of the greatest energy consumption hotspots in the whole system: recent studies suggest that open ponds provide the most sustainable means of cultivation infrastructure due to low energy requirements compared to more energy intensive photobioreactors. Much focus has also been placed on finding or developing strains of algae that are capable of yielding high oil concentrations combined with high productivity. Yet to cultivate such strains in open ponds is difficult because of microbial competition and limited radiation-use efficiency. To improve viability, the use of wastewater has been considered by many researchers as a potential source of nutrients with the added benefit of tertiary water treatment; however productivity rates are affected and optimal conditions can be difficult to maintain year round. This paper investigates the process streams that are likely to provide the most viable methods of energy recovery from cultivating and processing algal biomass. The key findings are the importance of a flexible approach that depends upon location of the cultivation ponds and the industry targeted. Additionally this study recommends moving toward technologies producing higher energy recoveries such as pyrolysis or anaerobic digestion as opposed to other studies that have focused on biodiesel production.

Cornelissen and colleagues present a detailed analysis of the supply potential and use of biomass in the context of a transition to a fully renewable global energy system by 2050 in chapter 3. They investigate bioenergy potential within a framework of technological choices and sustainability criteria, including criteria on land use and food security, agricultural and processing inputs, complementary fellings, residues, and waste. This makes their approach more comprehensive, more stringent in the applied sustainability criteria, and more detailed on both the supply potential and the demand side use of biomass than that of most other studies. They find that the potential for sustainable bioenergy from residues and waste, complementary fellings, energy crops, and algae oil in 2050 is 340 EJ a^{-1} of primary energy. This potential is then compared to the demand for biomass-based energy in the demand scenario related to this study, the Ecofys Energy Scenario [1]. This scenario, after applying energy efficiency and non-bioenergy renewable options, requires a significant contribution of bioenergy to meet the remaining energy demand; 185 EJ a^{-1} of the 340 EJ a^{-1} potential supply. For land use for energy crops, they find that a maximum of 2,500,000 km^2 is needed of a 6,730,000 km^2 sustainable potential. For greenhouse gas emissions from bioenergy, a 75%–85% reduction can be achieved compared to fossil references. They conclude that bioenergy can meet residual demand in the Ecofys Energy Scenario sustainably with low associated greenhouse gas emissions. It thus contributes to its achievement of a 95% renewable energy system globally by 2050.

Chapter 4, by Alam and colleagues, argues that fossil fuel energy resources are depleting rapidly, and most importantly the liquid fossil fuel will be diminished by the middle of this century. In addition, the fossil fuel is directly related to air pollution and land and water degradation. In these circumstances, biofuel from renewable sources can be an alternative to reduce our dependency on fossil fuel and assist to maintain the healthy global environment and economic sustainability. Production of biofuel from food stock generally consumed by humans or animals can be problematic and the root cause of worldwide dissatisfaction. Biofuels production from microalgae can provide some distinctive advantages such as their rapid growth rate, greenhouse gas fixation ability, and high production capacity of lipids. This paper reviews the current status of biofuel from algae as a renewable source.

Worldwide, algal biofuel research and development efforts have focused on increasing the competitiveness of algal biofuels by increasing the energy and financial return on investments, reducing water intensity and resource requirements, and increasing algal productivity. In chapter 5, Beal and colleagues present analyses in each of these areas—costs, resource needs, and productivity—for two cases: (1) an experimental case, using mostly measured data for a lab-scale system, and (2) a theorized highly productive case that represents an optimized commercial-scale production system, albeit one that relies on full-price water, nutrients, and carbon dioxide. For both cases, the analysis described herein concludes that the energy and financial return on investments are less than 1, the water intensity is greater than that for conventional fuels, and the amounts of required resources at a meaningful scale of production amount to significant fractions of current consumption (e.g., nitrogen). The analysis and presentation of results highlight critical areas for advancement and innovation that must occur for sustainable and profitable algal biofuel production that can occur at a scale that yields significant petroleum displacement. To this end, targets for energy consumption, production cost, water consumption, and nutrient consumption are presented that would promote sustainable algal biofuel production. Furthermore, this work demonstrates a procedure and method by which subsequent advances in technology and biotechnology can be framed to track progress.

Hunt and colleagues explore the surge of interest in bioenergy in chapter 6. This interest has been marked with increasing efforts in research and development to identify new sources of biomass and to incorporate cutting-edge biotechnology to improve efficiency and increase yields. It is evident that various microorganisms will play an integral role in the development of this newly emerging industry, such as yeast for ethanol and *Escherichia coli* for fine chemical fermentation. However, it appears that microalgae have become the most promising prospect for biomass production due to their ability to grow fast, produce large quantities of lipids, carbohydrates and proteins, thrive in poor quality waters, sequester and recycle carbon dioxide from industrial flue gases, and remove pollutants from industrial, agricultural and municipal wastewaters. In an attempt to better understand and manipulate microorganisms for optimum production capacity, many researchers have investigated alternative methods for stimulating their growth and metabolic behavior. One such novel approach is the use of electromagnetic fields for the stimulation of growth and metabolic cascades and controlling biochemical pathways. An effort has been made in this review to consolidate the information on the current status of biostimulation research to enhance microbial growth and metabolism using electromagnetic fields. It summarizes information on the biostimulatory effects on growth and other biological processes to obtain insight regarding factors and dosages that lead to the stimulation and also what kind of processes have been reportedly affected. Diverse mechanistic theories and explanations for biological effects of electromagnetic fields on intra- and extracellular environment have been discussed. The foundations of biophysical interactions such as bioelectromagnetic and biophotonic communication and organization within living systems are expounded with special consideration for spatiotemporal aspects of electromagnetic topology, leading to the potential of multipolar electromagnetic systems. The future direction for the use of biostimulation using bioelectromagnetic, biophotonic and electrochemical methods have been proposed for biotechnology industries in general with emphasis on an holistic biofuel system encompassing production of algal biomass, its processing, and conversion to biofuel.

Chapter 7 looks at how biomass can efficiently replace petroleum in the production of fuels for the transportation sector. Serrano-Ruiz and colleagues argue that one effective strategy for the processing of complex biomass feedstocks involves previous conversion into simpler compounds (platform molecules) that are more easily transformed in subsequent upgrading reactions. Lactic acid and levulinic acid are two of these relevant biomass derivatives that can easily be derived from biomass sources by means of microbial and/or chemical routes. The present paper intends to cover the most relevant catalytic strategies designed today for the conversion of these molecules into advanced biofuels (e.g. higher alcohols, liquid hydrocarbon fuels) that are fully compatible with the existing hydrocarbons-based transportation infrastructure. The routes described herein involve: (i) deoxygenation reactions that are required for controlling reactivity and for increasing energy density of highly functionalized lactic and levulinic acid combined with (ii) C C coupling reactions for increasing molecular weight of less-oxygenated reactive intermediates.

Jones and colleagues argue that some microalgae are particularly attractive as a renewable feedstock for biodiesel production due to their rapid growth, high content of triacylglycerols, and ability to be grown on non-arable land in chapter 8. Unfortunately, obtaining oil from algae is currently cost prohibitive in part due to the need to pump and process large volumes of dilute algal suspensions. In an effort to circumvent this problem, the authors have explored the use of anion exchange resins for simplifying the processing of algae to biofuel. Anion exchange resins can bind and accumulate the algal cells out of suspension to form a dewatered concentrate. Treatment of the resin-bound algae with sulfuric acid/methanol elutes the algae and regenerates the resin while converting algal lipids to biodiesel. Hydrophobic polymers can remove biodiesel from the sulfuric acid/methanol, allowing the transesterification reagent to be reused. They show that *in situ* transesterification of algal lipids can efficiently convert algal lipids to fatty acid methyl esters while allowing the resin and transesterification reagent to be recycled numerous times without loss of effectiveness.

Chapter 9 shows that biodiesel production from microalgae is being widely developed at different scales as a potential source of renewable energy with both economic and environmental benefits. Duong and colleagues argue that although many microalgae species have been identified and isolated for lipid production, there is currently no consensus as to which species provide the highest productivity. Different species are expected to function best at different aquatic, geographical, and climatic conditions. In addition, other value-added products are now being considered for commercial production, which necessitates the selection of the most capable algae strains suitable for multiple-product algae biorefineries.

Here the authors present and review practical issues of several simple and robust methods for microalgae isolation and selection for traits that may be most relevant for commercial biodiesel production. A combination of conventional and modern techniques is likely to be the most efficient route from isolation to large-scale cultivation.

In chapter 10, Liu and colleagues show that with fast development and wide applications of next-generation sequencing (NGS) technologies, genomic sequence information is within reach to aid the achievement of goals to decode life mysteries, make better crops, detect pathogens, and improve life qualities. NGS systems are typically represented by SOLiD/Ion Torrent PGM from Life Sciences, Genome Analyzer/HiSeq 2000/MiSeq from Illumina, and GS FLX Titanium/GS Junior from Roche. Beijing Genomics Institute (BGI), which possesses the world's biggest sequencing capacity, has multiple NGS systems including 137 HiSeq 2000, 27 SOLiD, one Ion Torrent PGM, one MiSeq, and one 454 sequencer. The authors have accumulated extensive experience in sample handling, sequencing, and bioinformatics analysis. In this paper, technologies of these systems are reviewed, and first-hand data from extensive experience is summarized and analyzed to discuss the advantages and specifics associated with each sequencing system. At last, applications of NGS are summarized.

Lopez and colleagues explore progress in genome sequencing in chapter 11. This progress is proceeding at an exponential pace, and several new algal genomes are becoming available every year. One of the challenges facing the community is the association of protein sequences encoded in the genomes with biological function. While most genome assembly projects generate annotations for predicted protein sequences, they are usually limited and integrate functional terms from a limited number of databases. Another challenge is the use of annotations to interpret large lists of "interesting" genes generated by genome-scale datasets. Previously, these gene lists had to be analyzed across several independent biological databases, often on a gene-by-gene basis. In contrast, several annotation databases, such as DAVID, integrate data from multiple functional databases and reveal underlying biological themes of large gene lists. While several such databases have been constructed for animals, none is currently available for the study of algae. Due to renewed interest in algae as potential sources of biofuels and the emergence of multiple algal genome sequences, a significant need has arisen for such a database to process the growing compendiums of algal genomic data. The Algal Functional Annotation Tool is a web-based comprehensive analysis suite integrating annotation data from several pathway, ontology, and protein family databases. The current version provides annotation for the model alga *Chlamydomonas reinhardtii*, and in the future will include additional genomes. The site allows users to interpret large gene lists by identifying associated functional terms and their enrichment. Additionally, expression data for several experimental conditions were compiled and analyzed to provide an expression-based enrichment search. A tool to search for functionally related genes based on gene expression across these conditions is also provided. Other features include dynamic visualization of genes on KEGG pathway maps and batch gene identifier conversion. The Algal Functional Annotation Tool aims to provide an integrated data-mining environment for algal genomics by combining data from multiple annotation databases into a centralized tool. This site is designed to expedite the process of functional annotation and the interpretation of gene lists, such as those derived from high-throughput RNA-seq experiments. The tool is publicly available at http://pathways.mcdb.ucla.edu webcite.

In the final chapter, Mani-Yazdi and colleagues explore the lack of sequenced genomes for oleaginous microalgae; they argue that this lack limits our understanding of the mechanisms these organisms utilize to become enriched in triglycerides. Here, they report the *de novo* transcriptome assembly and quantitative gene expression analysis of the oleaginous microalga *Neochloris oleoabundans*, with a focus on the complex interaction of pathways associated with the production of the triacylglycerol (TAG) biofuel precursor. After growth under nitrogen replete and nitrogen limiting conditions, they quantified the cellular content of major biomolecules including total lipids, triacylglycerides, starch, protein, and chlorophyll. Transcribed genes were sequenced, the transcriptome was assembled *de novo*, and the expression of major functional categories, relevant pathways, and important genes was quantified through the mapping of reads to the transcriptome. Over 87 million, 77 base pair high quality reads were produced on the Illumina HiSeq sequencing platform. Metabolite measurements supported by genes and pathway expression results indicated that under the nitrogen-limiting condition, carbon is partitioned toward triglyceride production, which increased fivefold over the nitrogen-replete control. In addition to the observed overexpression of the fatty acid synthesis pathway, TAG production during nitrogen limitation was bolstered by repression of the β-oxidation pathway, up-regulation of genes encoding for the pyruvate dehydrogenase complex that funnels acetyl-CoA to lipid biosynthesis, activation of the pentose phosphate pathway to supply reducing equivalents to inorganic nitrogen assimilation and fatty acid biosynthesis, and the up-regulation of lipases—presumably to reconstruct cell membranes in order to supply additional fatty acids for TAG biosynthesis. Their quantitative transcriptome study reveals a broad overview of how nitrogen stress results in excess TAG production in *N. oleoabundans*, and provides a variety of genetic engineering targets and strategies for focused efforts to improve the production rate and cellular content of biofuel precursors in oleaginous microalgae.

PART I

INTRODUCTION

CHAPTER 1

CURRENT STATUS AND PROSPECTS OF BIODIESEL PRODUCTION FROM MICROALGAE

XIAODAN WU, RONGSHENG RUAN, ZHENYI DU, and YUHUAN LIU

1.1 INTRODUCTION

In recent years, increasing consumption of conventional energy has caused serious concerns about energy security and environmental degradation. Therefore, renewable, non-polluting biomass energy has been receiving more and more attention from both the academic community and industries. Biomass derived from photosynthesis includes a variety of organisms, such as plants, animals and microbes. As photosynthetic microorganisms, microalgae can use and therefore remove nitrogen, phosphorus in wastewater, sequester CO_2 in the air, and synthesize lipids which can be converted into biodiesel. The declining supply of conventional fossil fuel and concern about global warming make microalgae-based biodiesel a very promising alternative. Although the potential and advantages of microalgae-based biodiesel over conventional biodiesel have been well recognized [1–3], broad commercialization of microalgae biodiesel has not yet to be realized, chiefly because of the techno-economic constraints, particularly in the areas of mass cultivation and downstream processing. The objectives of the present paper are to review recent development in microalgae production, especially in high density cultivation and downstream processing, and identify technological bottlenecks and strategies for further development.

1.2 THE SUPERIORITY OF MICROALGAE BIODIESEL

As an alternative feedstock for biodiesel production, microalgae have the following advantages over conventional oil crops such as soybeans: (1) microalgae have simple structures, but high photosynthetic efficiency with a growth doubling time as short as 24 h. Moreover, microalgae can be produced all year round. Some data in Table 1 [4] show microalgae are the only source of biodiesel that have the potential to completely displace fossil diesel. (2) The species abundance and biodiversity of microalgae over a broad spectrum of climates and geographic regions make seasonal and geographical restrictions much less of a concern compared with other lipid feedstocks. Microalgae may be cultivated on freshwater, saltwater lakes with eutrophication, oceans, marginal lands, deserts, etc. (3) Microalgae can effectively remove nutrients such as nitrogen and phosphorus, and heavy metals from wastewaters. (4) Microalgae sequester a large amount of carbons via photosynthesis, for example, the CO_2 fixation efficiency of Chlorella vulgaris was up to 260 $mg \cdot L^{-1} \cdot h^{-1}$ in a membrane photobioreactor [5]. Utilization of CO_2 from thermal power plants by large-scale microalgae production facilities can reduce a great deal of the greenhouse gas emissions blamed for global warming. (5) The production and use of microalgae biodiesel contribute near zero net CO_2 and sulfur to the atmosphere. (6) Microalgae can produce a number of valuable products, such as proteins, polysaccharides, pigments, animal feeds, fertilizers, and so on. In short, microalgae are a largely untapped biomass resource for renewable energy production.

TABLE 1: Comparison of some sources of biodiesel.

Crop	Oil yield ($L \cdot ha^{-1}$)
Corn	172
Soybean	446
Canola	1,190
Jatropha	1,892
Coconut	2,689
Oil palm	5,950
Microalgae (70% oil in biomass)	136,900
Microalgae (30% oil in biomass)	58,700

However, commercialization of microalgae biomass and biofuel production is still facing significant obstacles due to high production costs and poor efficiency. In face of these challenges, researchers are undertaking profound efforts to improve microalgae biomass production and lipid accumulation and lower downstream processing costs.

1.3 HIGH DENSITY CULTIVATION OF MICROALGAE

As simple photosynthetic organisms, microalgae can fix CO_2 and synthesize organic compounds, such as lipids, proteins and carbohydrates in large amounts over short periods of time. Traditional methods of microalgae cultivation based on photoautotrophic mode have many shortcomings, among which low cell density is a major issue giving rise to low productivity, harvesting difficulty, associated high costs, and hence poor techno-economic performance. Therefore, a signficant effort towards commercializing microalgae biomass production is to develop high density cultivation processes. Two approaches are being actively researched and developed: (1) metabolic pathways control; (2) cultivation system design.

1.3.1 METABOLIC PATHWAYS

Microalgae may utilize one or more of the three major metabolic pathways depending on light and carbon conditions: photoautotrophy, heterotrophy, and mixotrophy [6]. Most microalgae are capable of photoautotrophic growth. Photoautotrophic cultivation in open ponds is a simple and low-cost way for large-scale production; however the biomass density is low because of limited light transmission, contamination by other species or bacteria, and low organic carbon concentration [7]. Some microalgae can make use of organic carbons and O_2 to undergo rapid propagation through heterotrophic pathway. Heterotrophic cultivation has drawn increasing attention and it is regarded as the most practical and promising way to increase the productivity [8–10]. Currently, research on heterotrophic cultivation of microalgae is mainly focused on Chlorella. Cell densities as high as 104.9 $g \cdot L^{-1}$ (dry cell weight, *Chlorella pyrenoidosa*) have been reported [11]. Microalgae can adapt to different organic matters such as sucrose, glycerol, xylan, organic acids in slurry after acclimatization [12]. The ability of heterotrophic microalgae to utlize a wide variety of organic carbons provides an opportunity to reduce the overall cost of microalgae biodiesel production since these organic substrates can be found in the waste streams such as animal and municipal wastewaters, effluents from anaerobic digestion, food processing wastes, etc. On the basis of heterotrophic cultivation, researchers have carried out studies of mixotrophic cultivation which can greatly enhance the growth rate because it realizes the combined effects of photosynthesis and heterotrophy. After examining the biomass and lipid productivities characteristics of 14 microalgae, Park et al. [13] found that biomass and lipid productivities were boosted by mixotrophic cultivation. Andrade et al. [14] studied the effects of molasses concentration and light levels on mixotrophic growth of *Spirulina platensis*, and found the biomass production was stimulated by molasses, which suggested that this industrial by-product could be used as a low-cost supplement for the growth of this species. Bhatnagar et al. [15] found the mixotrophic growth of *Chlamydomonas globosa*, *Chlorella minutissima* and *Scenedesmus bijuga* resulted in 3–10 times more biomass production compared to that obtained under phototrophic growth conditions. The maximum lipid productivities of *Phaeodactylum tricornutum* in mixotrophic cultures with glucose, starch and acetate in medium were 0.053, 0.023 and 0.020 $g \cdot L^{-1} \cdot day^{-1}$, which were respectively 4.6-, 2.0-, and 1.7-fold of those obtained in the corresponding photoautotrophic control cultures [16].

1.3.2 CULTIVATION SYSTEMS

In order to achieve large-scale biodiesel production from microalgae, a cost effective cultivation system is of great significance. The cultivation systems include open and closed styles. The former, which simulates the growth environment in natural lakes, is just open-ponds characterized by simple and low cost structure and operations, low biomass concentration, and poor system stability. The closed culture systems are photobioreactors (PBR) of different configurations including tubular, flat plate, and column photobioreactors. Compared with open pond systems, the closed systems are usually more stable because it is easier to control the process conditions and maintain monoculture and allow higher cell density, but they have higher capital and operational costs. In both open and closed microalgae culture systems, light source and light intensity are critical to the performance of phototrophic growth of microalgae. With the development of optical trapping system, light delivery and lighting technologies, which improve the distribution and absorption of light, the advent of some new photobioreactors will improve the efficiency of photosynthesis [17]. In addition, gas-liquid mass transfer efficiency is another critical factor affecting CO_2 utilization and hence the phototrophic growth. Cheng et al. [18] constructed a 10 L photobioreactor integrated with a hollow fiber membrane module which increased the gas bubbles retention time from 2 s to more than 20 s, increasing the CO_2 fixation rate of *Chlorella vulgaris* from 80 to 260 $mg \cdot L^{-1} \cdot h^{-1}$.

1.4 HIGH LIPID CONTENT OF MICROALGAE

It is easy to increase the chemical composition of microalgae by changing the environmental conditions and other factors. Many studies indicate that the composition, including lipid content, of microalgae is closely related to the environmental conditions and medium composition. Improving the lipid content in microalgae is a focus of commercial production of microalgae biomass. Current studies on high lipid content of microalgae are focused mainly on selection of microalgae species, genetic modification of microalgae, nutrient management, metabolic pathways, cultivation conditions, and so on.

1.4.1 MICROALGAE SPECIES AND STRAINS

The lipid content of microalgae varies among different species and strains (Table 2 [19,20]). The lipid content of microalgae is usually in the range of 20% to 50% (dry base), and can be as high as 80% under certain circumstances. Selecting high lipid content and fast growing microalgae is an important step in the overall success of biodiesel production from microalgae. Traditional methods of screening microalgae for high lipid content rely on time-consuming and laborious lipid extraction process which involves cell wall disruption and solvent extraction of a reasonably large amount of microalgae cells. Recently, high-throughput screening techniques employing lipophilic fluorescent dye staining (such as Nile Red [21], BODIPY 505 [22]) and fluorescence microscopy or flow cytometry are being developed [23]. With these new techniques, the amount of sample and preparation time are greatly reduced because the lipid content of algal cells is measured in situ without the need for extraction. Lipid content and lipid productivity are two different concepts. The former refers to lipid concentration within the microalgae cells without consideration of the overall biomass production. However the latter takes into account both the lipid concentration within cells and the biomass produced by these cells. Therefore lipid productivity is a more reasonable indicator of a strain's performance in terms of lipid production.

1.4.2 GENETIC MODIFICATION

The screening of strains from local habitats and elsewhere should be considered as the first step in selecting high performance strains for biodiesel production. However, other approaches including genetic manipulation may be employed to optimize the lipid and biomass productivity of promising strains obtained through the screening process. The modern biotechnologies, such as genetic engineering, cell fusion, ribosome engineering, metabolic engineering, etc. are potential techniques to develop new algae strains with rapid growth and high lipid content [24,25]. Such techniques are the key to breakthroughs in microalgae biomass energy development. In the fatty acid biosynthesis pathway, acetyl-CoA carboxylase (ACCase) is the key rate-limiting enzyme that helps the substrates acetyl-CoA enter the carbon chain of fatty acids. Therefore, it is effective to enhance the expression of ACCase to promote the lipid synthesis in microalgae. Song et al. [26] successfully constructed a vector called pRL-489-ACC to realize the shuttle expression of the gene coding acetyl-CoA carboxylase in the fatty acid synthesis pathway. Zaslavskaia et al. [27] introduced a gene encoding a glucose transporter (glut1 or hup1) into Phaeodactylum tricor-

TABLE 2: Lipid content in the dry biomass of various species of microalgae.

Species Lipid content	(% dryweight)
Anabaena cylindrica	4–7
Botyococcus braunii	25–80
Chlamydomonas reinhardtii	21
Chlorella emersonii	28–32
Chlorella protothecoides	57.9
Chlorella pyrenoidosa	2
Chlorella vulgaris	14–22
Crypthecodinium cohnii	20
Cylindrotheca sp.	16–37
Dunaliella bioculata	8
Dunaliella primolecta	23
Dunaliella salina	6
Dunaliella tertiolecta	35.6
Euglena gracilis	14 20
Hormidium sp.	38
Isochrysis sp.	25–33
Monallanthus salina	>20
Nannochloris sp.	30–50
Nannochloropsis sp.	31–68
Neochloris oleoabundans	35–54
Nitzschia sp.	45–47
Phaeodactylum tricornutum	20–30
Pleurochrysis carterae	30–50
Porphyridium cruentum	9–14
Prymnesium parvum	22–38
Scenedesmus dimorphus	16–40
Scenedesmus obliquus	12–14
Schizochytrium sp.	50–77
Spirogyra sp.	11–21
Spirulina maxima	6–7
Spirulina platensis	4–9
Synechoccus sp.	11
Tetraselmis maculata	8
Tetraselmis sueica	15–23

nutumcan to allow the alga to grow on exogenous glucose in the absence of light. This represents progress of large-scale commercial production of microalgae with high lipid content by reducing limitations associated with light-dependent growth. In addition, phosphoenolpyruvate carboxylase (PEPC) is closely related to the fatty acid biosynthesis pathway because of the inhibition of PEPC activity redounding to catalyse acetyl-CoA to enter the fatty acid synthesis pathway. With successful clone of some PEPC gene in microalgae (such as *Anabaena sp.* PCC 7120 [28], *Synechococcus vulcanus* [29]) and detailed analysis of its sequence characteristics and structure, it will be possible to improve the lipid content of microalgae by regulation of PEPC expression by antisense technology [25].

1.4.3 NUTRIENT MANAGEMENT

Usually microalgae only synthesize small amounts of triacylglycerols (TAGs) under normal nutrient conditions, but can synthesize a large number of TAGs with a significant change in the fatty acid composition under stress conditions. Limiting nutrient availability such as nitrogen and phosphorus starvation during microalgae cultivation is a common method to induce lipid synthesis [30,31]. When the nitrogen is exhausted and becomes the limiting factor, microalgae will continue to absorb organic carbons, which are to be converted to lipids. The nutrient limitation also results in a gradual change in lipid composition, i.e., from free fatty acids to TAGs which are more suitable for biodiesel production [32]. Phosphorus is another important nutrient that influences algae growth and lipid accumulation. Khozin-Goldberg et al. [33] found that phosphate limitation could cause significant changes in the fatty acid and lipid composition of Monodus subterraneus. Some studies found that phosphorus deficiency led to reduced lipid content of *Nannochloris atomus* and *Tetraselmis sp.* [34]. Silicon is a necessary element for the growth of diatom. Roessler [35] found that silicon deficiency could induce lipid accumulation in *Cyclotella cryptica* by two distinct processes: (1) An increase in the proportion of newly assimilated carbons which are converted to lipids; (2) A slow conversion of previously assimilated carbon from non-lipid compounds to lipids. Unfortunately, higher lipid content achieved through nutrient limitation is usually at the expense of lower biomass productivity because nutrient deficiency limits cell growth. As mentioned above, lipid productivity, representing the combination of lipid content and biomass yield, is a more meaningful performance index to indicate the ability of lipid production of microalgae. Therefore, it is necessary to develop a nutrient management strategy which will first facilitate rapid biomass accumulation and then induce lipid accumulation in order to achieve maximum lipid productivity.

1.4.4 METABOLIC PATHWAY MANAGEMENT

The lipid content of algal cells is closely related to the metabolic pathways that the algae undergo. Photoautotrophic growth is usually characterized by lower rates and lower lipid content compared with heterotrophic growth in which algae use organic matter as carbon and energy source [36,37]. Efforts have been made to control metabolic activities during cultivation in order to maximize lipid accumulation. Wu et al. [38] added organic carbon (glucose) to and reduced inorganic nitrogen in the cultivation of *Chlorella protothecoides*, and found that the induced heterotrophic growth resulted in 55.2% lipid content, which was about four times that in photoautotrophic growth. Xiong et al. [39] developed a photosynthesis-fermentation model with double CO_2 fixation in both photosynthesis and fermentation stages which provided an efficient approach for the production of algal lipid. In this model, cultivation of *C. protothecoides* could realize 69% higher lipid yield on glucose achieved at the fermentation stage, and 61.5% less CO_2 released compared with typical heterotrophic metabolism.

1.4.5 CULTURE CONDITIONS

Lipid synthesis may also be induced under other stress culture conditions, such as extreme light [40], temperature [41], salinity [42], pH [43], and CO_2 conditions [44]. Therefore, during cultivation, process conditions must be closely monitored and/or appropriately adjusted for optimal lipid productivity.

1.5 PROCESSING OF MICROALGAE LIPID

1.5.1 MICROALGAE HARVESTING AND DEWATERING

Microalgae cells are small (typically in the range of Φ2–70 μm) and the cell densities in culture broth are low (usually in the range of 0.3–5 $g \cdot L^{-1}$). Harvesting microalgae from the culture broth and dewatering them are energy intensive and therefore a major obstacle to commercial scale production and processing of microalgae. Many harvesting technologies, such as centrifugation, flocculation, filtration, gravity sedimentation, floatation, and electrophoresis techniques have been tested [45]. The choice of harvesting technique depends on, in part, the characteristics of microalgae (such as their size and density) and the target products.

1.5.2 MICROALGAE LIPID EXTRACTION AND REFINING

Intracellular microalgae lipids can be extracted by a variety of methods, such as mechanical crushing extraction, chemical extraction, enzymatic extraction, supercritical carbon dioxide ($SCCO_2$) extraction [46], microwave extraction [47], etc. Microalgae lipids in the form of triglycerides or fatty acids can be converted to biodiesel through transesterification/(esterification for fatty acids) reactions after the extraction [48]. In order to achieve efficient reaction, the choice of catalyst is very important. The traditional liquid acid and alkali catalyst are called homogeneous catalysts because they act in the same liquid phase as the reaction mixture. Due to their simple usage and less time required for lipids conversion, the homogeneous catalysts dominate the biodiesel industry. However, the transesterification catalyzed by homogeneous catalysts needs high purity feedstock and complicated downstream processing [49], so high efficiency and low pollution catalysts such as solid acid catalysts, solid alkali catalysts, enzyme catalyst, supercritical catalyst systems and ionic liquid catalysts are receiving increasing attentions. Krohn et al. [50] studied the catalytic process using supercritical methanol and porous titania microspheres in a fixed bed reactor to catalyze the simultaneous transesterification and esterification of triglycerides and free fatty acids to biodiesel. The process was able to reach conversion efficiencies of up to 85%. Patil et al. [51] reported a process involving simultaneous extraction and transesterification of wet algal biomass containing about 90% of water under supercritical methanol conditions.

1.5.3 PYROLYSIS OF MICROALGAE LIPID

Thermochemical conversion covers different processes such as direct combustion, gasification, thermochemical liquefaction, and pyrolysis. In pyrolysis, heating rate affects reaction rate which in turn affects product composition. Traditional heating methods require expensive heating mechanisms to achieve rapid temperature rise with poor process control. Microwave assisted pyrolysis (MAP) has following advantages: fine grinding of biomass is not necessary; microwave heating is mature and scalable technology which is suitable for distributed biomass conversion. Due to insufficient understanding of the mechanism of pyrolysis and the lack of effective control of the pyrolysis process, pyrolytic bio-oils are complex mixture with low calorific value, high acidity, high oxygen volume, and poor stability. However, bio-oil from pyrolysis of microalgae appears to have higher quality than those from cellulosics [52].

In recent years, the role of catalyst and minerals in biomass pyrolysis was investigated. Lu et al. [53] reported that $ZnCl_2$ could catalyze fast pyrolysis of corn cob with the main products of furfural and acetic acid. Du et al. [54] used 1-butyl-3-methylimidazolium chloride and 1-butyl-3-methylimidazolium boron tetrafluoride as the catalysts in MAP of straw and sawdust.

1.6 THE BOTTLENECK OF COMMERCIAL DEVELOPMENT OF MICROALGAE BIODIESEL

Although the potential of microalgae lipid production is tremendous, no commercial development of microalgae based fuels has been achieved so far because of lack of price competitiveness versus petroleum diesel. The bottleneck that limits the development of microalgae biomass energy is that we do not have technologies to produce large quantities of low-cost and high lipid content microalgae biomass. In order to achieve large-scale and low-cost cultivation of microalgae, the following bottleneck issues should be addressed.

1.6.1 DEVELOPMENT OF HIGH PERFORMANCE MICROALGAE STRAINS

Breeding of high-quality microalgae with characteristics of high lipid productivity and strong adaptability is the key to realize both of high lipid content and great biomass. Excellent microalgae strains can be obtained by screening of a wide range of naturally available isolates, and the efficiency of those can be improved by selection and transformation. Considering the good results of mixotrophic cultivation, we should do some domestication making microalgae adapt to different organic matters such as sucrose, glycerol, xylan, organic acids in slurry. This will hopefully greatly reduce the cost of cultivation.

1.6.2 DEVELOPMENT OF COST-EFFECTIVE CULTIVATION SYSTEMS

Continuous production systems are a critical element in large scale commercial production of algal biomass. While we should continue to improve open pond operations, significant efforts should be invested in development of photobioreactors for high density cultivation of microalgae. Photobioreactors with a real time smart on-line monitoring system which can maintain optimal conditions for the growth of microalgae is very promising to realize high growth rate and cell density [55]. In order to address the issues with high capital and operational costs, innovative photobioreactor designs must be developed. Such designs may incorporate cost effective lighting techniques and renewable power combining solar energy, biogas, wind energy, waste heat.

1.6.3 WASTEWATER TREATMENT AND MICROALGAE CULTIVATION

Disposal of wastewaters from human activities and animal production is both an environmental and a financial issue. These nitrogen and phosphorus rich wastewaters have been proven suitable for microalgae growth. The feasibility of growing microalgae to remove inorganic nitrogen, phosphorus, metal elements and other pollutants in the wastewaters has long been recognized [56,57]. Therefore, a wastewater based microalgae production process has the due benefits of wastewater treatment and production of algal biomass with minimum external input of fresh water and nutrients. It has been argued that microalgae biodiesel production in conjunction with wastewater treatment is the area with the most plausible commercial potential in the short term [58,59]. Several applications in wastewater treatment have been reported in the literature. Chinnasamy et al. [60] found that both fresh water and marine algae showed good growth in carpet industry effluents and municipal sewage. Kim et al. [61] added fermented swine urine (3%) (v/v) to the medium to culture *Scenedesmus sp.*, and received the growth rate (3-fold), dry weight (2.6-fold). Wang et al. [62] investigated the effectiveness of using digested dairy manure as a nutrient supplement for cultivation of oil-rich green microalgae *Chlorella sp*, and found that the total fatty acid content of dry weight was increased from 9.00% to 13.7%, along with removal of ammonia, total nitrogen, total phosphorus, and COD by 100%, 75.7%–82.5%, 62.5%–74.7%, and 27.4%–38.4%, respectively. Li et al. [63] cultivated *Scenedesmus sp.* LX1 in secondary effluent, achieving high biomass yield (0.11 $g \cdot L^{-1}$, dry weight) and lipid content (31%–33%, dry weight).

1.6.4 CO_2 FIXATION AND MICROALGAE CULTIVATION

To significantly improve photoautotrophic cultivation, new microalgae strains and process conditions must be developed to enable fast growth and high lipid accumulation at high CO_2 levels [64,65]. If CO_2 is from flue gas, algae strains' high tolerance to SO_x, NO_x, and high temperature is desirable [66]. Morais et al. [67–69] isolated several microalgae from the waste treatment ponds of a coal fired thermoelectric power plant, and investigated their growth characteristics when exposed to different concentrations of CO_2. When cultivated with 6% and 12% CO_2, *Chlorella kessleri* showed a maximum biomass productivity at 6% CO_2 while *Scenedesmus obliquus* showed a maximum biomass productivity at 12% CO_2. They also found that *Spirulina sp., Scenedesmus obliquus*, and *Chlorella vulgaris* grew well when the culture medium contained up to 18% CO_2, and *Spirulina sp.* exhibited the highest rate among them. Chang et al. [70] found that some strains of Chlorella could grow in an atmosphere containing CO_2 up to 40%. Base on these studies, we can try to use some means to promote microalgae growth stimulated by CO_2 addition [71]. Such as pump CO_2 emission from power plants, industrial processes, or soluble carbonate through a sparger system of orifices evenly distributed over the bottom of the ponds when cultivating microalgae in large open-ponds.

1.6.5 DEVELOPMENT OF IMMOBILIZATION TECHNIQUE

The algal cells can be gathered into a corporation or fixed to the carrier by immobilization technology. Immobilized algae have numerous benefits as follows: high cell density, excellent ability to resist poison, high removal efficiency, good stability, easy product separation, etc. At present, immobilization carriers include mainly natural macromolecular materials (such as agar, sodium alginate, etc.) and synthetic polymeric gels (such as polyvinyl alcohol, polyacrylamide, etc.). The natural polymers have no toxicity and good mass transfer property, but their intensity is low. However, the synthetic organic polymer gels have high strength, but poor mass transfer property. Therefore, new immobilization carriers are expected to facilitate the development of algae immobilization technology.

High rate algal pond (HRAP) can achieve efficient removal of organic matter, heavy metals, nitrogen, phosphorus and other nutrients in wastewater by algae-bacteria symbiosis system. In the system, the microalgae provide oxygen and nutrition for bacteria; meanwhile bacteria provide useful growth-promoting substances for microalgae. Our team has developed a new type of microalgae immobilization carrier using the fungi [72]. The key of the technology is that fungi mediate pelletization of microalgae. After adding a certain number of fungal spores to the microalgae culture, algae-fungi symbiotic spheres formed accompanied by spore germination. The immobilized microalgae allow efficient recycling of livestock wastewater for algae cultivation and are easy to be harvested. In addition, the fungi can be converted to animal feed.

1.6.6 EXTENSION OF MICROALGAE CHAIN

In addition to lipids, microalgae can produce some value-added bioactive substances, such as polysaccharides, proteins and pigments. Therefore, microalgae have good prospects in the food, feed, pharmaceutical, and cosmetic industries. If a biorefinery approach that converts microalgae to wide range of products including biodiesel and value-added products, it will significantly improve the economic outlook of microalgae based fuels production.

1.7 CONCLUSIONS

Microalgae are a sustainable energy resource with great potential for CO_2 fixation and wastewater purification. This review discusses current status and anticipated future developments in the microalgae to biodiesel approach. For biodiesel production to have a significant impact on renewable fuels standards, technologies must be developed to enable large scale algae biomass production. The strategies for both high biomass and lipid productivities are discussed. Further efforts on microalgae biodiesel production should focus on reducing costs in large-scale algal biomass production systems. Combining microalgae mixotrophic cultivation with sequestration of CO_2 from flue gas and wastewater treatment, and taking the biorefinery approach to algal biomass conversion will improve the environmental and economic viability.

REFERENCES

1. Khan, S.A.; Rashmi; Hussain, M.Z.; Prasad, S.; Banerjee, U.C. Prospects of biodiesel production from microalgae in India. Renew. Sustain. Energy Rev. 2009, 13, 2361–2372.
2. Lim, S.; Teong, L.K. Recent trends, opportunities and challenges of biodiesel in Malaysia: An overview. Renew. Sustain. Energy Rev. 2010, 14, 938–954.
3. Stephens, E.; Ross, I.L.; King, Z.; Mussgnug, J.H.; Kruse, O.; Posten, C.; Borowitzka, M.; Hankamer, B. An economic and technical evaluation of microalgal biofuels. Nature Biotechnol. 2010, 28, 126–128.
4. Chisti, Y. Biodiesel from microalgae. Biotechnol. Adv. 2007, 25, 294–306.
5. Cheng, L.H.; Zhang, L.; Chen, H.L.; Gao, C.J. Carbon dioxide removal from air by microalgae cultured in a membrane-photobioreactor. Sep. Purif. Technol. 2006, 50, 324–329.
6. Chojnacka, K.; Noworyta, A. Evaluation of Spirulina sp. growth in photoautotrophic, heterotrophic and mixotrophic cultures. Enzyme Microb. Technol. 2004, 34, 461–465.
7. Greenwell, H.C.; Laurens, L.M.L.; Shields, R.J.; Lovitt, R.W.; Flynn, K.J. Placing microalgae on the biofuels priority list: A review of the technological challenges. J. R. Soc. Interface 2010, 7, 703–726.
8. Chen, F. High cell density culture of microalgae in heterotrophic growth. Trends Biotechnol. 1996, 14, 421–426.
9. Li, X.F.; Xu, H.; Wu, Q.Y. Large-scale biodiesel production from microalga Chlorella protothecoides through heterotrophic cultivation in bioreactors. Biotechnol. Bioeng. 2007, 98, 764–771.
10. Doucha, J.; Lívansky, K. Production of high-density Chlorella culture grown in fermenters. J. Appl. Phycol. 2012, 24, 35–43.
11. Wu, Z.Y.; Shi, X.M. Optimization for high-density cultivation of heterotrophic Chlorella based on a hybrid neural network model. Lett. Appl. Microbiol. 2006, 44, 13–18.
12. Heredia-Arroyo, T.; Wei, W.; Ruan, R.; Hu, B. Mixotrophic cultivation of Chlorella vulgaris and its potential application for the oil accumulation from non-sugar materials. Biomass Bioenergy 2011, 35, 2245–2253.
13. Park, K.C.; Whitney, C.; McNichol, J.C.; Dickinson, K.E.; MacQuarrie, S.; Skrupski, B.P.; Zou, J.; Wilson, K.E.; O'Leary, S.J.B.; McGinn, P.J. Mixotrophic and photoautotrophic cultivation of 14 microalgae isolates from Saskatchewan, Canada: Potential applications for wastewater remediation for biofuel production. J. Appl. Phycol. 2012, 24, 339–348.
14. Andrade, M.R.; Costa, J.A.V. Mixotrophic cultivation of microalga Spirulina platensis using molasses as organic substrate. Aquaculture 2007, 264, 130–134.
15. Bhatnagar, A.; Chinnasamy, S.; Singh, M.; Das, K.C. Renewable biomass production by mixotrophic algae in the presence of various carbon sources and wastewaters. Appl. Energy 2011, 88, 3425–3431.

16. Wang, H.Y.; Fu, R.; Pei, G.F. A study on lipid production of the mixotrophic microalgae Phaeodactylum tricornutum on various carbon sources. Afr. J. Microbiol. Res. 2012, 6, 1041–1047.
17. Masojidek, J.; Papacek, S.; Sergejevova, M.; Jirka, V.; Cerveny, J.; Kunc, J.; Korecko, J.; Verbovikova, O.; Kopecky, J.; Stys, D. A closed solar photobioreactor for cultivation of microalgae under supra-high irradiance: Basic design and performance. J. Appl. Phycol. 2003, 15, 239–248.
18. Cheng, L.H.; Zhang, L.; Chen, H.L.; Cao, C.J. Carbon dioxide removal from air by microalgae cultured in a membrane-photobioreactor. Sep. Purif. Technol. 2006, 50, 324–329.
19. Singh, A.; Nigam, P.S.; Murphy, J.D. Renewable fuels from algae: An answer to debatable land based fuels. Bioresour. Technol. 2011, 102, 10–16.
20. Singh, A.; Nigam, P.S.; Murphy, J.D. Mechanism and challenges in commercialisation of algal biofuels. Bioresour. Technol. 2011, 102, 26–34.
21. Elsey, D.; Jameson, D.; Raleigh, B.; Cooney, M.J. Fluorescent measurement of microalgal neutral lipids. J. Microbiol. Methods 2007, 68, 639–642.
22. Cooper, M.S.; Hardin, W.R.; Petersen, T.W.; Cattolico, R.A. Visualizing "green oil" in live algal cells. J. Biosci. Bioeng. 2010, 109, 198–201.
23. Mutanda, T.; Ramesh, D.; Karthikeyan, S.; Kumari, S.; Anandraj, A.; Bux, F. Bioprospecting for hyper-lipid producing microalgal strains for sustainable biofuel production. Bioresour. Technol. 2011, 102, 57–70.
24. Rosenberg, J.N.; Oyler, G.A.; Wilkinson, L.; Betenbaugh, M.J. A green light for engineered algae: Redirecting metabolism to fuel a biotechnology revolution. Curr. Opin. Biotechnol. 2008, 19, 430–436.
25. Song, D.H.; Fu, J.J.; Shi, D.J. Exploitation of oil-bearing microalgae for biodiesel. Chin. J. Biotechnol. 2008, 24, 341–348.
26. Song, D.H.; Hou, L.J.; Song, X.F.; Shi, D.J. Cloning of the acetyl coenzyme A carboxylase gene of Cyanophyta in fatty acid biosynthesis pathway and the construction of the shuttle expression vector. In Proceedings of the 14th National Algae Symposium, Hohhot, China, 12–15 August 2007.
27. Zaslavskaia, L.A.; Lippmeler, J.C.; Shih, C.; Ehrhardt, D.; Grossman, A.R.; Apt, K.E. Trophic conversion of an obligate photoautotrophic organism through metabolic engineering. Science 2001, 292, 2073–2075.
28. Luinenburg, I.; Coleman, J.R. Identification, characterization and sequence analysis of the gene encoding phosphoenolpyruvate carboxylase in Anabaena sp. PCC 7120. J. Gen. Microbiol. 1992, 138, 685–691.
29. Chen, L.M.; Omiya, T.; Hata, S.; Izui, K. Molecular characterization of a phosphoenolpyruvate carboxylase from a thermophilic cyanobacterium, Synechococcus vulcanus with unusual allosteric properties. Plant Cell Physiol. 2002, 43, 159–169.
30. Zhila, N.O.; Kalacheva, G.S.; Volova, T.G. Effect of nitrogen limitation on the growth and lipid composition of the green alga Botryococcus braunii Kütz IPPAS H-252 Russian. J. Plant Physiol. 2005, 52, 311–319.
31. Weldy, C.S.; Huesemann, M.H. Lipid production by Dunaliella salina in batch culture: Effects of nitrogen limitation and light intensity. J. Undergrad. Res. 2007, 7, 115–122.

32. Widjaja, A.; Chien, C.C.; Ju, Y.H. Study of increasing lipid production from fresh water microalgae Chlorella vulgaris. J. Taiwan Inst. Chem. Eng. 2009, 40, 13–20.
33. Khozin-Goldberg, I.; Cohen, Z. The effect of phosphate starvation on the lipid and fatty acid composition of the fresh water eustigmatophyte Monodus subterraneus. Phytochemistry 2006, 67, 696–701.
34. Reitan, K.I.; Rainuzzo, J.R.; Olsen, Y. Effect of nutrient limitation on fatty acid and lipid content of marine microalgae. J. Phycol. 1994, 30, 972–979.
35. Roessler, P.G. Effects of silicon dificiency on lipid composition metabolism in the diatom Cyclotella cryptica. J. Phycol. 1988, 24, 394–400.
36. Miao, X.L.; Wu, Q.Y. High yield bio-oil production from fast pyrolysis by metabolic controlling of Chlorella protothecoides. J. Biotechnol. 2004, 110, 85–93.
37. Wei, D.; Liu, L.J. Optimization of culture medium for heterotrophic Chlorella Protothecoides producing total fatty acids. Chem. Bioeng. 2008, 25, 35–40.
38. Xu, H.; Miao, X.L.; Wu, Q.Y. High quality biodiesel production from a microalga Chlorella protothecoides by heterotrophic growth in fermenters. J. Biotechnol. 2006, 126, 499–507.
39. Xiong, W.; Gao, C.F.; Yan, D.; Wu, C.; Wu, Q. Double CO_2 fixation in photosynthesis-fermentation model enhances algal lipid synthesis for biodiesel production. Bioresour. Technol. 2010, 101, 2287–2293.
40. Guedes, A.C.; Meireles, L.A.; Amaro, H.M.; Malcata, F.X. Changes in lipid class and fatty acid composition of cultures of Pavlova lutheri, in response to light intensity. J. Am. Oil Chem. Soc.2010, 87, 791–801.
41. Renaud, S.M.; Thinh, L.V.; Lambrinidis, G.; Parry, D.L. Effect of temperature on growth, chemical composition and fatty acid composition of tropical Australian microalgae grown in batch cultures. Aquaculture 2002, 211, 195–214.
42. Kotlova, E.R.; Shadrin, N.V. The role of membrane lipids in adaptation of Cladophora (Chlorophyta) to living in shallow lakes with different salinity. Botanicheskii Zhurnal 2003, 88, 38–45.
43. Guckert, J.B.; Cooksey, K.E. Triglyceride accumulation and fatty acid profile changes in Chlorella (Chlorophyta) during high pH-induced cell cycle inhibition. J. Phycol. 1990, 26, 72–79.
44. Chiu, S.Y.; Kao, C.Y.; Tsai, M.T.; Ong, S.C.; Chen, C.H.; Lin, C.S. Lipid accumulation and CO_2 utilization of Nanochloropsis oculata in response to CO_2 aeration. Bioresour. Technol. 2009, 100, 833–838.
45. Uduman, N.; Qi, Y.; Danquah, M.K.; Forde, G.M.; Hoadley, A. Dewatering of microalgal cultures: A major bottleneck to algae-based fuels. J. Renew. Sustain. Energy 2010, 2, 012701.
46. Halim, R.; Gladman, B.; Danquah, M.K.; Webley, P.A. Oil extraction from microalgae for biodiesel production. Bioresour. Technol. 2011, 102, 178–185.
47. Koberg, M.; Cohen, M.; Ben-Amotz, A.; Gedanken, A. Bio-diesel production directly from the microalgae biomass of Nannochloropsis by microwave and ultrasound radiation. Bioresour. Technol. 2011, 102, 4265–4269.
48. Johnson, M.B.; Wen, Z. Production of biodiesel fuel from the microalga Schizochytrium limacinum by direct transesterification of algal biomass. Energy Fuels 2009, 23, 5179–5183.

49. Borges, M.E.; Diaz, L. Recent developments on heterogeneous catalysts for biodiesel production by oil esterification and transesterification reactions: A review. Renew. Sustain. Energy Rev. 2012, 16, 2839–2849.
50. Krohn, B.J.; McNeff, C.V.; Yan, B.; Nowlan, D. Production of algae-based biodiesel using the continuous catalytic Mcgyan® process. Bioresour. Technol. 2011, 102, 94–100.
51. Patil, P.D.; Gude, V.G.; Mannarswamy, A.; Deng, S.G.; Cooke, P.; Munson-McGee, S.; Rhodes, I.; Lammers, P.; Nirmalakhandan, N. Optimization of direct conversion of wet algae to biodiesel under supercritical methanol conditions. Bioresour. Technol. 2011, 102, 118–122.
52. Du, Z.Y.; Li, Y.C.; Wang, X.Q.; Wan, Y.Q.; Chen, Q.; Wang, C.G.; Lin, X.Y.; Liu, Y.H.; Chen, P.; Ruan, R. Microwave-assisted pyrolysis of microalgae for biofuel production. Bioresour. Technol. 2011, 102, 4890–4896.
53. Lu, Q.; Wang, Z.; Dong, C.Q.; Zhang, Z.F.; Zhang, Y.; Yang, Y.P.; Zhu, X.F. Selective fast pyrolysis of biomass impregnated with $ZnCl_2$: Furfural production together with acetic acid and activated carbon as by-products. J. Anal. Appl. Pyrolysis 2011, 1, 273–279.
54. Du, J.; Liu, P.; Liu, Z.H.; Sun, D.G.; Tao, C.Y. Fast pyrolysis of biomass for bio-oil with ionic liquid and microwave irradiation. J. Fuel Chem. Technol. 2010, 5, 554–559.
55. Meireles, L.A.; Azevedo, J.L.; Cunha, J.P.; Malcata, F.X. On-line determination of biomass in a microalga bioreactor using a novel computerized flow injection analysis system. Biotechnol.Prog. 2002, 18, 1387–1391.
56. Aslan, S.; Kapdan, I. Batch kinetics of nitrogen and phosphorus removal from synthetic wastewater by algae. Ecol. Eng. 2006, 28, 64–70.
57. Shi, J.; Podola, B.; Melkonian, M. Removal of nitrogen and phosphorus from wastewater using microalgae immobilized on twin layers: An experimental study. J. Appl. Phycol. 2007, 19, 417–423.
58. Mulbry, W.; Kondrad, S.; Buyer, J. Treatment of dairy and swine manure effluents using freshwater algae: Fatty acid content and composition of algal biomass at different manure loading rates. J. Appl. Phycol. 2008, 20, 1079–1085.
59. Kong, Q.X.; Li, L.; Martinez, B.; Chen, P.; Ruan, R. Culture of microalgae Chlamydomonas reinhardtii in wastewater for biomass feedstock production. Appl. Biochem. Biotechnol. 2010, 160, 9–18.
60. Chinnasamy, S.; Bhatnagar, A.; Hunt, R.W.; Das, K.C. Microalgae cultivation in a wastewater dominated by carpet mill effluents for biofuel applications. Bioresour. Technol. 2010, 101, 3097–3105.
61. Kim, M.K.; Park, J.W.; Park, C.S.; Kim, S.J.; Jeune, K.H.; Chang, M.U.; Acreman, J. Enhanced production of Scenedesmus spp. (green microalgae) using a new medium containing fermented swine wastewater. Bioresour. Technol. 2007, 98, 2220–2228.
62. Wang, L.; Li, Y.C.; Chen, P.; Min, M.; Chen, Y.F.; Zhu, J.; Ruan, R. Anaerobic digested dairy manure as a nutrient supplement for cultivation of oil-rich green microalgae Chlorella sp. Bioresour. Technol. 2010, 101, 2623–2628.
63. Li, X.; Hu, H.Y.; Yang, J. Lipid accumulation and nutrient removal properties of a newly isolated freshwater microalgae, Scenedesmus sp. LX1, growing in secondary effluent. New Biotechnol. 2010, 27, 59–63.

64. Yoo, C.; Jun, S.Y.; Lee, J.Y.; Ahn, C.Y.; Oh, H.M. Selection of microalgae for lipid production under high levels carbon dioxide. Bioresour. Technol. 2010, 101, 571–574.
65. Jiang, L.L.; Luo, S.J.; Fan, X.L.; Yang, Z.M.; Guo, R.B. Biomass and lipid production of marine microalgae using municipal wastewater and high concentration of CO2. Appl. Energy 2011, 88, 3336–3341.
66. Lee, J.N.; Lee, J.S.; Shin, C.S.; Park, S.C.; Kim, S.W. Methods to enhance tolerances of Chlorella KR-1 to toxic compounds in flue gas. Appl. Biochem. Biotechnol. 2000, 84–86, 329–342.
67. Morais, M.G.; Costa, J.A.V. Isolation and selection of microalgae from coal fired thermoelectric power plant for biofixation of carbon dioxide. Energy Convers. Manag. 2007, 48, 2169–2173.
68. Morais, M.G.; Costa, J.A.V. Carbon dioxide fixation by Chlorella kessleri, C. vulgaris, Scenedesmus obliquus and Spirulina sp. cultivated in flasks and vertical tubular photobioreactors. Biotechnol. Lett. 2007, 29, 1349–1352.
69. Morais, M.G.; Costa, J.A.V. Biofixation of carbon dioxide by Spirulina sp. and Scenedesmus obliquus cultivated in a three-stage serial tubular photobioreactor. J. Biotechnol. 2007, 129, 439–445.
70. Chang, E.H.; Yang, S.S. Some characteristics of microalgae isolated in Taiwan for biofixation of carbon dioxide. Bot. Bull. Acad. Sini. 2003, 44, 43–52.
71. Park, J.B.K.; Craggs, R.J. Wastewater treatment and algal production in high rate algal ponds with carbon dioxide addition. Water Sci. Technol. 2010, 61, 633–639.
72. Zhou, W.G.; Cheng, Y.L.; Li, Y.; Wan, Y.Q.; Liu, Y.H.; Lin, X.Y.; Ruan, R. Novel fungal pelletization assisted technology for algae harvesting and wastewater treatment. Appl. Biochem. Biotechnol. 2012, 167, 214–228.

This chapter was originally published under the Creative Commons Attribution License. Wu, X., Ruan, R., Du, Z., and Liu, Y. Current Status and Prospects of Biodiesel Production from Microalgae. Energies 2012, 5(8), 2667-2682.

CHAPTER 2

SUSTAINABLE FUEL FROM ALGAE: CHALLENGES AND NEW DIRECTIONS

DOUGLAS AITKEN and BLANCA ANTIZAR-LADISLAO

2.1 INTRODUCTION

In the current climate there are many reasons for considering alternative fuel sources and algae has been heralded as a potential "silver bullet", however, after initial excitement it appears that the concept will not be commercially viable for at least 10 years. The technology to produce fuel from algae is currently available and vehicles have been powered by this feedstock in a number of cases [1]. As a commercially viable alternative to fossil fuels, however, the technologies are not yet there, the energy balance of producing the fuel is high, the economics cannot compete and the overall sustainability is in doubt. The concept of algal fuels is one that has been with us since the 1950s [2]. Funding for this area of research has fluctuated roughly in line with rising and falling crude oil prices. One of the main contributors in pioneering algal fuels was Professor W.J. Oswald who designed systems to cultivate algae on a large scale in the 1950s and 60s [3, 4]. He developed the concept to remove nutrients from wastewater and provide a useful biomass for food or fuel. At the time however the viability of the concept appeared unachievable and as oil prices dropped so did funding for such projects. The price of oil however is continuing to rise with little sign of slowing down and it is now important to focus once again on improving the viability of fuel from algal feedstock. This paper will investigate where we have come since the first research on algal cultivation and energy recovery was initiated, what is currently hindering the commercial application of the concept and where we need to go from here. Continued research will allow us to recover the maximum potential from algal biomass which will provide an increasingly important resource for us in the future as predicted by Professor Oswald [5].

The aim of this paper is to investigate the current state of algal bio-energy production and consider what conventional and novel processes can be employed to provide a more sustainable approach. A focus is placed on the energetic consumption and the environmental acceptability and where combining value added benefits such as wastewater treatment and carbon mitigation through flue gas utilisation may be beneficial.

2.2 BACKGROUND

2.2.1 THE BEGINNING FOR ALGAL BIOFUELS

With predictions of an ever-increasing global population in the 1940s and 1950s, many researchers were considering how feeding such vast number of people would be possible. Traditionally livestock is fed using arable crops however researchers believed that algae could play a large part in providing a high protein food source for livestock and thus a method for feeding the global population [5]. At University California at Berkley, Professor Oswald began designing pond systems to cultivate freshwater algae on a large scale. The idea was to design a low impact system (i.e., low energy requirements and environmental impact) which provided conditions allowing high productivity of the cultivated algae. Oswald's work also focussed upon combining algal cultivation with wastewater treatment providing a co-benefit [4]. The algae could therefore provide a means of improving the water quality of raw or partially treated effluent as well as providing livestock feed.

The biomass produced from the cultivation process was not restricted to livestock feed and studies were performed assessing the amount of biogas the algal biomass was capable of providing [2]. Algae was deemed a potentially valuable substrate for biogas production and various strains have been tested for their suitability up to the present day [6-8]. Further investigations led to algal biomass being assessed for alternative fuel types. Due to the high oil content of many algae species [9-14] biodiesel was considered a valuable fuel which could be extracted and processed from algal biomass. The concept of producing biodiesel from microalgae was developed considerably by the US Department of Energy's Aquatic Species Program: Biodiesel from Algae [15]. The program ran from 1978 to 1996 and was focused upon producing biodiesel from microalgae fed with CO_2 from flue gases. The program was born out of a requirement for energy security as the US relied heavily upon gasoline for transport fuel, disruption to supplies could have significant repercussions to the economy. The program provided excellent contributions to the area of algal cultivation for biofuel but when funds were diverted to alternative fuel research the program was phased out in 1996 [15].

Recently the interest in biofuels from algae has dramatically increased as a result of increased fossil fuel prices and the need to find an alternative energy source due to the threat of climate change. Areas of studies include optimising biofuel yields, methods of reducing energy consumption, investigating alternative products and assessing environmental impacts.

2.2.2 ALGAE AND WASTEWATER TREATMENT

All autotrophic algal species require a source of nutrients. The most important nutrients, (i.e., those that are needed in greatest concentrations) are nitrogen and phosphorous, but many other nutrients and trace metals are also necessary for optimal growth [16]. There are many media recipes designed to provide complete nutrition for most species of algae. Nutrient rich effluents however are often capable of providing almost all of the nutrients required by certain algal species and [17, 18] consequent cultivation in effluent provides two significant benefits. Firstly, direct uptake of these nutrients and metals, produces cleaner water. Secondly, the algae generate oxygen which aids aerobic bacterial growth leading to additional metal and nutrient assimilation. In the 1950s experiments were carried out by Oswald and his colleagues investigating the symbiotic relationship between algae and bacteria for wastewater treatment in oxidation ditches [3, 19-22]. The experiments which were undertaken used the algae *Euglena sp.* due to their natural presence in ditches under examination. Oswald and his colleagues discovered that the bacteria and algae in the oxidation ditch develop a symbiotic relationship producing a more stable and less hazardous effluent [21]. The economic potential of the algal cells for livestock feed was identified, and the merits of using a faster growing species of algae in effluent specifically for the purpose of livestock feed were discussed [19]. Particular attention in the late 1950s was given to the design of wastewater treatment ponds and relationships between oxygen production, biological oxygen demand (BOD) removal and light use efficiency over specific periods of time as a function to a variety of species, depths of pond, treatment time, loading rates to identify optimal operational conditions [3]. The importance of algae in a heavily loaded oxidation pond to provide the necessary oxygen for sludge oxidation was highlighted, and this early research remains of particular interest as it identified optimal conditions for effluent treatment and demonstrated that oxidation ponds using the symbiotic relationship can achieve considerable BOD removal (>85%) [3].

Further important work was carried out on the use of algae in wastewater treatment in the 1970s and 1980s. Of particular interest was the research lead by Professor Shelef, with a focus on the growth of dominant species of algae in open ponds using raw sewage as the main source of nutrients [23]. His research indicated that *Micractinium* and *Chlorella* dominated in most cases, with retention times of around three to six days. Using an influent with concentrations of total suspended solids (TSS) ca. 340 mg/L and BOD ca. 310 mg/L, a considerable reduction to levels ca. 60 mg/L TSS and below 20mg/L BOD was reported. Additionally phosphate was reduced to very low levels and around 10–40 mg/L ammonia remained. This resulted in low levels of organic contamination which allowed the use of the treated wastewater for irrigation, and the residual nutrients provided a source of fertiliser as an added value.

In the 1980's the concept of growing algae was further developed however the focus turned to utilising the biomass for fuel production due to the energy crisis in the United States at the time. Oswald continued his research and was joined by Dr. Benemann and together they investigated the potential for the cultivation of algae on a large scale for fuel production. During this time most research moved away from wastewater treatment and more towards high productivity of biomass and high fuel yields with work conducted on the use of flue gas as a source of carbon dioxide to increase productivity and provide a method of carbon mitigation. Due to the continued search however for a more economically and environmentally viable system, once again research has turned to the potential benefits of growth in wastewater. Several different wastewater types have been investigated, most common are domestic sewage [24-27], agricultural wastewater (swine and cattle) [26, 28-31] and several industrial wastewaters, e.g., carpet manufacturing [32] and distillation [33]. Previous research suggests that cultivation in the tertiary treatment stages of wastewater treatment may provide the ideal conditions for reasonable algal growth due to high residual nutrient loading and prior removal of organic contaminants [25]. In such a scenario algae can provide an effective means of nutrient polishing. Various strains of algae have been shown to effectively remove nitrogen and phosphorous forms in a number of synthetic and actual wastewaters (Table 1).

The table displayed in Table 1 suggests that there is considerable potential for nutrient removal in wastewaters with high nutrient loading using a variety of species of common algae and mixtures of locally dominant algae. Much of the research reported has however been conducted at a lab scale in photo-bioreactors which provide controlled conditions (e.g., temperature, light and species control) and therefore will improve the productivity of the algae and thus the uptake rate of nutrients. There are however exceptions where polycultures of dominant algae have been grown both using an outdoor turf scrubber [34] and high rate algal ponds [35] as the culture methods. In the cases where biomass has been cultivated on turf scrubbers or in ponds the system has developed using species which come to dominate as species selection is not possible. Despite the lack of controllability of species dominance and environmental factors, these studies still demonstrate good N and P removal alongside reasonable productivity rates of biomass.

With many industries looking for cost effective means of nutrient removal from wastewaters, the cultivation of algal biomass can provide a suitable method which may return a positive balance of energy as opposed to conventional processes which generally require an energetic input.

2.2.3 THE POSSIBILITY OF CARBON MITIGATION

Cultivation of algal biomass could provide a method of carbon mitigation through CO_2 uptake from flue gases during photosynthesis. Providing algae can utilise industrial gases, there is the potential to remove CO_2, which would otherwise be emitted. The mitigation of CO_2 from flue gas using algae would be ideal in an industrial scenario, as targets for reducing greenhouse gas emissions are becoming tighter [44]. The improvement of biomass yields by introducing a concentrated source of CO_2 has been reported [45, 46], however, there are many barriers yet to overcome; for example concentration of CO_2 in flue gas may be too high for many strains of algae resulting in toxicity, and/or the presence of other toxins in the gas may adversely affect productivity, and/or gas transport cost to algal biomass growth reactors or ponds may be unviable. Nevertheless, as mentioned, there is certainly potential for flue gases to play a part in an algal biomass cultivation system.

The atmosphere provides a CO_2 concentration of 0.038% for the growth of algae; theoretically, with a higher concentration available, higher productivity is possible [47]. Early studies found *Chlorella sp.* to be highly suitable for cultivation in flue gases due to its capacity to be grown with the injection of gas containing a CO_2 concentration of 15% [48], a concentration similar to that of most flue gases [46]. Experimentation conducted within the US aquatic species programme [15] using flue gases as a source of CO_2 indicated that local strains of algae dominated with a high CO_2 use efficiency. Single algal biomass productivity rates as high as 50 g/m^2/day were recorded, although attempts to achieve consistently high productivity rates failed during a long-term experiment for one year, provably due to low ambient temperatures [15]. In 2002, research was conducted

Table 1. Nutrient removal efficiencies of algae in wastewaters.

Algae	Method	Nremoval (%)	Premoval (%)	Productivity	Refs
Synthetic wastewater					
S. obliquus	PBR[1]	70 (NO_3)	94 (PO_4)	-	[36]
C.vulgaris	PBR	50 (NH_4)	78 (PO_4)	-	[37]
Scenedesmus sp.	PBR	66 (NO_3)	-	39.3 mg/L/day	[38]
Municipal wastewater					
Polyculture	PBR	96 (NH_4)	99 (PO_4)	24.4 mg (Lipid)/L/day	[26]
Scenedesus	PBR	99 (NH_4)	99 (PO_4)	250 mg/L/day	[39]
Cyanobacteria	PBR	88.3 (TKN)	64.8 (PO_4)	10.9 g/m^2/day	[40]
Chlorella sp.	PBR	74.7-82.4 (NH_4)	83.2-90.6 (P)	0.343-0.948/day	[41]
Micractinium sp. Desmodesmus sp.	HRAP	65 (NH_4)	-	8 g VSS/m^2/day	[35]
Swine and dairy manure					
Polyculture (filamentous)	Turf scrubber	95 (N)	77 (P)	9.4 g/m^2/day	[42]
C. sorokiniana	PBR	94–100 (NH_4)	70–90 (PO_4)	-	[43]
Polyculture	Turf scrubber	51–83 (N)	62–91 (P)	8.3-25.1 g/m^2/day	[34]

[1]S.obliquus *immobilised in chitosan beads*

by the National Renewable Energy Laboratory (NREL) and the US Department of Agriculture investigating uptake of CO_2 from synthetic and flue gas sources and its commercial and environmental viability. The technical feasibility and economic viability of integrating a micro-algal cultivation system with a coal fired power plant was investigated [49], using a bench scale system as a test rig. An artificial flue gas (12% CO_2; 5.5% O_2; 423 ppm SO_2; 124 ppm NO_x) based on the composition of a North Dakota power station boiler was produced and sparged into a bio-reactor tank. Two strains of algae were cultivated, *Monoraphidium* and *Nannochloropsis*, both of which grew successfully under the administered conditions. It was reported that growth rates of the microalgae varied between 15 to 25 g/m^2/day and contained 41% protein, 26% lipid and 33% carbohydrate [49].

Research using real flue gases for CO_2 uptake and cultivation of algal biomass has also been conducted [45, 46, 50, 51]. For example, *Chlorella sp.* was cultivated using a photobioreactor system approach and the productivity of *Chlorella sp.* was investigated in presence of a flue gas (6–8% CO_2) from a natural gas boiler and in presence of a control gas, which resulted in higher productivity in the flue than in the control gas, of 22.8 ± 5.3 g/m^2/day [52]. As a result it was suggested that 50% of the flue gas could be decarbonised using that system [51]. Similar studies conducted with *Chlorella vulgaris* using a photobioreactor system approach and flue gas from a municipal waste incinerator indicated that this strain was tolerant to a concentration of 11% (v/v) CO_2 as well as to the flue gas, with a higher biomass productivity in the flue [46]. Both studies suggest that the presence of potential contaminants in the flue had little adverse impact upon the algae. Examples of various strains of algae cultivated with the addition of CO_2 with their productivity rates are summarized in Table 2. Existing research indicates that an improved growth of algal biomass has been obtained using artificial and flue gases with CO_2 concentration up to approximately 12% (Table 2). Above this concentration it appears that productivity is reduced, most likely due to acidity caused by the high CO_2 levels. It is suggested that, although most strains would benefit from an increased concentration of CO_2, testing is required to identify optimal CO_2 concentrations as this appears to vary between strains. Similarly only a limited number of flue gas sources have been investigated; if the concept is likely to be taken up across many different industries, a variety of flue gases will need to be tested.

In summary, research to date suggests mitigation of CO_2 using algae cultivation is promising providing the gases are at a low concentration and contain low levels of contamination and operational conditions (e.g., pH, temperature, light) are controlled. Additionally the cost and energy input of transporting the gases to the ponds or bioreactors must be balanced by the benefits that the extra CO_2 and the carbon mitigation provide. In a study investigating the potential for using power plant flue gas [54], it was estimated that an electricity consumption of 15.1 kWh/day was necessary for the direct injection of CO_2 to one hectare of cultivation pond. The energy benefits as a result of the injection would therefore need to be at least 15.1 kWh/d/ha and the CO_2 savings greater than what would be emitted by conducting injection.

TABLE 2: Examples of algal biomass cultivated in a source of CO2 and biomass productivity.

Algae Species	Gas	CO2 (%)	Productivity	Refs.
Chlorella sp.	Air	Air	0.68 /day	[53]
Chlorella sp.	Synthetic	2	1.45 /day	[53]
Chlorella sp.	Synthetic	5	0.90 /day	[53]
Chlorella sp.	Synthetic	10	0.11 /day	[53]
C. vulgaris	Flue gas (MSW incinerator)	10–13	2.50 g/L/day	[46]
Spirulina sp.	Synthetic	Air	0.14 g/L/day	[50]
Spirulina sp.	Synthetic	6	0.22 g/L/day	[50]
Spirulina sp.	Synthetic	12	0.17 g/L/day	[50]
S. Obliquus	Synthetic	Air	0.04 g/L/day	[50]
S. Obliquus	Synthetic	6	0.10 g/L/day	[50]
S. Obliquus	Synthetic	12	0.14 g/L/day	[50]

2.2.4 COMPARISON OF OPEN PONDS AND PHOTO-BIOREACTORS

The two main methods of infrastructure considered suitable for cultivation of algae are open (raceway) ponds or photo-bioreactors (PBRs) [55], and are compared in Table 3. Raceway ponds are similar to oxidation ditches used in wastewater treatment systems being large, open basins of shallow depth and a length at least several times greater than that of the width. Raceway ponds are typically constructed using a concrete shell lined with polyvinyl chloride (PVC) with dimensions ranging from 10 to 100 m in length and 1 to 10 m in width with a depth of 10 to 50 cm [55]. Oswald considered the open pond to be the most viable method of combining algal cultivation and wastewater treatment in the 1950s [56].

Photobioreactors are more commonly used for growing algae for high value commodities or for experimental work at a small scale. Recently, however, they have been considered for producing algal biomass on a large scale as they are capable of providing optimal conditions for the growth of the algae [55]. A closed reactor allows species to be protected from bacterial contamination, shallow tubing allows efficient light utilisation, bubbling CO_2 provides high efficiency carbon uptake and water loss is minimised. PBRs provide very high productivity rates compared to raceway ponds. In their life-cycle assessment (LCA) study, Jorquera et al. [55] estimated volumetric productivity to be at least eight times higher in flat-plate and tubular PBRs. The reason why PBRs however have not become widespread is due to the energy and cost intensity of production and operation. PBRs require a far higher surface area for the volume of algal broth compared to alternative infrastructure. Much higher volumes of material are therefore required which in turn requires a higher capital energy input and increases environmental impacts [57]. During operation algal biomass must be kept in motion to provide adequate mixing and light utilisation. These increase productivity but also require additional energy for pumping. So far in comparison to raceway ponds the benefits of PBRs do not outweigh the necessary energy requirements identified in the LCA study published by Jorquera et al. [55]. A net energy ratio (i.e., energy produced/energy consumed) of 8.34 has been reported for raceway ponds as compared to a net energy ratio of 4.51 and 0.20 for flat-plate and tubular photobioreactors, respectively [55]. It is likely that ponds will continue to provide the most effective infrastructure for algal cultivation due to their low impact design and low energy input requirement. PBRs will continue to be important however, for laboratory work, developing cultures and producing biomass with high economic value. As research continues it may also be possible to develop infrastructure that will provide the benefits of both PBRs and open ponds together.

TABLE 3: Comparison of raceway ponds and photo-bioreactors.

	Raceway Pond	Photobioreactor	Refs
Estimated productivity (g/m^2/day)	11	27	[55]
Advantages	Low energy	High productivity	[55]
	Simple technology	High controllability	
	Inexpensive	Small area required	
	Well researched	Concentrated biomass	
Disadvantages	Low productivity	High energy	
	Contamination	Expensive	
	Large area required	Less researches	
	High water use		
	Dilute biomass		

2.2.5 BIOMASS PROCESSING

2.2.5.1 HARVESTING

As the biomass cannot be utilised efficiently at low concentrations in media, the first step in the biomass processing stage is harvesting the algae for subsequent processing. The method of harvesting used depends very much upon the type of algae which is under cultivation. Microalgae require more intensive harvesting methods in comparison to macroalgae, because of their cell size. Depending upon circumstances, often a series of harvesting methods is required to produce a final biomass below a desired moisture content. Common methods of harvesting of algae are: microfiltration, flocculation, sedimentation, flotation and centrifugation [58].

One of the most effective methods of harvesting is filtration using micro-filters. This method of filtration generally uses a rotary drum covered with a filter to capture the biomass as the influent passes through from the centre outwards [59]. Initial harvesting tests in the 1960s tested micro-filters but found that the majority of algal cells simply passed through most of the filter types [60]. It was later suggested that micro-filtration was suitable for strains of algae with a cell size greater than around 70 μm and was not suitable for those species with cell sizes lower than 30 μm [61]. The size of the opening in the filter mesh dictates what percentage of biomass is captured likewise with the size of the biomass cells. The pore size also affects how much pressure is required to facilitate the flow of water through the filter which will in turn affect the energy consumption [62]. The concentration of the algae in suspension also influences the efficiency of removal as highly concentrated biomass will foul the filter very quickly causing reduced performance and a requirement for backwashing and thus further energy consumption. If filtration is to be used it is essential that the method suits the species of algae which is being harvested, otherwise the filtration will be ineffective and provide low yields of biomass. If the cultivated algal species allows for filtration (e.g., *Spirulina, Spirogyra, Coelastrum*), the filtration method can prove very efficient and cost effective method of harvesting. Mohn [63] for example found that gravity filtration using a microstainer and vibrating screen both provided good initial harvesting of *Coelastrum* up to a total suspended solid of 6% with low energy consumption (0.4 kWh/m^3). Mohn [63] also investigated pressure filtration of *Coelastrum* which provided even higher total solids of concentrate up to 27%, although requiring more than twice the energy. Clearly inexpensive and low energy harvesting of biomass is possible with filtration, providing the dominant algae being harvested is of a suitable cell size and optimal concentration level.

Sedimentation and flotation have also been proven as viable options for harvesting algal biomass with no requirement for specific cell size. Both sedimentation and flotation rely on biomass density to facilitate the process, both processes are aided by flocculation and flotation is aided additionally with bubbling. As a method of biomass removal, sedimentation was considered a viable process in the 1960s due to its prominence in wastewater treatment and its low energy requirement [60]. Due to the low specific gravity of algae, the settlement process is, however, slow but, under certain conditions, the self-flocculation of some strains of algae is possible. Nutrient and carbon limitation and pH adjustment appear to be methods of auto-flocculation of algae which may provide a low-cost solution to the initial harvesting process [64]. Recent studies have focussed upon bio-flocculation which occurs as a result of using several bacteria or algal strains to flocculate with the desired algal biomass to allow settlement. Gutzeit et al. [65] found that gravity sedimentation was possible using bacterial-algae flocs developed in wastewater for the removal of nutrients, and reported that the flocs of *Chlorella vulgaris* were stable and settled quickly. Other approaches investigated the combined use of autoflocculating microalgae (*A. falcatus, Scenedesmus obliquus* and *T. suecica*) to allow for flocculation of non-flocculating oil-accumulating algae (*Chlorella vulgaris* and *Neochloris oleoabundans*) [66], which resulted in a faster sedimentation as well as a higher percentage of biomass harvested. This method of harvesting appears viable due to its low energetic inputs but also because it does not rely on chemicals, thus allowing the water to be discharged or recycled without further treatment. However, it should be noted that this method of flocculation may not be suitable for all types of algae, and thus further research is required in this area.

Conventional methods of flocculation using flocculants common to wastewater treatment such as alum, ferric chloride, ferric sulphide, chitosan among other commercial products are likely to provide a more consistent and effective solution to flocculation. Much research has been conducted upon the removal of algae using flocculants with varying degrees of success (Table 4). For example, a complete removal of freshwater microalgae, Chlorella and Scenedesmus, using 10 mg/L of polyelectrolytes while 95% removal using 3 mg/L of polyelectrolites has been reported [60]. A comparative study where alum and ferric chloride were use as flocculants for three species of algal biomass (*Chlorella vulgaris, I. galbana* and *C. stigmatophora*) indicated the low dosages of alum (25 mg/L) and ferric chloride (11 mg/L) were sufficient for optimal removal of *Chlorella vulgaris*, while higher dosages of alum and ferric chloride were required for the removal of marine cultures *I. galbana* (225 mg/L alum; 120 mg/L ferric chloride) and *C. stigmatophora* (140 mg/L alum; 55 mg/L ferric chloride) [67]. Additionally it has been reported that the combined use of chitosan at low concentrations (2.5 mg/L) and ferric chloride provided much quicker flocculation of the algal cells, *Chlorella vulgaris, I. galbana* and *C. stigmatophora,* and reduced the requirement of ferric chloride [43]. The use of chitosan as a flocculant for the removal of freshwater algae (*Spirulina, Oscillatoria* and *Chlorella*) and brackish algae (*Synechocystis*) has been investigated [43], and chitosan has been found to be a very effective flocculant, at maximum concentrations of 15 mg/L removing about 90% of algal biomass at pH 7.0. The use of conventional and polymeric flocculants for the removal of algal biomass in piggery wastewater has been recently investigated [43]: ferric chloride and ferric sulphate were found to be effective flocculants at high doses (150–250 mg/L) providing removal rates greater than 90%; polymeric flocculants required less dosing (5–50 mg/L), although provided lower biomass recoveries; chitosan performed poorly at both low and high dosages for each of the algal species types with a maximum removal of 58% at a dose of 25 mg/L for a consortium of *Chlorella.*

2.2.5.2 SEDIMENTATION

Sedimentation of algal biomass is a further method of biomass removal but generally requires prior flocculation for high removal efficiencies. Sedimentation can be carried out with some species without flocculation, but removal efficiency is generally considered poor. Flocculation can be used to increase cell dimensions allowing improved sedimentation. If carried out in conjunction with flocculation, a sedimentation tank can provide a reliable solution for biomass recovery [69].

TABLE 4: Maximum removal rates of various flocculants for the removal of algal biomass.

Flocculant	Algae	Removal (%)	Dosage (mg/L)	Media type	Refs.
$FeCl_3$	*Chlorella*	98	250	Piggery wastewater	[43]
	S. obliquus	95	100		
	Chlorococcum sp.	90	150		
$Fe_2(SO_4)_3$	*Chlorella*	90	250		
	S. obliquus	98	150		
	C. sorokiniana	98	250		
Chitosan	*Spirulina, Oscillatoria, Chlorella*	>90	15	Nutrient media	[68]
Polyelectrolyte					
(Puriflocs 601 & 602)	*Chlorella, Scenedesmus*	95	3	Sewage	[60]

2.2.5.3 FLOTATION

Flotation was a method of harvesting considered in the 1960s [60][70] however the recoverability of biomass was generally found to be poor with a wide range of reagents tested. It has been reported that using dissolved air flotation mixed algal species could be harvested up to a slurry of 6% total solids; using electro-flotation, which creates air bubbles through electrolysis which then attach to the algal cells, mixed algal species could be harvested up to a slurry of 5%, but this approached required a significant energy input; using dispersed air flotation which uses froth or foam to capture the algal cells resulted also in similar results [69]. Existing research indicates that flotation offers a quicker alternative to sedimentation following algal flocculation, but more energy is required and thus cost is higher whilst providing a final product with lower total solids content.

2.2.5.4 CENTRIFUGATION

Probably the most effective method of biomass removal, with very high recovery rates, is centrifugation. As with the other alternative methods, centrifugation was considered a feasible option in early algal biomass dewatering work in the 1960s. Golueke and Oswald [4] investigated various means of dewatering algae further to provide a biomass with a sufficiently low moisture content. One of the methods they looked at was centrifugation and three of the four centrifuges that they tested proved to be extremely effective producing a maximum removal of 79% and a biomass with solids content of 11.5% and maximum of 18.2%. Further research was conducted by Mohn [63] in the area of harvesting algal biomass using centrifugation and he focussed on suitability of algal strains, cost and energy use. In accordance with Golueke and Oswald, Mohn found centrifuges to be very effective for the removal of Scenedesmus and Coelastrum, particularly the Westfalia self-cleaning plate separator and the Westfalia nozzle centrifuge [63]. The centrifuges provided biomass with total solids content of 2–22% with a minimum energy consumption of 0.9 kWh per m^3. Table 5 provides an overview of Mohn's findings indicating the possible harvesting methods, effectiveness, energy requirements and reliability of several harvesting methods. Mohn's results suggest filtration provides the best harvesting strategy in terms of high concentration of solids with low energy requirements [63].

TABLE 5: Harvesting methods, effectiveness and energy requirements.

Harvesting Method	Algae species	% TSS of Concentrate	Concentration Factor	Energy (kWh)	Reliability	Refs.
Gravity filtration	Coelastrum	6	60	0.4	Good	[63]
Pressure filtration	Coelastrum	22–27	245	0.88	Very high	[63]
Centrifuge (Westfalia self-cleaning)	Scenedesmus, Coelastrum proboscideum	12	120	1	Very good	[63]
Centrifuge (Westfalia screw)	Scenedesmus, Coelastrum proboscideum	22	11	8	Very good	[63]

Despite centrifugation being an effective method of concentrating biomass, the energy requirements are much higher than that of filtration. However clearly the choice of harvesting depends heavily upon the biomass type, if the cell size is large enough, then filtration is likely to be the most effective and economically viable option. Otherwise it is likely that a process stream involving flocculation, sedimentation, flotation or centrifugation is necessary. There is little parallel between the effectiveness of common flocculants for harvesting algae in research conducted. It can be observed that there are many effective flocculants for algae removal however suggested optimal dosages vary significantly between studies. Ferric chloride can be considered a viable option potentially combined with chitosan to improve yield and reduce time and material input. Further research is necessary for individual scenarios to choose the most effective method of flocculation and consequent harvesting.

2.2.6 FUELS

Following harvesting, the algal biomass requires conversion to fuel through a variety of techniques to extract and process the sought-after products within the cells. The necessary processing depends upon the desired fuel [71]. The three fuel types that will be covered here are those that are currently considered the most suitable for energy recovery from algae: being biodiesel, bioethanol and biogas. Each of these biofuel types require a different process stream and diagrams are presented in Figures 1–3.

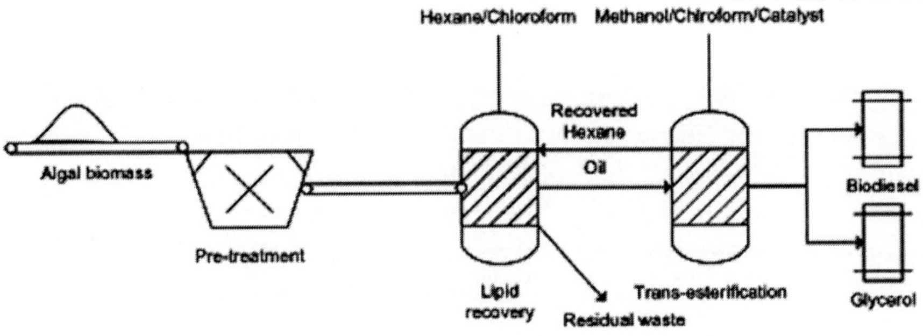

FIGURE 1: Simplified process diagram for biodiesel production.

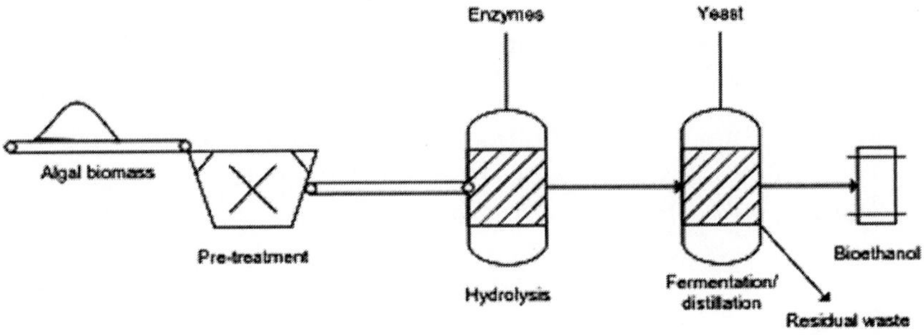

FIGURE 2: Simplified process diagram of bioethanol production from algal biomass.

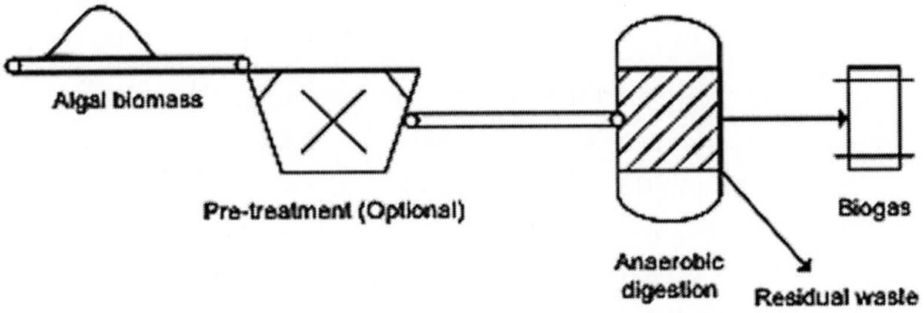

FIGURE 3: Simplified process diagram of biogas production from algal biomass.

2.2.6.1 BIODIESEL

Biodiesel is the most common fuel type researched as a method of recovering energy from algae due to the high oil content of many algae strains [11, 13, 14]. The production of biodiesel initially requires the extraction of the lipid content of the algal cells. Most researchers follow a standard protocol written by Bligh and Dyer in 1959 [72] which uses chloroform and methanol as the extraction technique. Prior to lipid extraction the cells must be disrupted to allow access to the oils within the cell. Disruption can be achieved by homogenisation, bead beating, mechanical pressing, microwave treatment, acid/alkali treatment, sonication, lyophilisation and autoclaving among others.

Lee et al. [73] produced a study investigating the various methods of cell disruption and corresponding lipid extraction efficiencies. They found that for each algal strain (*Botryococcus spp. Chlorella vulgaris* and *Scenedesmus spp.*) microwave treatment provided the highest lipid yield. In terms of productive strains, *Botryococcus spp.* provided the highest yield using microwave treatment at 28.6% lipid recovery from the biomass. Bead-beating however, almost matched this value. Each of the disruption methods (autoclaving, bead-beating, microwaving, sonication and osmotic shock) produced lipid yields higher than a no-disruption technique.

The next step of the process is the lipid extraction and most studies extract the lipid content of the biomass using a modified version of Bligh and Dyer's method [72]. This requires the addition of methanol and chloroform, typically in proportions of approximately 1:1 methanol to chloroform mixed with the sample also at a ratio of about 1:1 methanol/chloroform mixture to sample [73]. Once the reaction is complete the oil can be separated using a centrifuge or funnelling method as the densities of the materials differ. Methanol, chloroform and a catalyst (acid or base) are then mixed with oil to allow trans-esterification to occur. The two products from the reaction are methyl esters (biodiesel) and glycerol. The produces are biphasic and thus can be easily separated.

Research in the area is now looking at the possibility of improving extraction of oils from wet biomass which eliminates the energy consumption required for drying of the biomass. It is generally considered that removal of oil from dry biomass is most efficient and practical [74]. Johnson and Wen [75] investigated the use of both wet and freeze-dried algal biomass (*S. limacinum*) for the production of biodiesel. The researchers found that wet biomass produced 20% less fatty acid methyl-esters than the dried biomass, lowering the biodiesel value. Further research has been conducted by Patil et al. [74] who conducted experiments producing fatty acid methyl-esters from wet biomass via a supercritical methanol method. The process required only one step for extraction and trans-esterification with addition of methanol at ratio of 1:9, biomass to methanol respectively, a temperature of 255 °C and reaction time of 25 min. The results showed a Fatty Acid Methyl Ester (FAME) recovery of around 88% from *Nannochloropsis* biomass. The research suggests that high recovery is possible without the energy intensive process of drying and separate lipid extraction. Similarly positive results of direct extraction from wet biomass were produced from Wahlen et al. [76] who experimented with direct biodiesel production from various freshwater green algae strains, cyanobacteria and mixed wild algae. More research is required to assess the potential of recovering biodiesel from wet algae in a single stage process yet the concept appears promising. Energy costs of the process may be higher but this could well be outweighed by the reduced energy cost from drying of the biomass as was calculated by Lardon et al. [77] in their LCA of biodiesel from microalgae. This LCA study compared methods of cultivating and processing algal biomass for maximum energy recovery, they investigated the energy consumption associated with producing 1 kg of biodiesel. In their study they found that drying required 81.8 MJ of heat and 8.52 MJ of electricity per kg of biodiesel with no heating requirement for wet biomass. Oil extraction required higher energy consumption for wet biomass than dry but the final energy balance for wet biomass was a high positive value (105 MJ/kg biodiesel) compared to the negative balance for dry biomass (−2.6 MJ/kg biodiesel).

2.2.6.2 BIOETHANOL

An alternative or addition to the production of biodiesel is the production of bio-ethanol from the carbohydrates and starches in the algal cells. Depending upon the strain and composition of the algal species significant yields of ethanol can be produced from algal biomass [78-81]. Strains with filamentous cells such as *Spirulina* and *Spirogyra* are considered most promising due to the higher percentage of carbohydrate in their make-up. The conventional process of producing bioethanol using hydrolysis and fermentation is well understood for many feedstocks but optimal conversion has not yet been achieved for algal biomass. Similarly to lipid extraction, the first stage in the process is the disruption of the biomass cells which can be carried out using numerous techniques including bead-beating, autoclaving, microwaving and acid or alkali treatment. Once the cells have been disrupted the carbohydrates and starches can be converted into sugars using enzymatic or acid hydrolysis. Following hydrolysis the sugars are then be fermented with yeast (typically *S. cerevisiae* or *S. bayanus*) which will provide a broth of up to 17% (v/v) ethanol depending upon the concentration of sugars (AB Mauri, personal correspondence). The next step to produce bioethanol is to distil the broth to produce an ethanol concentration of around 98% (v/v) then further refinement of the ethanol produces a fuel which can be used as an additive to conventional engines or up to a maximum of 85% in specialised E85 engines [82].

As the concept of converting algal biomass into bioethanol is relatively under-researched most studies have simply focussed upon investigating what ethanol recoveries are possible. In an early study by Hirano et al. [79] a variety of freshwater and marine algae was selected for testing. *Chlorella vulgaris* was found to contain a high proportion of starch (37%) and a recovery of 65% of ethanol from the starch was obtained using enzymatic hydrolysis followed by fermentation with *S. cerevisiae*. An overall recovery of 24% from the biomass was therefore obtained. Using the strain *Chlorococum spp.*, a conversion efficiency of about 38% of the ethanol was obtained [78], which can be considered promising however this was an optimal value and no consideration was given to the energy requirement of processing. What is interesting from this research is that when the lipid content of the biomass was recovered prior to fermentation, ethanol yields were far higher [78]. This suggests that biomass could provide both diesel and ethanol, maximising potential recoveries. Nguyen et al. [83] found in several studies that yields of up to 29% ethanol recovery efficiency were possible using *Chlamydomonas reinhardtii*. The studies mentioned above prove that high ethanol yields from algal biomass are possible but further studies are necessary to assess the viability in terms of energy balance, economics and environmental impacts.

Alternative methods of ethanol production have been investigated which focus upon intracellular ethanol production in which algae produce ethanol under dark, anaerobic conditions. The species which are capable of the process are cyano-bacteria and include the species: C*hlamydomonas reinhardtii, Oscillatoria limosa, Microcystis, Cyanothece, Cicrocystis aeruginosa* and *Oscillatoria spp.* [84]. The process requires the algae to be cultivated in a closed environment with the addition of CO_2 under which conditions, it is believed that concentrations of between 0.5 and 5% ethanol can be produced. Hirano et al. [79] investigated this phenomenon using *Chlamydomonas reinhardtii* and Sak-1 isolated from salt water, and a maximum yield of 1% , w/w produced by *C. reinhardtii* was reported. The ethanol-water mix can then be extracted and treated further to produce highly concentrated ethanol for fuel use. The benefits of the process are that no other organisms (e.g., enzymes and yeast) are required for hydrolysis/fermentation and the algae remains unaffected and can continue to grow without a requirement for harvesting. The energy requirements are likely to be lower than those necessary for conventional fermentation of biomass however the two methods need to be directly compared. In their study Luo et al. [84] show that the whole process provides a positive energy balance with the greatest surplus of energy when the maximum ethanol concentration is produced. Additionally the greenhouse gas emissions compare well to emissions via gasoline production but to reach 20% of the emissions from gasoline (a government aim) would require further reductions in the process chain.

Bioethanol production from algal biomass is still very much in its infancy, the concept is proven but the viability is not. Further life-cycle analyses are required to understand the potential of the concept. Post lipid processing and intracellular ethanol production look promising as energy consumption is minimised, further research will establish viability.

2.2.6.3 BIOGAS

A simpler method of energy recovery may be facilitated by anaerobic digestion of algal biomass providing a promising source of bio-energy in the form of biogas. The process was considered a potential source of useful energy recovery from algal cultivation near the start of modern research [2]. Anaerobic digestion is a process that has been used for hundreds of years to provide a source of energy from low value organic matter with minor energetic inputs. In the case of algal biomass, all the carbohydrates, proteins and fats can be converted into methane and carbon dioxide, although some components provide greater methane yields than others. It follows therefore that there is slightly less necessity to cultivate particular strains of algae for increased yields.

Table 6, taken from a study by Sialve et al. [85], displays the methane potential of each biomass component. Research has been conducted investigating the potential of various strains of algal biomass and Sialve et al. [85] used the methane potential to calculate yields for a number of strains. Their results can be viewed in Table 7 which compares theoretical results with experimental results from literature. Table 7 suggests that the values of methane yield can vary between species due to compositional make-up and that the yield depends very much upon the growth conditions as this can have a great impact upon the composition of the biomass. Comparing the actual yields with the theoretical yields shows a realistic conversion efficiency loss of about 50% in the majority of cases.

It is important therefore that in further studies investigating potential yields, exaggerated or over-optimistic yields are not used as these may not reflect real performance.

TABLE 6: Methane potential from biomass substrate [85].

Substrate	L CH_4 /g VS
Proteins	0.851
Lipids	1.014
Carbohydrates	0.415

As opposed to direct conversion, anaerobic digestion can alternatively be used to recover energy from the waste biomass following extraction of the more valuable components from the biomass cells. In their life-cycle assessment, Lardon et al. [77] calculated that the only feasible way of producing a positive energy balance of algal biodiesel was to recover further energy using anaerobic digestion of the residual waste. In fact, in normal culture conditions, they found that the energy produced from anaerobic digestion would be greater than that from extracted biodiesel. In their investigation of biogas from algae, Sialve et al. [85] suggest that at lipid contents below 40% it is unlikely to be worth recovering the lipids using current methods and the biomass should simply be digested to recover the maximum energy yield. In their LCA study of algae digestion Collet et al. [86] found the environmental impacts of biogas from algae to be poor in comparison to algal biodiesel using results from the study conducted previously by Lardon et al. [77]. The study compared the results for 1MJ of energy produced in a combustion engine. The difference in impacts was mainly due to electricity consumption and assuming anaerobic digestion is also applied following the biodiesel extraction in the biodiesel scenario. The figures used in the mentioned study provided high values of energy consumption which contrast with those used in other studies [87], the impacts may therefore not be as adverse as suggested. Collet et al. [86] concluded that the impacts can be improved with reduced energy consumption and a combined process of lipid extraction and anaerobic digestion may provide the optimal solution.

The biogas produced through anaerobic digestion differs from biodiesel and bio-ethanol in that it is not a fuel that can be used directly for combustion in vehicle engines. There are two options for biogas, one is combustion within a co-generator to produce electricity with possible heat recovery. The alternative is to refine the biogas removing the CO_2 and the methane can then be used as a fuel within a gas engine [90]. Further energy is required to upgrade the gas to a useable transport fuel and this is often ignored in studies with the energetic content of the gas is only considered. Further research is necessary to investigate the impact of downstream processing if comparison to the alternative biofuel types as a transport fuel is desired.

Anaerobic digestion is one of the methods of recovering energy that seems to provide a positive net energy balance due to the low inputs required [77]. The results may however be optimistic as real yields are much

TABLE 7: Theoretical and actual methane from different algal species.

Algae Species	Proteins (%)	Lipids (%)	Carbohydrates(%)	CH_4 (L/g VS) (Theoretical) [85]	CH4 (L/g VS) (Experimental)	**Refs.**
Euglena gracilis	39–61	14–20	14–18	0.52–0.8	-	[85]
Chlamydomonas reinhardtii	48	21	17	0.69	0.59	[88]
Chlorella pyrenoidosa	57	2	26	0.8	0.17–0.32 (*Chlorella-Scenedesmus*)	[2]
Chlorella vulgaris	51–58	14–22	12–17	0.63–0.79	0.24	[89]
Dunaliella salina	57	6	32	0.68–0.74	0.44–0.45 (*Dunaliella*)	[85]
Spirulina maxima	60–71	6–7	13–16	0.63–0.74	0.31–0.32 (*Spirulina*)	[85]
Spirulina platensis	46–63	4–9	8–14	0.47–0.69	0.31–0.32 (*Spirulina*)	[85]
Scenedesmus obliquus	50–56	12–14	10–17	0.59–0.69	0.17–0.32 (*Chlorella-Scenedesmus*)	[2]

lower than theoretical calculated yields. Additionally the biogas may require further processing to be useful as a fuel and this will affect the energy consumption and environmental impacts. Nevertheless the process is capable of recovering energy from all strains of algae regardless of the composition and therefore can be very useful as part of a flexible approach.

2.3 LIMITATIONS

Despite having been researched for over 50 years now, there are still only a few companies that are growing algae for fuel on a large or commercial scale. The economics of producing algae for fuel do not currently justify the intensity of the numerous processing stages and current practicalities. Cultivating algae with high productivity year round is a challenging task unless grown in controlled conditions, however, this itself, is not a viable solution. Attempts have been made to cultivate pure strains of algae in environmental conditions but with little success. In most cases local strains of algae come to dominate, out-competing the selected strain. This section reviews the limitations currently facing biofuel production from algal feedstock.

2.3.1 OPEN POND CULTIVATION AND SPECIES CONTROL

As mentioned above, open pond cultivation is currently considered the most viable option for large scale cultivation of algal biomass for energy. According to LCA research conducted by Jorquera et al. [55], open ponds provide a much higher net energy ratio in comparison to PBRs, both tubular and flat plated. Jorquera et al. [55] investigated the amount of energy consumed and produced using the three different cultivation methods for a yield of 100,000 kg of biomass annually. Productivities and energy yields in PBRs were found about three times higher than ponds, due to differences in efficiency. The energy consumption of PBRs to produce the equivalent amount of algal biomass compared to cultivation in ponds was however considerably higher (ca. ten times for tubular PBRs). The high energy consumption of the PBRs is mainly due to the air pumping, water pumping and the caloric content of the equipment used. On the whole, it can be assumed that despite low productivity requirements, open ponds provide a higher biomass yield for the energy consumed. One limitation of this reported research [55] is that it assumed that the cultivation of algae and the estimated productivity in the pond is possible all year round.

It has been previously reported that, under environmental conditions, wild strains of algae are likely to dominate and the strain of algae will also change depending upon the season [91]. Before discussing controlled conditions further, it is important to stress that certain strains of algae may be controlled by their requirement for extreme conditions. *Spirulina* and *Dunaliella,* for example, require a high pH and raised salinity to survive, most invasive species would be intolerant of these conditions [91]. The majority of species, however, require less extreme conditions and competition by native algae (and possibly other living microorganisms) remains a problem. During the summer seasons the dominant algae will be those that thrive in higher temperatures and conversely in winter those species that survive colder weather will dominate. Tseng et al. [92] found that at temperatures between 17 to 22 °C, *Chlorella vulgaris* dominated whereas at higher temperatures from 22 to 27 and even up to 32 °C, *Scenedesmus ellipsoideus, S. dimorphus* and *Wastella botryoides* were dominant. Alternative observations were, however, made by Professor Shelef finding that in Israel, *Chlorella* and *Micratinium* were dominant in the summer whereas *Euglena* and *Scenedesmus* were the common species in winter [23]. Clearly the location of cultivation has a large impact upon which species will dominate in each season. What is most important, however, is that when growing algae in environmental conditions, selectivity of strain is not currently possible.

Recently research has investigated methods of species control in cultivation ponds. For example, two high-rate algae ponds have been compared, one which included recycling of algae and one which did not [91]. The species under investigation was *Pediastrum spp.* and algal biomass was collected every day and settled in algal settling cones. One litre of biomass was returned to one of the ponds and not the other. The pond with recycling provided over 90% dominance of *Pediastrum sp.* in one year whilst the non-recycled pond provided only 53% dominance. It is suggested that this method of recycling may provide a useful method of species control in open pond cultivation, however it may not be successful for every strain required for cultivation. Further research is required to assess whether recycling will be beneficial for any considered strains.

2.3.2 WATER RESOURCE SCARCITY

With water becoming a scarce commodity, intensive use of water for bio-energy cannot be considered sustainable if water extraction is affecting agriculture, domestic use or causing environmental impacts. Being an aquatic species, algae require more water than terrestrial bio-energy plants and when cultivated in open ponds there are great water losses mainly through evaporation. According to Williams and Laurens [93] the dissociation of one mol of water occurs for every mol of CO_2 required in the photosynthetic process. In their study of water use in algal cultivation Murphy and Allen [94] calculate that 33.2 m^3/m^2 of water per year is required to cultivate algae in a raceway pond in the United States. It is possible to recycle much of the water that is drained from the ponds during harvesting but there will be losses in harvesting the algae; freshwater must therefore be sourced. In the same study it was reported that the management of the water will require seven times the amount of energy that can be produced from biodiesel extracted from the algae [94]. The majority of countries around the world are becoming increasingly water stressed and therefore using extra freshwater in biofuel production is not sustainable. The use of wastewater as an alternative to freshwater provides an ideal solution however freshwater would still be necessary for downstream processing of the biomass or for dilution of highly concentrated wastewater.

2.3.3 ENERGY CONSUMPTION

Energy consumption in the production process is deemed the largest obstacle to algal biofuel production and a positive energy balance is a necessity but difficult to achieve. There are now several LCA studies which have investigated the amount of energy consumed in each of the necessary processes and comparing this to the energy recovery potential. Some studies have suggested that a positive energy balance is possible others suggest the contrary. Lardon et al. [77] found that if algae was cultivated purely for biodiesel, a positive balance would be unattainable. However if the residual biomass were to be anaerobically digested, a positive balance could be achieved in the scenario of growing algae in low nitrogen media and processing wet biomass [86].

The majority of other life-cycle analyses conducted recently suggest that algal biofuel can be produced with a positive energy balance though possibly not as positive as some alternative biofuels. Clarens et al. [87] modelled the growth of algae in raceway ponds and compared the energy consumption and environmental impacts of the fuel produced to fuel from corn, canola and switchgrass. In terms of energy consumption it was found that algal biodiesel required a far higher input, at least four times as much) as the next highest, and a sensitivity analysis revealed that the energy consumption was mainly a result of fertiliser use and carbon dioxide production [88]. Sander and Murthy [95] conducted a LCA comparing the difference between harvesting methods of filter pressing and centrifugation, and a positive energy balance for both methods was reported, with a higher net energy yield for the filter press (almost double that of the centrifuge). The mentioned study did not provide details of the modelled strain nor likely productivity rates; it assumed that the algae contained 30% lipids, which would be difficult to achieve for an outdoor cultured strain; and did consider year round production, which would also be a challenge. In a study by Stephenson et al. [96] air-lift tubular photobioreactors and raceway ponds were compared in terms of energy consumption and yield, and their results suggested that the majority of energy was consumed in the cultivation stage (i.e., the cultivation stage in bio-reactors required approximately 10 times more energy than raceway ponds), which is in contrast to previous studies [77][87]. This study has similarities with that of Jorquera et al. [55] who showed raceway ponds provided a far greater energy balance than bioreactors. A high energy consumption in the bio-reactors was attributed to the manufacture of the PVC material and circulation of the culture [55]. The majority of energy consumed from cultivation in raceway ponds was due to circulation using a paddlewheel. It was suggested that for raceway cultivation the anaerobic digestion of the residual biomass could offset the energy required from cultivation, however this was not the case for the tubular photo-bioreactors [55].

In a further study conducted by Clarens et al. [97], different process chains and how these affect the energy balance or Energy Return on Investment (EROI) were compared, particularly the study looked at the various end-products from the algae (i.e., anaerobic digestion (AD) to electricity, biodiesel and AD to electricity, biodiesel and combustion to electricity and direct combustion) as well as source options for CO_2 (i.e., virgin CO_2, carbon capture, flue gas) and nutrients (wastewater supplementation). In each case direct combustion of the biomass to electricity produced the highest EROI and the best option was direct combustion of the biomass with direct compression of flue gas providing a source of CO_2 [97]. The EROI for this scenario was 4.10, a similar scenario using flue gas and wastewater supplementation provided an EROI of 4.09. According to Clarens et al. [97] a value greater than 3 is considered sustainable, comparing well to canola (2.73), but not to switchgrass (15.90). Table 8 provides a summary of the best energy balances produced through the main studies conducted investigating energy recovery from algae.

Current production of biodiesel from algae without some other form of energy recovery will usually give a negative energy balance. To overcome this, it is necessary to include another form of energy recovery such as anaerobic digestion or combustion. It may even be far more beneficial to ignore biodiesel and to recover energy directly from anaerobic digestion or combustion as their input requirements are significantly lower. Energy reduction measures in each process will further improve the viability of biofuel from algae whatever the process stream used. Further work is required to find the optimal recovery method that can compete with the energy balance of conventional biofuels.

2.3.4 FERTILISERS

To achieve the highest productivity, it is necessary to add a source of nutrients to produce an effective medium. The fertiliser requirements can be calculated using the stoichiometric requirements of the algae. In their LCA, Clarens et al. [87] used to triangular distributions to calculate minimum, maximum and most likely dosing rates for nitrogen and phosphorous. These dosing rates were found to be just under two times the stoichiometric requirement providing a surplus but with the excess allowing other reactions to remove the surplus. The fertilisers used were assumed to have been sourced from urea and superphosphate. Stephenson et al. [96] estimated a nitrogen requirements of 59 kg per ton of biodiesel produced which contrasts significantly with the 6 kg per ton estimated by Lardon et al. [77]. Collet et al. [86] in their LCA study assumed a nitrogen dosing of 221 kg per day which equates to 5.74 kg per ton of biomass. The study assumed the biomass if processed to biogas so to compare the studies it is possible to calculate the nitrogen requirement for the energy produced. In which case the requirements in the study by Stephenson et al. [96] is 1.56 kg N/GJ, for the study by Lardon et al. [77] it is 0.16 kg N/GJ and for the study by Collet et al. [86] the value is 0.65 kg/GJ (assuming the energy content of biodiesel and biogas is 37.8 MJ/kg and 0.036 MJ/L respectively). Clearly each study makes different assumption and thus the fertiliser requirement estimates vary, it's most likely that higher dosing is required to provide an abundance of nutrients thus avoiding nutrient limitation.

The major drawback to the use of fertilisers is their energy input, cost and environmental impact. Lardon et al. [77] found fertilisers to be one of the major contributors to energy consumption and to the negative energy balance of the whole process system. When the low nitrogen scenario was considered, the energy consumption was far lower. Clarens et al. [87] came to similar conclusions as nutrient-derived energy consumption accounted for the greatest energy use in algae production. Similar observations were made by Shirvani et al. [98]. Not only do fertilisers negatively affect the energy balance, but they also provide a significant source of environmental impacts to the system. Fertiliser production requires a high energy input from both electricity and fossil fuels, both of which are high emitters of greenhouse gases. Some studies have ignored the impact of fertilisers however results in the study by Clarens et al. [87] suggested that using alternative sources of nutrients (wastewater) could in fact uptake CO_2 and return a positive energy balance in the best case (using source-separated urine as a nutrient source).

TABLE 8: A comparison of LCA results of energy balances calculated in algae-biofuel studies.

LCA Study	Energy Balance	LCA Method	Comments	Refs.
Algae-biodiesel	0.95	Well to fuel	Not taking into account wastewater treatment or CO_2 from flue gas, both of these contributing the most energy use, cultivation in ponds	[87]
Algae-biodiesel	6.7	Well to pump	Co-product allocation provides greatest energy recovery, wastewater assumed to provide nutrients, harvesting greatest energy consumer, cultivation in ponds	[95]
Algae-biodiesel	1.34	Well to fuel	Wet biomass processing and low nitrogen addition for high lipid content, anaerobic digestion of oil cake essential for positive energy balance, cultivation in ponds	[77]
Algae-biodiesel	3.05	Cultivation	Considers just the cultivation stage and energy content of the oil in the biomass, cultivation in ponds	[55]
Algae-bioethanol	5	Well to wheel	80% heat exchange efficiency	[84]
Algae-bioelectricity (combustion)	4.10	Well to wheel	Use of flue gas for CO_2	[97]

2.3.5 CARBON DIOXIDE

Increased concentrations of carbon dioxide (above atmospheric concentration) have been proven to improve the productivity [45-47] of algal cultivation. Production of synthetic CO_2 however is too energy-intensive to generate and a source of waste carbon dioxide is required. Many studies have proven the advantages of using CO_2 injection combined with algal cultivation [45, 47, 50, 51, 99]. As producing CO_2 synthetically is not sustainable, it is necessary for an existing source of CO_2 to be situated near to the algae growth ponds. Researchers have considered the plant flue gas from coal-fired power stations as an ideal source of CO_2 [48, 50] and flue gases have been shown to be successful as a source of CO_2. Nevertheless barriers would need to be overcome to implement the concept in a scaled-up system. It is evident from literature that CO_2 concentrations that are too high (above 15%) will cause a decrease in biomass productivity and potentially death of the cells. This may limit the number of possibilities for use of flue gas, although it must be noted that generally flue gases contain CO_2 concentrations lower than this [5]. It is not only the CO_2 that could be lethal to the cells: other toxins may also negatively impact the biomass. SO_2 can have a great impact upon the biomass and the pH of the water and high SO_2 concentration cause the pH to drop to very low levels. pH can be adjusted using NaOH but this requires additional materials and energy. In addition, the temperature of flue gas is generally above that of normal culture conditions and is likely to be too high to allow biomass growth. Cooling would be necessary to reduce the temperature to an acceptable level thus requiring water and additional energy for pumping.

Clearly there are many issues related to the use of waste flue gas as a source of CO_2 that must be addressed to allow implementation on a larger scale. It may be the case that transporting and treating flue gas prior to injection would require too much energy compared to the benefit that could be gained.

2.3.6 ENVIRONMENTAL IMPACTS

As with any production process, algal biofuel will undoubtedly have an impact on the environment relating to land use, water use, atmospheric emissions and terrestrial/water emissions. One of the key aims of biofuel production is to produce a fuel with fewer environmental impacts than conventional fossil fuels [100]. The intensive processing of the biomass, however, could result in a fuel with greater environmental impacts.

When considering environmental impacts of a product, many factors are taken into account. One of the main impact categories which is considered is the greenhouse gas emissions (GHG) in kg CO_2 equivalent, effectively the benchmark for how "green" a product is. In the best case, a biofuel can have a negative greenhouse gas emission in that during its production more carbon dioxide is taken up than is released during production and use of the fuel. Many studies have been carried out assessing the greenhouse gas emissions of various fuel types from different feedstocks. Recent studies which have investigated the production of algal biofuel have found that, under most circumstances, algal biofuels are likely to have a net positive greenhouse gas emissions [77, 87, 96]. This is in contrast with many other biofuels produced from conventional first and second generation feedstocks which are produced uptaking more greenhouse gases than are emitted in the process [101-103]. A comparison of carbon dioxide emissions from algal biofuel and alternative feedstocks is shown in Table 9. The table shows the CO_2 emissions per MJ of energy recovered as biofuel. The LCA method is included showing at which point the study stopped i.e., at fuel production (well to fuel) or at combustion (well to wheel). The data displayed in table 8 exhibits how poorly algal biofuel currently performs when compared to alternative feedstocks whether they are processed to bioethanol or biodiesel. One of the studies finds algal biodiesel to provide a negative GHG balance [95], nevertheless this is in contrast to the majority [77, 87, 96]. The different termination points of the study make comparison more difficult as predictably there are GHG emissions associated with the transport and combustion of the fuel.

The majority of greenhouse gases in algal biofuel production are emitted as a result of energy production. Clarens et al. [87] for example demonstrated that CO_2 procurement demands 40% of total energy consumption and 30% of GHG emissions. Any electricity required will create GHG emissions at the point of generation. In their more recent study Clarens et al. [97] compared the greenhouse gas emissions of two scenarios: algal biodiesel with bioelectricity generated from residual biomass and just bioelectricity generated from the biomass. The results were compared with biodiesel and bioelectricity from canola and bioelectricity from switchgrass. The energy from algae scenarios both performed well with direct bioelectricity from algae producing the least GHG emissions. The process stream configuration greatly affects the energy requirements. The greater the number of processes (particularly those including lipid extraction and digestion) required more energy and thus also produced greater greenhouse gas emissions.

GHG emissions may be the most common impact category yet there are many others that also require consideration including eutrophication potential, global warming potential, land use and human toxicity. In their life-cycle analysis Lardon et al. [77] investigated the environmental impacts of their algal-biofuel best-case scenario (low N, wet processing) to alternative feedstocks (rapeseed, soybean, palm, diesel). In some areas the algal biofuel performed well (such as in land use and eutrophication) however it did not compare well for the majority of categories, particularly for photochemical oxidation, ionizing radiation, marine toxicity, acidification and abiotic depletion. In the study published by Clarens et al. [87] a fewer number of categories were investigated but the results are similar for eutrophication and land use, both of which are favourable in comparison to corn, canola and switchgrass. Clearly improvements need to be made to minimise the adverse impacts that would be caused by the production and combustion of algal biofuels. These impacts are unlikely to ever be non-existent but it is important that the concept can perform favourably in comparison to alternatives regarding environmental impacts.

TABLE 9: GHG emissions from various biofuels from different feedstocks.

Feedstock	Biofuel	Cultivation	LCA Method	GHG Emissions (CO_2e) kg CO_2/MJ	Refs.
Algae	Biodiesel	PBR	Well to wheel	0.32	[96]
Algae	Biodiesel	Raceway pond	Well to fuel	0.057	[87][96][77]
			Well to wheel	0.18	[95]
			Well to pump	0.2	
			Well to fuel	−0.021	
Canola	Biodiesel	Agricultural	Well to fuel	−0.05	[87]
Soy bean	Biodiesel	Agricultural		0.030	[101]
Corn	Bioethanol	Agricultural	Well to fuel	−0.082	[87]
Switchgrass	Bioethanol	Agricultural	Well to fuel	−0.076	[87] [102]
			Well to fuel	−0.024	
Poplar	Bioethanol	Agricultural	Well to fuel	−0.024	[102]

2.4 A SUSTAINABLE VISION

To be considered sustainable, as a fuel source, it is essential that the overall process provide a positive energy balance with minimal environmental and social impacts whilst maintaining economic viability. Improving the energy balance is likely to improve the other areas. For example, reducing energy consumption requires less electricity generation which in turn will reduce environmental impacts whilst lessening production cost. If optimal process configurations can be designed for the production of algal biofuels maximising energy yield whilst minimising consumption the concept, could in the future, become a method of producing a sustainable fuel. As a result, many current studies are focussing upon grand scale systems centred around large power plants for CO_2 with potential utilisation of wastewater if available for nutrient provision/water treatment. Although research is heading in the right direction by reducing energy consumption through combining wastewater treatment for nutrient provision and carbon abatement with the use of flue gases, perhaps the future lies in more flexible, localised solutions which are adaptable to unique conditions.

2.4.1 INTEGRATED AND LOCALISED SOLUTIONS

In many industrial processes there is often a source of effluent as well as flue gases. Wastewater treatment plants, farms with AD plants, breweries, distilleries and oil refineries all have the potential to offer both materials. The basic requirements of a system to cultivate algae are an area for infrastructure, a source of nutrients (most importantly N and P), a source of concentrated carbon dioxide, freshwater and a consumer for the products obtained. Table 10 displays a number of areas where algal cultivation may be appropriate and the advantages and disadvantages of such a system. Table 10 shows a variety of potential industries where cultivation of algae could be possible. The majority of these industries provide a wastewater stream with sufficient nutrient loading for the growth of algae. Oil refinery wastewater may have a nutrient concentration too low for optimal growth nevertheless if necessary additional fertiliser could be supplemented. The flue gases found in each of the industries are likely to contain a CO_2 concentration of up to 15% and this value is considered to be near the maximum level that will allow algal growth before it becomes toxic [49]. Each of the flue gases mentioned is likely to boost algal growth whilst sequestering carbon simultaneously.

Depending upon the industry, there are likely to be many problems to overcome. High nutrient loadings (farm effluent, distillery effluent) would lead to poor treatment or toxicity and therefore require dilution with freshwater. To dilute such high concentrations would require significant sources of freshwater possibly needing expensive transportation costs and environmental issues if located in a water stressed area. Wastewaters, particularly those from chemical industries such as oil refining and bioethanol production, could potentially contain toxic contaminants. Similarly flue gases may contain toxins that could affect the growth of the algal biomass. It is evident that there are many different opportunities for the implementation of algal cultivation in industry. Nevertheless it is not possible to have one fixed solution. Every scenario will have different wastewater characteristics, available water and land, varying flue gas characteristics, energy needs and problems related to implementation. Every approach to implementation may be different but the concept allows flexibility.

2.4.2 ALGAL SPECIES

Selecting an algal strain because of its beneficial properties is unlikely to be the most successful method of recovering energy from its cultivation. As recent studies have shown, producing high value biofuel from algae may not be the most effective means of energy recovery. Instead it seems that anaerobic digestion or combustion may be more appropriate. Given this situation, it may be more important to utilise strains of algae that are most suited to individual scenarios (wastewater type, climate etc.). It is also likely in many locations with climatic variation in seasons that the species of algae dominating will change as temperatures and amount of sunlight vary. There are examples of such species change in the literature, Professor Shelef in his study of algal cultivation in raw wastewater in open ponds found that in Spring *Micratinium* dominated, in Summer *Chlorella* was most common and in autumn and winter *Euglena* became dominant [23].

The alternative to allowing various strains to dominate naturally is to select a strain that is capable of tolerating extreme conditions or recycling the favoured algae. *Spirulina* is a species of algae renowned for good biomass control due to high pH requirements [111, 112]. Conditions could be manipulated to promote the growth of species such as *Spirulina* by adjusting pH in wastewater streams. In a study conducted by Olguin et al. [113]

Spirulina was cultivated in piggery wastewater and seawater. In the study, continuous cultivation of *Spirulina* was achieved with no issues relating to contamination. Calculations would be necessary to understand whether or not promoting specific strain dominance would be worthwhile from an energy recovery perspective. It may be more productive to simply allow a naturally dominant strain to develop requiring fewer inputs. As studied by Park et al. [91], it is also possible to recycle algae improving dominance of selected strains. This may provide a robust method of selectivity and could allow for improved productivity with little input required.

TABLE 10: Applicability of various industries for implementation of algal cultivation.

Industry	Total N (mg/L)	Total P (mg/L)	Flue Gas Source	Advantages	Disadvantages
WWTP[a]	15[b] (NH_4)	11.5[b] (PO_4)	AD co-generator	Provides tertiary treatment	Land requirement
				Abatement of CO_2 from co-digester	Contamination of wastewater could affect algae
				AD of biomass available	
Farm	1210[c]	303[c]	AD co-generator	Treatment of excess nutrients	Potentially no CO_2 source
	5600[d]	1600[d]	Composting facility	Treated biomass for feed	High nutrient loading may require dilution
				Available land	
Brewery/ distillery	56.5[e] (NH_4)	177–215[e]	Fermentation process	Wastewater treatment	Land area requirement
	51[f]	57–325.8[h] (PO_4)	Boiler flue gas	Biomass for co-generator produced	Low pH wastewater
	560–834[g] (TKN)			Sustainability targets	
	3–106[h] (NH_3)				
Oil refinery	8[i] (NH_3)	0.1[i]	Flue gases	Abatement of GHGs	Wastewater/flue gas may be too toxic
				Sustainability targets	Low nutrient loading

a Wastewater treatment plant; b Secondarily treated wastewater [104]; c Raw dairy manure [105]; d Raw swine manure [106]; e Bioethanol distillery [102], f Distillery stillage [107]; g Grape distillery [108]; h Brewery wastewater [109]; i [110].

2.4.3 CULTIVATION METHODS

As discussed, the only potentially sustainable cultivation method currently available is the use of open ponds. This is because of their lower energy requirements compared to PBRs. Open ponds require a far greater area of land for the mass of biomass produced and area requirements need to be considered for individual cases. The necessary area will be dependent upon the volume of wastewater that requires treatment, pond depth, nutrient loading, discharge limits and hydraulic retention times (HRT).

If the focus of algal cultivation is for wastewater treatment the treatment efficiency will have a great impact upon the pond area required. Treatment of water with algae depends upon the productivity of the algae, the higher the productivity the greater the nutrients assimilated. Hydraulic retention times of around 10 days are most common [55, 96] and as the HRT increases, the area required increases proportionally. It may be important to minimise the area requirements by reducing cultivation time but nevertheless if the wastewater discharged from the system is above the required limits then the time is likely to be too short. The HRT of each system will depend upon the influent nutrient loading and limits of discharge and therefore must be calculated accordingly.

2.4.4 LOW ENERGY HARVESTING

Harvesting of the algal biomass is one of the greatest energy consumers in the process chain for algal biofuel production. Many options exist for extracting the algae with each having their own advantages and disadvantages. Low energy harvesting is favoured but options are limited by cell sizes. If the algae being harvested are of a large 70 μm [61] the algae can be filtered. Ideally, if the conditions allow, gravity filtration is possible and this requires very little energy input. This would be the optimal solution economically and environmentally due to low energy requirements. Alternatively, should cell size allow, the biomass can be pressure filtered requiring slightly more energy but providing a higher removal efficiency. In most cases it is likely that flocculation would be required to allow the biomass to settle or float more readily. Conventional flocculants appear favourable in terms of harvesting yield yet may cause issues downstream due to contamination. Ideally if flocculation is necessary, bio-flocculation could be carried out using bacteria or other algae strains to obtain flocs, but further research is required to fully understand in which conditions this is possible. Alternatively organic flocculants such as chitosan could provide a more sustainable option but, again, efficiency of biomass removal using chitosan requires further study. It is likely that if flocculation is necessary, a combination of flocculants would be required for the most sustainable solution. Following flocculation, sedimentation or flotation of the biomass should be a successful method of harvesting without significant energy input.

Centrifugation of algal biomass could be necessary if biomass of high solids content is required. Due to the energy requirement, centrifugation should, where possible, be avoided but due to the high and rapid recovery it could provide a necessary step. The least energy-intense processes for harvesting algae are sedimentation/flotation and gravity filtration as there is little energetic input. Ideally biomass would be filtered or settled using sedimentation due to low inputs however with the majority of strains this may not be practical without prior treatment. Flocculation with the least intensive and damaging flocculants should be used if necessary and centrifugation a last option for dewatering.

2.4.5 SUGGESTED CONVERSION TECHNIQUES

Following biomass harvesting it is then necessary to extract the maximum energy possible from the biomass to provide the best return. The three main fuel types that have received the majority of research related to algal biomass are biodiesel, biogas and bioethanol. Due to the potentially high oil content of certain strains of algae, the ease of extraction and value of the end product biodiesel has received the most attention. Algal strains such as *Chlorella* are noted for being able to produce up to 70% oil content within their cell walls [14]. This scenario however requires very specific conditions (low nitrogen and no contamination) and would be very difficult to obtain in practice. It has been suggested by Sialve et al. [85] that it would not be economically viable to extract lipids from algae containing an oil yield any less than 40%, and therefore for the majority of algal species anaerobic digestion would provide the highest positive energy balance due to low input requirements. Similarly in their environmental study Clarens et al. [97] found that direct combustion of biomass to produce electricity provided the highest energy return on investment when compared to anaerobic digestion, biodiesel production plus anaerobic digestion and biodiesel production plus direct combustion.

Given the high energy consumption required to produce biodiesel from algae, the process does not seem beneficial to be used for a flexible system where the algae cultivated are likely to be a mix of species with low lipid content. Bioethanol has potential and in such a system the algae is likely to contain a high proportion of convertible carbohydrates but the energy balance of such a process is untested and is unlikely to yield great efficiencies in the near future and is likely to require high inputs (enzymes and yeast). It seems far more likely that anaerobic digestion or combustion of the biomass will provide the maximum energy recovery. Another benefit of such a concept is that facilities to carry out the digestion or combustion are likely to be already operating on site with no requirement for new development.

2.4.6 RESOURCE CONSERVATION AND RECYCLING

The proposed concept utilises a variety of waste streams from industry thus saving energy and environmental impacts by avoiding manufacture of raw materials. Furthermore as a method of wastewater treatment, energy and associated impacts will be saved by avoiding alternative methods of treatment. The energy recovered through anaerobic digestion or combustion can be returned to the system, powering the units which require an energy source (paddlewheel or centrifugation). The waste heat can be used to dry the biomass if required or alternatively used to heat the ponds if the temperature falls below optimal conditions. The residual waste from the energy recovery system can be fed back into the treatment ponds supplying additional nutrients if required or alternatively sold as a fertiliser.

2.4.7 CURRENT STATE OF CONCEPT

It is well known that many strains of algae are capable of growing in wastewater and by doing so providing a form of treatment. Outdoor productivities are, however, difficult to find in the literature as most studies are performed in the laboratory and therefore conditions are less realistic. Likely strains to dominate in specific scenarios and locations are not known and therefore it is hard to speculate what type of strain would be dominant. It is therefore also difficult to know what harvesting technique would be most appropriate for the strain cultivated and how much energy could be expected to be recovered from conversion to biodiesel, bioethanol, biogas or from combustion.

Each of the processes studied have been tested and are considered practically viable. Each of the harvesting methods considered are currently used for harvesting algal biomass regardless of their energy use and overall viability. Conversion techniques have been shown to be feasible, again regardless of how viable they are in real situations. What are missing are pilot-scale studies of the whole system to give valuable information about applicability to different situations.

2.4.8 WHERE TO GO FROM HERE

Much of the research reported so far is based on laboratory work and speculation while testing a full system would allow realistic life-cycle assessments to be carried out investigating similar systems in a number of industrial scenarios. Data related to energy consumption and yield would prove the viability of the concept. Setting up pilot scale infrastructure within most industries with suitable wastewater would be a simple undertaking with great research benefits. The ponds would need to be inoculated with a mix of local strains and the dominance of those strains monitored. Suitable harvesting techniques would need to be tested for the algal mix, cultivated within the ponds to identify the most effective and sustainable method for each case. Further research is required to optimize energy recovery from conversion techniques which provide the maximum energy yield, most likely anaerobic digestion or combustion.

2.5 CONCLUSIONS

The review of the current state of knowledge and technology suggests that it is unlikely that there is one solution to biofuel recovery from algal biomass. Production of energy from algae is most likely to be successful on a case by case basis based on applicability to the particular industry and the site under question. The majority of wastewaters from common industries have shown capacity to support the cultivation of various strains of algae. Allowing natural domination of algal strains means that algae which are most effective for that particular situation should develop. If a preferred algal strain is required, the pond could be seeded with the algae and recycled continuously to promote growth.

The biomass processing stages can use existing technologies which are tested for many strains of algae. Harvesting can be optimised for each individual scenario. Optimal recovery of energy for maximum efficiency is likely to be similar for each industry. Literature suggests that recovery through anaerobic digestion or combustion provides the highest energy return for mixed strains. As the strains are likely to be mixed and varying there is little point in designing systems for specialised biofuels (biodiesel or bioethanol) which require specific biomass characteristics. Therefore a system which is flexible for numerous industries is possible. Pilot scale tests of such systems will be essential for implementation to optimise systems individually.

REFERENCES

1. Wolfson, W., Synthetic Biology Transforms Green Goo to Black Gold. Chemistry & Biology, 2009. 16(3): p. 237-238.
2. Golueke, C.G., W.J. Oswald, and H.B. Gotaas, Anaerobic Digestion of Algae. Applied Microbiology, 1957. 5(1): p. 47-55.
3. Oswald, W.J., H.B. Gotaas, C.G. Golueke, and W.R. Kellen, Algae in Waste Treatment. Sewage and Industrial Wastes, 1957. 29(4): p. 437-455.

4. Golueke, C.G. and W.J. Oswald, Power from Solar Energy - Via Algae-Produced Methane. Solar Energy, 1963. 7(3): p. 86-92.
5. Oswald, W.J., My sixty years in applied algology. Journal of Applied Phycology, 2003. 15(2-3): p. 99-106.
6. Saxena, V.K., S.M. Tandon, and K.K. Singh, Anaerobic-Digestion of Green Filamentous Algae and Waterhyacinth for Methane Production. National Academy Science Letters-India, 1984. 7(9): p. 283-284.
7. Cecchi, F., G. Vallini, P. Pavan, A. Bassetti, and J. Mataalvarez, Management of Macroalgae from Venice Lagoon through Anaerobic Co-Digestion and Co-Composting with Municipal Solid-Waste (Msw). Water Science and Technology, 1993. 27(2): p. 159-168.
8. Ras, M., L. Lardon, S. Bruno, N. Bernet, and J.P. Steyer, Experimental study on a coupled process of production and anaerobic digestion of Chlorella vulgaris. Bioresource Technology, 2011. 102(1): p. 200-206.
9. Vijayaraghavan, K. and K. Hemanathan, Biodiesel Production from Freshwater Algae. Energy & Fuels, 2009. 23: p. 5448-5453.
10. Weyer, K.M., D.R. Bush, A. Darzins, and B.D. Willson, Theoretical Maximum Algal Oil Production. Bioenergy Research, 2010. 3(2): p. 204-213.
11. Rodolfi, L., G.C. Zittelli, N. Bassi, G. Padovani, N. Biondi, G. Bonini, and M.R. Tredici, Microalgae for Oil: Strain Selection, Induction of Lipid Synthesis and Outdoor Mass Cultivation in a Low-Cost Photobioreactor. Biotechnology and Bioengineering, 2009. 102(1): p. 100-112.
12. Lee, J.Y., C. Yoo, S.Y. Jun, C.Y. Ahn, and H.M. Oh, Comparison of several methods for effective lipid extraction from microalgae. Bioresource Technology, 2010. 101: p. S75-S77.
13. Demirbas, A., Production of Biodiesel from Algae Oils. Energy Sources Part a-Recovery Utilization and Environmental Effects, 2009. 31(2): p. 163-168.
14. Chisti, Y., Biodiesel from microalgae. Biotechnology Advances, 2007. 25(3): p. 294-306.
15. Sheehan, J., T. Dunahay, J. Benemann, and P. Roessler, A Look Back at the U.S. Department of Energy's Aquatic Species Program—Biodiesel from Algae. 1998: Golden, Colorado.
16. Chen, M., H.Y. Tang, H.Z. Ma, T.C. Holland, K. Ng, and S.O. Salley, Effects of nutrients on growth and lipid accumulation in the green algae dunaliella tertiolecta. Bioresource Technology, 2011. 102: p. 1649-1655.
17. Sturm, B.S.M., E. Peltier, V. Smith, and F. deNoyelles, Controls of microalgal biomass and lipid production in municipal wastewater-fed bioreactors. Environmental Progress & Sustainable Energy, 2012. 31(1): p. 10-16.
18. Ruiz-Marin, A., L.G. Mendoza-Espinosa, and T. Stephenson, Growth and nutrient removal in free and immobilized green algae in batch and semi-continuous cultures treating real wastewater. Bioresource Technology, 2010. 101(1): p. 58-64.
19. Ludwig, H.F., W.J. Oswald, H.B. Gotaas, and V. Lynch, Algae Symbiosis in Oxidation Ponds .1. Growth Characteristics of Euglena-Gracilis Cultured in Sewage. Sewage and Industrial Wastes, 1951. 23(11): p. 1337-1355.
20. Oswald, W.J. and H.B. Gotaas, Photosynthesis in the Algae - Discussion. Industrial and Engineering Chemistry, 1956. 48(9): p. 1457-1458.

21. Oswald, W.J., H.B. Gotaas, H.F. Ludwig, and V. Lynch, Algae Symbiosis in Oxidation Ponds .2. Growth Characteristics of Chlorella-Pyrenoidosa Cultured in Sewage. Sewage and Industrial Wastes, 1953. 25(1): p. 26-37.
22. Oswald, W.J., H.B. Gotaas, H.F. Ludwig, and V. Lynch, Algae Symbiosis in Oxidation Ponds .3. Photosynthetic Oxygenation. Sewage and Industrial Wastes, 1953. 25(6): p. 692-705.
23. Shelef, G., Combined systems for algal wastewater treatment and reclamation and protein production. 1981: Technion: Haifa, Israel.
24. Orpez, R., M.E. Martinez, G. Hodaifa, F. El Yousfi, N. Jbari, and S. Sanchez, Growth of the microalga Botryococcus braunii in secondarily treated sewage. Desalination, 2009. 246(1-3): p. 625-630.
25. Delanoue, J., G. Laliberte, and D. Proulx, Algae and Waste-Water. Journal of Applied Phycology, 1992. 4(3): p. 247-254.
26. Woertz, I., A. Feffer, T. Lundquist, and Y. Nelson, Algae Grown on Dairy and Municipal Wastewater for Simultaneous Nutrient Removal and Lipid Production for Biofuel Feedstock. Journal of Environmental Engineering-Asce, 2009. 135(11): p. 1115-1122.
27. Patel, A., J. Zhu, and G. Nakhla, Simultaneous carbon, nitrogen and phosphorous removal from municipal wastewater in a circulating fluidized bed bioreactor. Chemosphere, 2006. 65(7): p. 1103-1112.
28. Wilkie, A.C. and W.W. Mulbry, Recovery of dairy manure nutrients by benthic freshwater algae. Bioresource Technology, 2002. 84(1): p. 81-91.
29. Shalaru, V.M., V.W. Shalaru, T.I. Dudnicenco, and M.D. Ichim, The use of algae in wastewater treatment from animal farms. Phycologia, 2005. 44: p. 93-93.
30. Pizarro, C., W. Mulbry, D. Blersch, and P. Kangas, An economic assessment of algal turf scrubber technology for treatment of dairy manure effluent. Ecological Engineering, 2006. 26(4): p. 321-327.
31. Park, K.Y., B.R. Lim, and K. Lee, Growth of microalgae in diluted process water of the animal wastewater treatment plant. Water Science and Technology, 2009. 59(11): p. 2111-2116.
32. Chinnasamy, S., A. Bhatnagar, R. Claxton, and K.C. Das, Biomass and bioenergy production potential of microalgae consortium in open and closed bioreactors using untreated carpet industry effluent as growth medium. Bioresource Technology, 2010. 101(17): p. 6751-6760.
33. Travieso, L., F. Benitez, E. Sanchez, R. Borja, M. Leon, F. Raposo, and B. Rincon, Performance of a Laboratory-scale Microalgae Pond for Secondary Treatment of Distillery Wastewaters. Chemical and Biochemical Engineering Quarterly, 2008. 22(4): p. 467-473.
34. Mulbry, W., S. Kondrad, C. Pizarro, and E. Kebede-Westhead, Treatment of dairy manure effluent using freshwater algae: Algal productivity and recovery of manure nutrients using pilot-scale algal turf scrubbers. Bioresource Technology, 2008. 99(17): p. 8137-8142.
35. Craggs, R., D. Sutherland, and H. Campbell, Hectare-scale demonstration of high rate algal ponds for enhanced wastewater treatment and biofuel production. Journal of Applied Phycology, 2012. 24(3): p. 329-337.

36. Fierro, S., M.D.P. Sanchez-Saavedra, and C. Copalcua, Nitrate and phosphate removal by chitosan immobilized Scenedesmus. Bioresource Technology, 2008. 99(5): p. 1274-1279.
37. Aslan, S. and I.K. Kapdan, Batch kinetics of nitrogen and phosphorus removal from synthetic wastewater by algae. Ecological Engineering, 2006. 28(1): p. 64-70.
38. Voltolina, D., H. Gomez-Villa, and G. Correa, Nitrogen removal and recycling by Scenedesmus obliquus in semicontinuous cultures using artificial wastewater and a simulated light and temperature cycle. Bioresource Technology, 2005. 96(3): p. 359-362.
39. Di Termini, I., A. Prassone, C. Cattaneo, and M. Rovatti, On the nitrogen and phosphorus removal in algal photobioreactors. Ecological Engineering, 2011. 37(6): p. 976-980.
40. Su, Y.Y., A. Mennerich, and B. Urban, Municipal wastewater treatment and biomass accumulation with a wastewater-born and settleable algal-bacterial culture. Water Research, 2011. 45(11): p. 3351-3358.
41. Wang, L.A., M. Min, Y.C. Li, P. Chen, Y.F. Chen, Y.H. Liu, Y.K. Wang, and R. Ruan, Cultivation of Green Algae Chlorella sp in Different Wastewaters from Municipal Wastewater Treatment Plant. Applied Biochemistry and Biotechnology, 2010. 162(4): p. 1174-1186.
42. Kebede-Westhead, E., C. Pizarro, and W.W. Mulbry, Treatment of swine manure effluent using freshwater algae: Production, nutrient recovery, and elemental composition of algal biomass at four effluent loading rates. Journal of Applied Phycology, 2006. 18(1): p. 41-46.
43. de Godos, I., C. Gonzalez, E. Becares, P.A. Garcia-Encina, and R. Munoz, Simultaneous nutrients and carbon removal during pretreated swine slurry degradation in a tubular biofilm photobioreactor, in Applied Microbiology and Biotechnology. 2009. p. 187-194.
44. Guinee, J.B., R. Heijungs, and E. van der Voet, A greenhouse gas indicator for bioenergy: some theoretical issues with practical implications. International Journal of Life Cycle Assessment, 2009. 14(4): p. 328-339.
45. Chiu, S.Y., C.Y. Kao, C.H. Chen, T.C. Kuan, S.C. Ong, and C.S. Lin, Reduction of CO2 by a high-density culture of Chlorella sp in a semicontinuous photobioreactor. Bioresource Technology, 2008. 99(9): p. 3389-3396.
46. Douskova, I., J. Doucha, K. Livansky, J. Machat, P. Novak, D. Umysova, V. Zachleder, and M. Vitova, Simultaneous flue gas bioremediation and reduction of microalgal biomass production costs. Applied Microbiology and Biotechnology, 2009. 82(1): p. 179-185.
47. Brune, D.E., T.J. Lundquist, and J.R. Benemann, Microalgal Biomass for Greenhouse Gas Reductions: Potential for Replacement of Fossil Fuels and Animal Feeds. Journal of Environmental Engineering-Asce, 2009. 135(11): p. 1136-1144.
48. Maeda, K., M. Owada, N. Kimura, K. Omata, and I. Karube, Co2 Fixation from the Flue-Gas on Coal-Fired Thermal Power-Plant by Microalgae. Energy Conversion and Management, 1995. 36(6-9): p. 717-720.
49. Stepan, D.J., R.E. Shockey, T.A. Moe, and R. Dorn, Subtask 2.3 - carbon dioxide sequestering using microalgal systems. 2002.

50. de Morais, M.G. and J.A.V. Costa, Isolation and selection of microalgae from coal fired thermoelectric power plant for biofixation of carbon dioxide. Energy Conversion and Management, 2007. 48(7): p. 2169-2173.
51. de Morais, M.G. and J.A.V. Costa, Biofixation of carbon dioxide by Spirulina sp and Scenedesmus obliquus cultivated in a three-stage serial tubular photobioreactor. Journal of Biotechnology, 2007. 129(3): p. 439-445.
52. Doucha, J., F. Straka, and K. Livansky, Utilization of flue gas for cultivation of microalgae (chlorella sp) in an outdoor open thin-layer photobioreactor. J Appl Phycol, 2005. 17: p. 403-412.
53. Chiu, S.Y., C.Y. Kao, C.H. Chen, T.C. Kuan, S.C. Ong, and C.S. Lin, Reduction of CO2 by a high density culture of chlorella sp in a siemicontinuous photobioreactor. Bioresource Technol, 2008. 99: p. 3389-3396.
54. Kadam, K., Microalgae Production from Power Plant Flue Gas: Enivornmental Implications on a Life Cycle Basis. 2001, NREL: Golden.
55. Jorquera, O., A. Kiperstok, E.A. Sales, M. Embirucu, and M.L. Ghirardi, Comparative energy life-cycle analyses of microalgal biomass production in open ponds and photobioreactors. Bioresource Technology, 2010. 101(4): p. 1406-1413.
56. Oswald, W.J., H.B. Gotaas, H.F. Ludwig, and V. Lynch, Algae symbiosis in oxidation ponds. 3. Photosyntehtic oxygenation. Sewage Ind Wastes, 1953. 25: p. 692-705.
57. Soratana, K. and A.E. Landis, Evaluating industrial symbiosis and algae cultivation from a life cycle perspective. Bioresource Technology, 2011. 102(13): p. 6892-6901.
58. Grima, E.M., E.H. Belarbi, F.G.A. Fernandez, A.R. Medina, and Y. Chisti, Recovery of microalgal biomass and metabolites: process options and economics. Biotechnology Advances, 2003. 20(7-8): p. 491-515.
59. Benemann, J., B. Koopman, J. Wieissman, D. Eisenberg, and R. Goebel, Development of microalgae harvesting and high-rate pond technologies in California. 1980: Elsevier/North-Holland Biomedical Press.
60. Golueke, C.G. and W.J. Oswald, Harvesting and processing sewage-grown planktonic algae. Pollution Control Federation, 1965. 37: p. 471-498.
61. Brennan, L. and P. Owende, Biofuels from microalgae-A review of technologies for production, processing, and extractions of biofuels and co-products. Renewable & Sustainable Energy Reviews, 2010. 14(2): p. 557-577.
62. Uduman, N., Y. Qi, M.K. Danquah, and A.F.A. Hoadley, Marine microalgae flocculation and focused beam reflectance measurement. Chemical Engineering Journal, 2010. 162(3): p. 935-940.
63. Mohn, F.M., Experiences and strategies in the recovery of biomas from mass cultures of microalgae. Algal biomass, ed. G. Shelef and C.J. Soeder. 1980, Amsterdam, The Netherlands: Elsevier.
64. Spilling, K., J. Seppala, and T. Tamminen, A Potential Low-Cost Method for Harvesting Microalgae Using High Ph and Regulation of Particle Encounter Rate. Phycologia, 2009. 48(4): p. 123-123.
65. Gutzeit, G., D. Lorch, A. Weber, M. Engels, and U. Neis, Bioflocculent algal-bacterial biomass improves low-cost wastewater treatment. Water Science and Technology, 2005. 52(12): p. 9-18.

66. Salim, S., R. Bosma, M.H. Vermue, and R.H. Wijffels, Harvesting of microalgae by bio-flocculation. Journal of Applied Phycology, 2011. 23(5): p. 849-855.
67. Sukenik, A., D. Bilanovic, and G. Shelef, Flocculation of microalgae in rackish and se waters. Biomass 1988. 15: p. 187-199.
68. Divakaran, R. and V.N.S. Pillai, Flocculation of algae using chitosan. Journal of Applied Phycology, 2002. 14(5): p. 419-422.
69. Shelef, G., A. Sukenick, and M. Green, Microalgae harvesting and processing: A literature review. 1984: Solar Energy Research Institute. US Department of Energy.
70. Golueke, C.G. and W.J. Oswald, Harvesting and processing sewage-grown planktonic algae. Water Pollut Control Federation, 1965. 37: p. 471-498.
71. Antizar-Ladislao, B. and J.L. Turrion-Gomez, Decentralized Energy from Waste Systems. Energies, 2010. 3(2): p. 194-205.
72. Bligh, E.G. and W.J. Dyer, A Rapid Method of Total Lipid Extraction and Purification. Canadian Journal of Biochemistry and Physiology, 1959. 37(8): p. 911-917.
73. Lee, J.Y., C. Yoo, S.Y. Jun, C.Y. ahn, and H.M. Oh, Comparison of several methods for effective lipid extraction from microalgae. Bioresource Technol, 2010. 101: p. S75-S77.
74. Patil, P.D., V.G. Gude, A. Mannarswamy, S.G. Deng, P. Cooke, S. Munson-McGee, I. Rhodes, P. Lammers, and N. Nirmalakhandan, Optimization of direct conversion of wet algae to biodiesel under supercritical methanol conditions. Bioresource Technology, 2011. 102(1): p. 118-122.
75. Johnson, M.B. and Z.Y. Wen, Production of Biodiesel Fuel from the Microalga Schizochytrium limacinum by Direct Transesterification of Algal Biomass. Energy & Fuels, 2009. 23: p. 5179-5183.
76. Wahlen, B.D., R.M. Willis, and L.C. Seefeldt, Biodiesel production by simultaneous extraction and conversion of total lipids from microalgae, cyanobacteria, and wild mixed-cultures. Bioresource Technology, 2011. 102(3): p. 2724-2730.
77. Lardon, L., A. Helias, B. Sialve, J.P. Stayer, and O. Bernard, Life-Cycle Assessment of Biodiesel Production from Microalgae. Environmental Science & Technology, 2009. 43(17): p. 6475-6481.
78. Harun, R., M.K. Danquah, and G.M. Forde, Microalgal biomass as a fermentation feedstock for bioethanol production. Journal of Chemical Technology and Biotechnology, 2010. 85(2): p. 199-203.
79. Hirano, A., R. Ueda, S. Hirayama, and Y. Ogushi, CO2 fixation and ethanol production with microalgal photosynthesis and intracellular anaerobic fermentation. Energy, 1997. 22(2-3): p. 137-142.
80. Hirayama, S., R. Ueda, Y. Ogushi, A. Hirano, Y. Samejima, K. Hon-Nami, and S. Kunito, Ethanol production from carbon dioxide by fermentative microalgae. Advances in Chemical Conversions for Mitigating Carbon Dioxide, 1998. 114: p. 657-660.
81. Choi, S.P., M.T. Nguyen, and S.J. Sim, Enzymatic pretreatment of Chlamydomonas reinhardtii biomass for ethanol production. Bioresource Technology, 2010. 101(14): p. 5330-5336..
82. Hromadko, J., J. Hromadko, P. Miler, V. Honig, and P. Sterba, Use of Bioethanol in Combustion Engines. Chemicke Listy, 2011. 105(2): p. 122-128.

83. Nguyen, M.T., S.P. Choi, J. Lee, J.H. Lee, and S.J. Sim, Hydrothermal Acid Pretreatment of Chlamydomonas reinhardtii Biomass for Ethanol Production. Journal of Microbiology and Biotechnology, 2009. 19(2): p. 161-166.
84. Luo, D.X., Z.S. Hu, D.G. Choi, V.M. Thomas, M.J. Realff, and R.R. Chance, Life Cycle Energy and Greenhouse Gas Emissions for an Ethanol Production Process Based on Blue-Green Algae. Environmental Science & Technology, 2010. 44(22): p. 8670-8677.
85. Sialve, B., N. Bernet, and O. Bernard, Anaerobic digestion of microalgae as a necessary step to make microalgal biodiesel sustainable. Biotechnology Advances, 2009. 27(4): p. 409-416.
86. Collet, P., A. Helias, L. Lardon, M. Ras, R.A. Goy, and J.P. Steyer, Life-cycle assessment of microalgae culture coupled to biogas production. Bioresource Technology, 2011. 102(1): p. 207-214.
87. Clarens, A.F., E.P. Resurreccion, M.A. White, and L.M. Colosi, Environmental Life Cycle Comparison of Algae to Other Bioenergy Feedstocks. Environmental Science & Technology, 2010. 44(5): p. 1813-1819.
88. Mussgnug, J.H., V. Klassen, A. Schluter, and O. Kruse, Microalgae as substrates for fermentative biogas production in a combined biorefinery concept. Journal of Biotechnology, 2010. 150(1): p. 51-56.
89. Ras, M., L. Lardon, S. Bruno, N. Bernet, and J.P. Steyer, Experimental study on a copuled process of production and anaerobic digestion of chlorella vulgaris. Biorestrouce Technol, 2011. 102: p. 200-206.
90. Fredriksson, H., A. Baky, S. Bernesson, A. Nordberg, O. Noren, and P.A. Hansson, Use of on-farm produced biofuels on organic farms - Evaluation of energy balances and environmental loads for three possible fuels. Agricultural Systems, 2006. 89(1): p. 184-203.
91. Park, J.B.K., R.J. Craggs, and A.N. Shilton, Recycling algae to improve species control and harvest efficiency from a high rate algal pond. Water Research, 2011. 45(20): p. 6637-6649.
92. Tseng, K.F., J.S. Huang, and I.C. Liao, Species Control of Microalgae in an Aquaculture Pond. Water Research, 1991. 25(11): p. 1431-1437.
93. Williams, P.J.L. and L.M.L. Laurens, Microalgae as biodiesel & biomass feedstocks: Review & analysis of the biochemistry, energetics & economics. Energy & Environmental Science, 2010. 3(5): p. 554-590.
94. Murphy, C.F. and D.T. Allen, Energy-Water Nexus for Mass Cultivation of Algae. Environmental Science & Technology, 2011. 45(13): p. 5861-5868.
95. Sander, K. and G.S. Murthy, Life cycle analysis of algae biodiesel. International Journal of Life Cycle Assessment, 2010. 15(7): p. 704-714.
96. Stephenson, A.L., E. Kazamia, J.S. Dennis, C.J. Howe, S.A. Scott, and A.G. Smith, Life-Cycle Assessment of Potential Algal Biodiesel Production in the United Kingdom: A Comparison of Raceways and Air-Lift Tubular Bioreactors. Energy & Fuels, 2010. 24: p. 4062-4077.
97. Clarens, A.F., H. Nassau, E.P. Resurreccion, M.A. White, and L.M. Colosi, Environmental Impacts of Algae-Derived Biodiesel and Bioelectricity for Transportation. Environmental Science & Technology, 2011. 45(17): p. 7554-7560.

98. Shirvani, T., X.Y. Yan, O.R. Inderwildi, P.P. Edwards, and D.A. King, Life cycle energy and greenhouse gas analysis for algae-derived biodiesel. Energy & Environmental Science, 2011. 4(10): p. 3773-3778.
99. Packer, M., Algal capture of carbon dioxide; biomass generation as a tool for greenhouse gas mitigation with reference to New Zealand energy strategy and policy. Energy Policy, 2009. 37(9): p. 3428-3437.
100. Antizar-Ladislao, B. and J.L. Turrion-Gomez, Second-generation biofuels and local bioenergy systems. Biofuels Bioproducts & Biorefining-Biofpr, 2008. 2(5): p. 455-469.
101. Hu, Z.Y., P.Q. Tana, X.Y. Yan, and D.M. Lou, Life cycle energy, environment and economic assessment of soybean-based biodiesel as an alternative automotive fuel in China. Energy, 2008. 33(11): p. 1654-1658.
102. Adler, P.R., S.J. Del Grosso, and W.J. Parton, Life-cycle assessment of net greenhouse-gas flux for bioenergy cropping systems. Ecological Applications, 2007. 17(3): p. 675-691.
103. Reiinders, L. and M.A.J. Huijbregts, Life cycle greenhouse gas emissions, fossil fuel demand and solar energy conversion efficiency in European bioethanol production for automotive purposes. Journal of Cleaner Production, 2007. 15(18): p. 1806-1812.
104. Orpez, R., M.E. Martinex, G. Hodaifa, F. El Yousfi, N. Jbari, and S. Sanchez, Growth of the microalga botryococcus braunii in secondarily treated sewage. Desalination, 2009. 246: p. 625-630.
105. Wilkie, A.C. and W.W. Mulbry, Recovery of dairy manure nutrients by benthic freshwater algae. Bioresource Technol, 2002. 84: p. 81-91.
106. Karakashev, D., J.E. Schmidt, and I. Angelidaki, Innovative process scheme for removal of organic matter, phosphorus and nitrogen from pig manure. Water Research, 2008. 42(15): p. 4083-4090.
107. Douskova, I., F. Kastanek, Y. Maleterova, P. Kastanek, J. Doucha, and V. Zachleder, Utilization of distillery stillage for energy generation and concurrent production of valuable microalgal biomass in the sequence: biogas-cogeneration-microalgae-products. Energy Conversion and Management, 2010. 51(3): p. 606-611.
108. Musee, N., M.A. Trerise, and L. Lorenzen, Post-treatment of distillery wastewater after UASB using aerobic techniques. South African Journal of Enology and Viticulture, 2007. 28(1): p. 50-55.
109. Raposo, M.F.D., S.E. Oliveira, P.M. Castro, N.M. Bandarra, and R.M. Morais, On the Utilization of Microalgae for Brewery Effluent Treatment and Possible Applications of the Produced Biomass. Journal of the Institute of Brewing, 2010. 116(3): p. 285-292.
110. Moreno, C., N. Farahbakhshazad, and G.M. Morrison, Ammonia removal from oil refinery effluent in vertical upflow macrophyte column systems. Water Air and Soil Pollution, 2002. 135(1-4): p. 237-247.
111. Kim, C.J., Y.H. Jung, S.R. Ko, H.I. Kim, Y.H. Park, and H.M. Oh, Raceway cultivation of Spirulina platensis using underground water. Journal of Microbiology and Biotechnology, 2007. 17(5): p. 853-857.

112. Celekli, A., M. Yavuzatmaca, and H. Bozkurt, Modeling of biomass production by Spirulina platensis as function of phosphate concentrations and pH regimes. Bioresource Technology, 2009. 100(14): p. 3625-3629.
113. Olguin, E.J., S. Galicia, G. Mercado, and T. Perez, Annual productivity of Spirulina (Arthrospira) and nutrient removal in a pig wastewater recycling process under tropical conditions. Journal of Applied Phycology, 2003. 15(2-3): p. 249-257.

Aitken, D., and Anitzar-Ladislo, A. Sustainable Fuel from Algae: Challenges and New Directions. In: Gikonyo B., ed. Advances in Biofuel Production: Algae and Aquatic Plants, Waretown, NJ and Oakville, Ontario, Canada: Apple Academic Press, Inc. Adapted by the authors from Aitken, D., and Anitzar-Ladislo, A. Achieving a Green Solution: Limitations and Focus for Sustainable Algal Fuels, in Energies 2012;5(5): 1613-1647. Used with permission.

PART II

BIOFUEL FOR TODAY

CHAPTER 3

THE ROLE OF BIOENERGY IN A FULLY SUSTAINABLE GLOBAL ENERGY SYSTEM

STIJN CORNELISSEN, MICHÈLE KOPER, and YVONNE Y. DENG

3.1 INTRODUCTION

To reduce human dependence on fossil fuels and to reduce climate change, we need to make a switch to a fully renewable energy system with no or low associated greenhouse gas emissions. In our Ecofys Energy Scenario modelling, we describe the transition to such a system towards 2050 reaching 95% renewable energy without a major reduction in activity levels [1]. In the Ecofys Energy Scenario we describe how energy efficiency options and non-bioenergy renewable options can accommodate the majority of this transition. However, after applying these options, there is still a large demand that needs to be met with sustainable bioenergy options. This is illustrated in Fig. 1.

From Fig. 1 it becomes clear that the majority of this remaining demand consists of energy carriers that cannot be easily provided by renewable options other than bioenergy. These include:

- Transport fuels where energy storage density is often a crucial factor; especially:
  - Long distance road transport
  - Aviation
  - Shipping
- Industrial fuels where electric or solar heating is insufficient; especially:
  - Long distance road transport
  - Applications that require a specific energy carrier, e.g. a gaseous fuel or solid fuel. One example is the steel industry where the structural strength of a solid fuel is required.

In this study we answer the question: can we meet the remaining energy demand in a fully renewable system like the Ecofys Energy Scenario using only sustainable bioenergy options?

In literature, there have been several attempts to quantify the potential of biomass available for energy supply with varying degrees of sustainability constraints. Estimates can differ within a very large range, depending on whether the study takes a holistic view at land management and how stringent the applied sustainability criteria are. Not many of these studies look at the end uses for this biomass potential in detail [2], [3], [4], [5], [6], [7], [8], [9], [10] and [11]. Other studies postulate the use of biomass to fill a demand need, but do not always specify in detail where this biomass would come from [12], [13] and [14]. None of these studies is as comprehensive, as stringent in the applied sustainability criteria and as detailed on both the supply potential and the demand side use of biomass as the study we present here.

This is represented by our detailed bioenergy potential modelling approach that acknowledges that:

- Bioenergy requires a thorough analytical framework to analyse sustainability, as cultivation, harvesting and processing of biomass and use of bioenergy have a large range of associated sustainability issues.
- Bioenergy encompasses energy supply for a multitude of energy carrier types, e.g. heat, electricity and transport fuels, using a multitude of different energy sources. Therefore a detailed framework of conversion routes is needed.

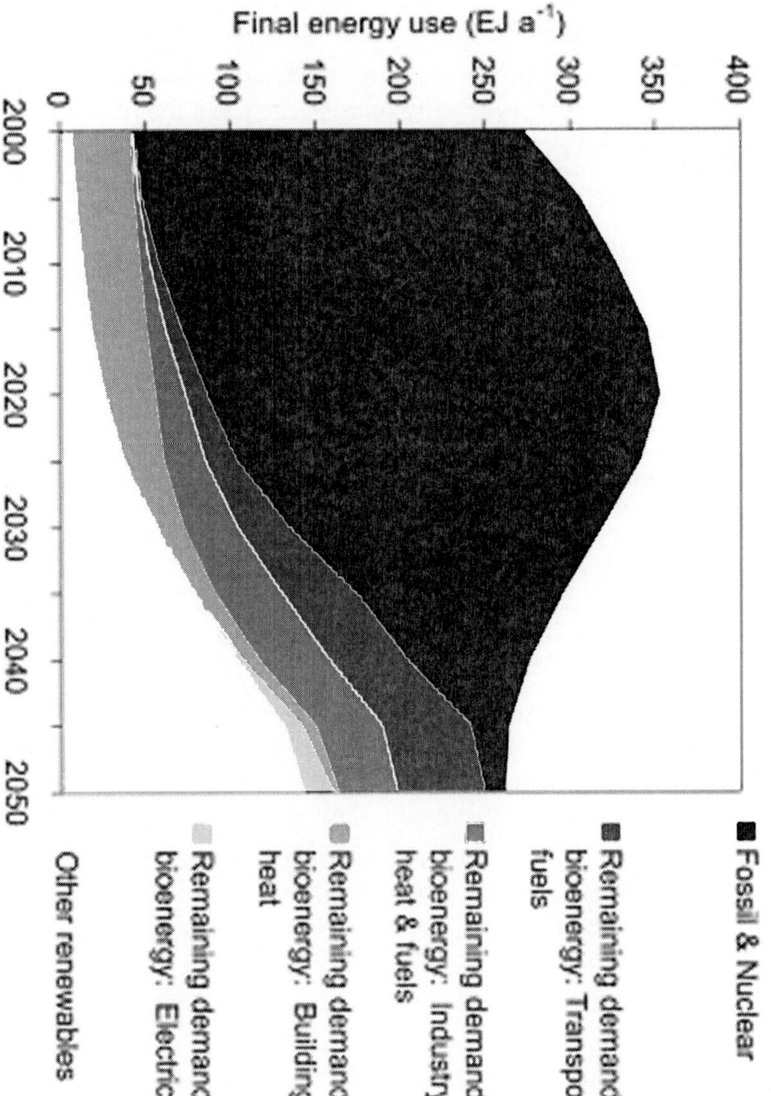

FIGURE 1: The need for bioenergy to fill the demand gap. Remaining demand in Ecofys Energy Scenario after applying energy efficiency, other renewable energy options and a small amount of remaining fossil fuels. This demand needs to be met with bioenergy options.

The outline of this approach is described in Section 2. For clarity, Section 3 presents detail on the steps of the followed methodology alongside the results obtained from the analysis. In Section 4 we discuss the sensitivity of our results and compare them to bioenergy potential values found in literature. In Section 5 we draw conclusions from our work and answer whether sustainable bioenergy options can meet the remaining demand in the Ecofys Energy Scenario.

3.2 APPROACH

From the demand modelling and supply modelling of all non-bioenergy options in the demand scenario, we find a remaining energy demand that needs to be met with bioenergy options. We applied technological choices (Section 2.1) and sustainability criteria (Section 2.2) to establish how this demand could be met.

3.2.1 BIOENERGY CONVERSION ROUTES

Because biomass can provide energy supply in a multitude of different energy carrier types, often in the same conversion route, the biomass use was channelled through a multitude of bioenergy routes, taking into consideration residues resulting from some of these routes. This approach is illustrated in a simplified diagram in Fig. 2.

In order to keep the projection robust, the key principle in selecting the supply and technology options was to only use options that are currently available or for which only incremental technological development is needed. One exception, where technological change of a more radical character is needed, is the inclusion of oil from algae as a supply option. To allow for development still needed in algae growing and harvesting, we included the use of significant amounts of algae oil from 2030 onwards only.

Another important assumption is made on the traditional use of biomass. Currently, about 35 EJ of primary biomass is used in traditional applications. This consists primarily of woody biomass and agricultural

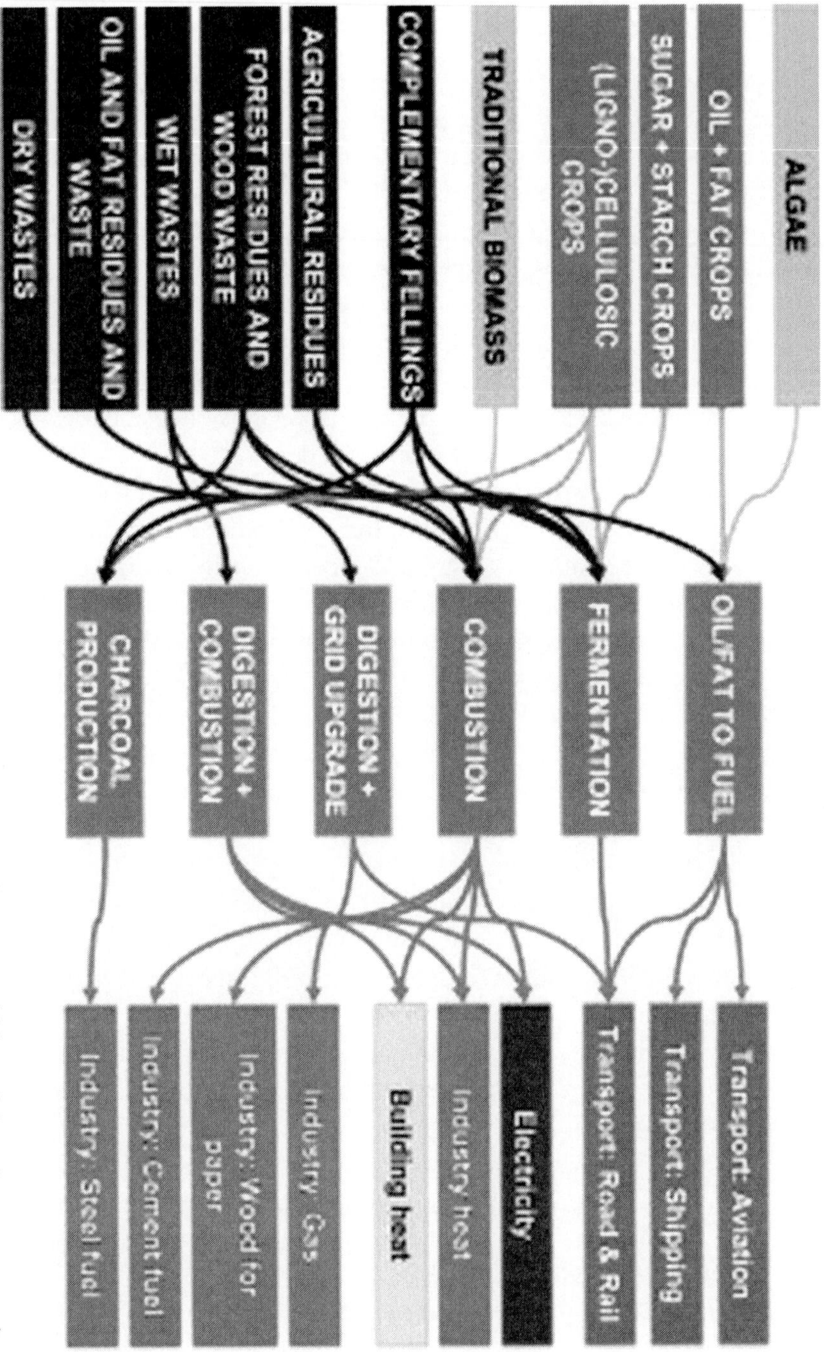

FIGURE 2: Bioenergy approach. Shown are the different biomass input types and conversion routes which result in various energy carrier types that meet demand.

residues harvested for home heating and cooking in developing countries. Towards 2050, the Ecofys Energy Scenario will supply energy for these demands through a route alternative to traditional biomass use. The traditional use of biomass is therefore phased out over time. A proportion of this phased out biomass is used within this work in a sustainable manner in other bioenergy routes, see also Section 3.3.2.

Within this study, the different bioenergy routes displayed in Fig. 2 are prioritised according to their feedstock as follows:

1. Traditional biomass: As this biomass is currently in use, it is used first in the Ecofys Energy Scenario to reflect the current situation. Over time, the contribution of this category becomes less as it is completely phased out towards 2050.
2. Sustainable residues and waste: Sustainable residues and waste, originating from agriculture, forestry and the food processing industry for example, are used to meet as much demand as possible.
3. Sustainable complementary fellings: This category consists of woody biomass gained from sustainable harvesting of additional forest growth and of the sustainable share of the biomass currently used in traditional uses. It is used to fill remaining demand in lignocellulosic routes, as much as possible.
4. Sustainable energy crops: Energy crops are used to fill as much of the remaining energy demand as possible while staying within their sustainable potential. Energy crops include oil crops, starch and sugar crops and (ligno) cellulosic crops.
5. Sustainable algae: Algae are used to yield oil to fill the remaining demand in the oil routes. Algae are used last because their growing and harvesting is currently not a proven technology on a commercial scale. However, although they are prioritised below sustainable energy crops, their use can be necessary before the entire potential for cropland for energy cropping is used. This is caused by the fact that we based our land availability on suitability for growing different crop types and only a share of the cropland potential is suitable for growing oil crops. Section 3.1 provides more detail on this methodology.

We assumed conversion efficiencies for each route as displayed in Table 1.

TABLE 1: Conversion efficiencies of bioenergy routes used in the Ecofys Energy Scenario. Values reflect the ratio between energy contents of inputs and outputs.

Technology	Biomass type	Conversion efficiency	Comments
Oil/fat to fuel	Oil from oil crops	88%	Efficiency is fuel output compared to oil input
			Additional inputs per MJ fuel: 0.14 MJ of heat, 0.01 MJ electricity
			Additional outputs per MJ fuel: ~1.5 MJ of residues
	Oil from algae	80%	Efficiency is fuel output compared to oil input and includes processing heat and electricity inputs
			Residues from algae are recycled in cultivation step
Fermentation	Starch or sugar	80%	Efficiency is fuel output compared to sugar/starch input
			Additional inputs per MJ fuel: 0.25 MJ of heat, <0.01 MJ electricity
			Additional outputs per MJ fuel: ~1.4 MJ of residues
	(Ligno) cellulose	39%	Efficiency is fuel output compared to (ligno) cellulose input and includes processing heat and electricity input
			Additional outputs per MJ fuel: ~1.0 MJ of residues
Combustion	Industrial direct fuel	100%	The conversion efficiencies are effectively included in the industrial demand numbers
			Valid for wood fuel for paper and cement kiln fuel
	Dry waste from municipal solid waste	78% (building heat)	Low efficiencies assumed because of suboptimal fuel, suboptimal combustion process and necessary flue gas cleaning
		73% (industry heat)	
		30% (electricity)	
	Other combustible biomass	95% (building heat)	Based on current best practices
		90% (industry heat)	
		40% (electricity)	

TABLE 1: *Cont.*

Technology	Biomass type	Conversion efficiency	Comments
Digestion + upgrading to gas grid quality	All wet wastes	52%	Based on 67% digestion efficiency
			Reduced by losses because of gas cleaning and compression
			End carrier is clean biogas which is equal to natural gas
Digestion + combustion	All wet wastes	60% (building heat)	Based on 67% digestion efficiency
			Combustion efficiency to end carrier equal to combustion of "other combustible biomass". Exception: 45% electric efficiency because gas engine can be used.
		57% (industry heat)	
		23% (electricity)	
Charcoal production	All woody biomass	40%	Based on current best practices

3.2.2 SUSTAINABILITY OF BIOENERGY

The bioenergy supply must use sustainable sources and create high greenhouse gas emission savings compared to fossil references in order for the Ecofys Energy Scenario, to which this study is related, to be considered sustainable. To ensure this we have derived a set of sustainability criteria to assess the sustainable bioenergy potentials from residues, wastes, complementary fellings, energy crops and algae that we apply in this study. An overview of these criteria is presented in Table 2. For reader convenience, the details and application of these criteria are described together with their results in the Section 3 as they are closely related to each other.

TABLE 2: Sustainability criteria used for bioenergy in this study.

Topic	Subtopic	Criteria applied to ensure sustainability topic is addressed
Land use and food security	Current land use	Exclusion of conversion of current forested, protected and urban land and agricultural cropland
	Agricultural water use	Exclusion of areas not suitable for rain-fed agriculture
	Biodiversity protection	Partially contained in current land use criterion
		Additional exclusion of land with high biodiversity valu
	Human development	Partially contained in current land use criterion
		Additional exclusion of land for human development
	Food security	Partially contained in current land use criterion
		Additional exclusion of land for meeting growth in food demand
Agricultural and processing inputs	Processing water use	Closed loop for processing water in biofuel production
	Agricultural nutrient use	N fertiliser production from sustainable energy and feedstock
		P and K fertiliser use: closed loop approach
Complementary fellings	Sustainable use of additional forest growth	Exclusion of protected, inaccessible and undisturbed forest areas
		Exclusion of non-commercial species
		Exclusion of wood needed for industrial fibre purposes
	Use of sustainable share of traditional biomass	Exclusion of 70% of the current traditionally used biomass
Residues and waste	Availability of residues	Exclusion of residues that are not available due to other uses and sustainability reasons
	Sustainable waste use	Additional recycling
		Exclusion of waste from non-renewable sources

3.3 SUSTAINABLE BIOMASS AVAILABILITY

3.3.1 SUSTAINABILITY OF BIOENERGY: LAND USE AND FOOD SECURITY

Our work explicitly prioritises a number of land uses over land use for bioenergy cropping. In addition, we restrict bioenergy cropping to land suitable for rain-fed cultivation of energy crops in order not to require irrigation as an agricultural input.

Therefore the following land is not used for bioenergy cropping in this study:

- Land used for supplying food, feed and fibre; taking into account future population growth and a diet change scenario
- Land used for protection of biodiversity and high carbon stock forest ecosystems
- Land used for human development by expanding the built environment
- Land not or marginally suitable for rain-fed cultivation of energy crops

We performed an assessment of land potential for rain-fed cultivation of energy crops based on this land use prioritisation. This assessment was based on data from a recent IIASA study [15]. Section 2.7 in that report provides an assessment of production potential for different bioenergy crops. We used the source data of this study and additional own analyses to perform the assessment using a stepwise approach. In this stepwise approach we started with the total global land mass excluding Antarctica and applied a number of exclusions to account for the sustainability criteria for land use in our work. This resulted in a 6,730,000 km^2 (673 million hectares) potential for energy crops in this study as shown in Fig. 3.

The assessed potential is located on grassland and non-densely vegetated woodland. Most land of these types is currently used as low-intensity grazing lands for livestock. It can be made available for other purposes through a combination of limiting future demand for livestock products and intensifying livestock systems with a low current intensity. As a reference,

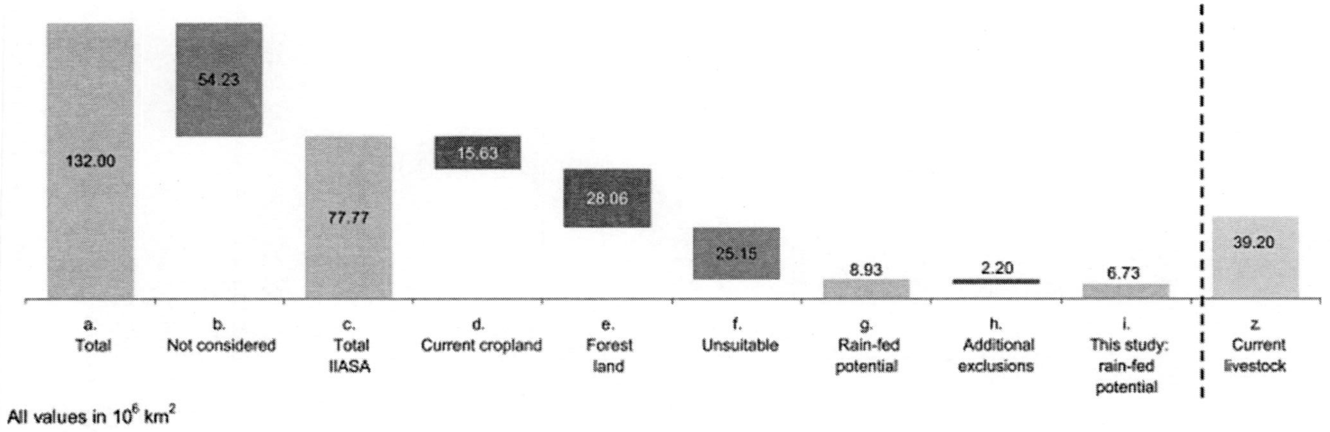

FIGURE 3: Results of land potential assessment for rain-fed cultivation of energy crops. a. Total global land mass (excluding Antarctica); b. Excluded: protected land, barren land, urban areas, water bodies; c. Total land considered in the IIASA study; d. Excluded: current agricultural cropland; e. Excluded: unprotected forested land; f. Excluded: not suitable for rain-fed agriculture; g. Potential for rain-fed agriculture; h. Excluded: additional land for biodiversity protection, human development, food demand; i. Potential for energy crops found in this study; z. Current land used to support livestock (for reference only; overlaps with other categories).

Figure options

the total land currently used to support livestock is included in Fig. 3. The value of 39,200,000 km^2 (3920 million hectares) is based on estimates of the agricultural cropland used to grow animal feed crops [15] and the land used as permanent meadow or pasture [16].

The applied exclusions were as follows:

1. Exclusion of all current protected land, barren land, urban land and inland water bodies as they cannot be used for agriculture. According to the IIASA data, this totals 54,230,000 km^2 (5423 million hectares).
2. Exclusion of current agricultural cropland to safeguard current food production. According to the IIASA data, this totals 15,630,000 km^2 (1563 million hectares).
3. Exclusion of conversion of all current unprotected forested land to protect forest biodiversity and forest carbon stocks. According to the IIASA data, this totals 28,060,000 km^2 (2806 million hectares).
4. Exclusion of all land that is not or marginally suitable for rain-fed agriculture to ensure only rain-fed bioenergy cropping. We derived from the IIASA data that this totals 25,150,000 km^2 (2515 million hectares). This is based on a conservative assumption where we used the highest per-crop value for availability of moderately suitable, suitable and very suitable land among the different crops analysed in the IIASA study as the value for all crops combined. This means that we assume that there might be a 100% spatial overlap of the given per-crop potentials. The used availability of moderately suitable, suitable and very suitable land is therefore the minimum availability, meaning the applied exclusion of not or marginally suitable land is the maximum exclusion possible based on the data set.
5. Exclusion of additional land for future requirements for biodiversity protection, human development and food demand. These amounted to an additional total of 2,200,000 km^2 (220 million hectares). This number was based on Ecofys analyses discussed in more detail in Sections 3.1.1 through 3.1.3.

3.3.1.1 BIODIVERSITY PROTECTION

The German Advisory Council on Global Change (WBGU) performed a study on the conservation of the biosphere [17]. One of the conclusions of this work was that between 10 and 20% of the world land mass should be protected to preserve the different functions of the biosphere, such as climate regulation, and its biodiversity.

Recent statistics provided by the World Database on Protected Areas state that 14% of the land mass is currently protected [18].

Therefore, to be at the upper limit of the range put forward by the WBGU, an additional 6% of the global land mass should be protected. Although it is not known where this 6% is located, we assume in our calculations that meeting this requirement will also reduce the land potential for rain-fed cultivation of energy crops by 6%. The reduction then totals 540,000 km^2 (54 million hectares). This reduction is additional to the exclusion of currently protected land based on the IIASA study.

3.3.1.2 HUMAN DEVELOPMENT

Hoogwijk performed a study on potentials of renewable energy sources, including an assessment of the increase in land use for the built environment [3]. Current land use for the built environment was estimated to be 2% of the total global land mass excluding Antarctica. United Nations projections estimate this land use to be 4% in 2030 [19].

We therefore assume that land use for the built environment will increase from the current 2% to 4% in 2030. We extrapolated this figure, using the population growth numbers used throughout the Ecofys Energy Scenario, to a 5% land use in 2050. The growth from current to 2050 land use for the built environment therefore requires excluding 3% of global land mass, excluding Antarctica, for this purpose.

Next, we have assumed that all of this expansion will take place on unprotected grass- and woodland because expansion into other land types is either not possible, not acceptable or much less likely. 3% of the global land mass, excluding Antarctica, amounts to 12% of the unprotected grass and woodland. We have therefore reduced the land potential for rain-fed cultivation of energy crops by 12% for human development.

The reduction then totals 1,040,000 km^2 (104 million hectares). This reduction is additional to the exclusion of urban areas based on IIASA data.

3.3.1.3 FOOD DEMAND

The need for agricultural cropland to meet future food demand is a highly debated issue. We took a pragmatic approach to assess whether or not the current agricultural cropland is able to sustain future food demand growth. The premise of our approach was the assumption that, in 2005, food supply equalled food demand; both were indexed at 100%. For this analysis, we did not analyse current or future food distribution patterns that might lead to local food shortages. We only assumed that there is no shortage of food production at the global level for which we would need to set aside additional land and that there is no overuse of cropland at the global level which could be taken out of food production without affecting supply.

We forecast the evolution of food demand and supply to 2050 as follows:

We extrapolated the growth in food demand using the following stepwise approach:

1. We started with current per capita calorie values of ~10 MJ $caput^{-1}$ (~2400 kilocalories $caput^{-1}$) of plant product [20] and 1.46 MJ $caput^{-1}$ (350 kilocalories $caput^{-1}$) and 3.97 MJ $caput^{-1}$ (950 kilocalories $caput^{-1}$) animal product in non-OECD and OECD regions [21], respectively. Animal product calories were converted into crop equivalents with conversion factors based on the crop feed intake necessary to produce them. The basis for these factors were feed intakes of ~17 kg kg^{-1}, ~2.4 kg kg^{-1} and ~1.7 kg kg^{-1} of feed per produced amount of meat, eggs and dairy respectively. For meat, this factor was derived from literature values for feed efficiencies per animal type [2] and current distribution of consumption of

meat per animal type [22], which is a mixed diet of bovine, ovine, pig and poultry products. For eggs, literature values from [23] were used. For dairy, literature values from [2], [24] and [25] were used. The feed is assumed to have an energy content of 19 MJ kg^{-1} of dry matter [2].

2. We calculated a "business-as-usual" (BAU) per capita diet in the period 2005–2050 differentiated between OECD and non-OECD countries. This was done based on existing diet projections [22].
3. We then assumed that total animal product consumption worldwide will be constrained to a growth of no more than ~65% between 2005 and 2050, which means that the average animal product consumption per capita (in crop equivalents) increases by about 10% over the same timeframe, given population projections. In practice, it could be desirable to divide the global 10% increase across regions in a non-equal manner, e.g. a reduction in animal product consumption in OECD countries and a significant increase in non-OECD countries.
4. We then multiplied the constrained per capita diets with population growth numbers used in [1] to get a total growth in food demand in crop equivalents. This was indexed against 2005.

We extrapolated the growth in food supplied by the current agricultural cropland by using a yield increase of 1% per year. This value is an intermediate value in a range of yield increase projections of 0.4–1.5 % found in literature ([7], [8], [22], [26] and [27]). The impact of climate change on yield projections was not explicitly considered in this analysis. However, by choosing the intermediate value of yield increase projections we have tried to be moderate in our assumptions. This yield increase was applied to the indexed value in 2005 of 100%.

The results for the extrapolated food demand and food supply are presented in Fig. 4. For reference, Fig. 4 also contains the indexed yield development of coarse grains over the last 50 years, which has been higher than the 1% assumed in this work.

From the graph in Fig. 4 it can be observed that, based on our assumptions the current agricultural cropland is projected to be able to supply the entire food demand in 2050. However, in intermediate years this is not

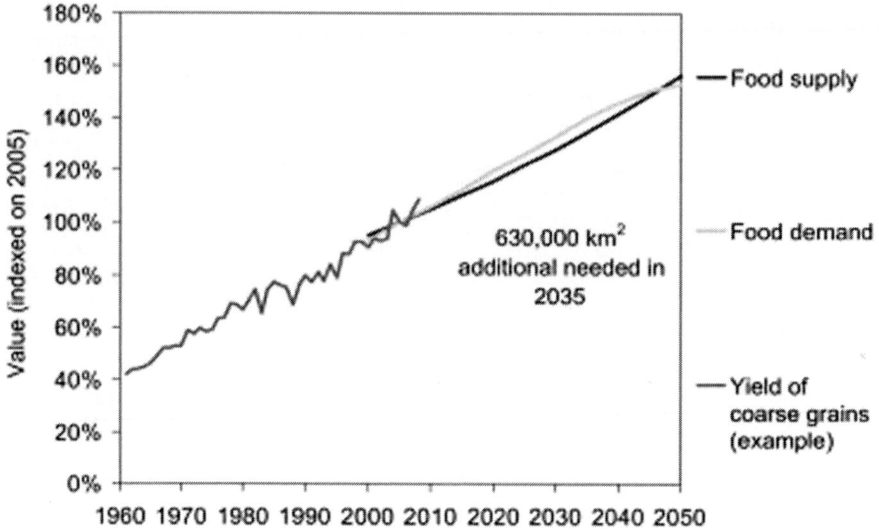

FIGURE 4: Calculation of land exclusions in our study based on global food security.

always the case. We have calculated that the maximum shortage of food from current agricultural cropland occurs in 2035 and amounts to about 4% of current agricultural cropland. This totals 630,000 km^2 (63 million hectares).

Although the identified 630,000 km^2 (63 million hectares) is the largest amount of additional land needed for meeting food demand in any given year to 2050, we choose to exclude it from the potential over the entire period in our analysis. This reduction is additional to the exclusion of current agricultural cropland based on IIASA data.

3.3.2 SUSTAINABILITY OF BIOENERGY: AGRICULTURAL AND PROCESSING INPUTS

To guarantee sustainable agriculture and processing of energy crops, we included a sustainable framework for the inputs required for them.

The yield projections for energy crops in this study are based on rainfed agricultural systems where nutrients are added to the land.

3.3.2.1 AGRICULTURAL WATER USE

Regarding agricultural water use, this means that no irrigation is used for the energy crops in this study. Energy crop yields are scaled in accordance with the land's suitability for rain-fed agriculture to reflect this. This means that most of the yields in this work are at around 50–70% of the maximum yield currently obtained in high input agricultural systems.

Table 3 gives ranges for these yields, as they differ per region. The number given in Table 3 is in primary biomass yield of the main product.

TABLE 3: Yields of energy crops used in this study in primary biomass yield of the main product.

Crop type	Range of yields across the regions (GJ ha−1)	Comments
Oils + fats	25–35 (~0.5–1 tonne ha^{-1} of oil)	Equates to ~22–31 GJ ha^{-1} of transport fuel
		Number includes only primary oil yields; agricultural and fuel processing residues are included elsewhere
		Marker crops: rapeseed, soybeans and oil palm
Sugar + starch	62–121 (~4–7 tonne ha^{-1} of starch or sugar)	Equates to ~49–95 GJ ha^{-1} of transport fuel
		Number includes only primary starch/sugar yields; agricultural and fuel processing residues are included elsewhere
		Marker crops: sugar cane and maize
		Highest yields in South America due to suitability for sugar cane
(Ligno) cellulosic crops	160–230 (~8–12 tonne ha^{-1} of dry matter)	Equates to ~61–88 GJ ha^{-1} of transport fuel
		Number includes all primary biomass yields; fuel processing residues are included elsewhere

3.3.2.2 AGRICULTURAL NUTRIENT USE

Regarding agricultural nutrient use, we envisage closed nutrient loops where possible. In addition, we included in the demand side of the Ecofys Energy Scenario, to which this study is related [1]:

- All heat energy input required to produce N fertiliser for bioenergy cropping.
- All electricity required to produce hydrogen for N fertiliser for bioenergy cropping. This mechanism can also be used as storage for supply-driven renewable electricity sources (see [1]).

3.3.2.3 PROCESSING WATER USE

We assessed, from literature, expert opinions and our previous experience, that a closed loop approach for biofuel processing water is currently commercially available.

3.3.3 SUSTAINABILITY OF BIOENERGY: COMPLEMENTARY FELLINGS

We have analysed the sustainable potential for the harvesting of woody biomass from forests for energy purposes, taking into consideration the demand for wood for other uses.

3.3.3.1 ADDITIONAL FOREST GROWTH

Sustainable additional forest growth is defined here as growth that is currently not harvested and that:

- is not needed for future growth in demand for industrial roundwood, (e.g. for construction or paper production)
- can be harvested in an ecologically sound way.

The potential for sustainable additional forest growth was primarily based on a study by Smeets [2]. According to the study, the world's technical potential for additional forest growth would be ~64 EJ of woody biomass in 2050. However, the ecologically constrained potential is found to be ~8 EJ. The difference lies in the exclusion of all protected, inaccessible and undisturbed areas from the ecological potential. This means that only areas of forest classified as 'disturbed and currently available for wood supply' are included. A further sustainability safeguard is the use of only commercial species in the gross annual increment, rather than all available species.

Because the calculations by Smeets are partially based on an older source from 1998 [28], an additional calculation was done for a selection of six countries (Brazil, Russia, Latvia, Poland, Argentina and Canada). This was considered necessary because in some of these countries, the area of 'disturbed' forest could have considerably changed in the time period from 1998 until today.

In the additional calculation, the amount of sustainable complementary fellings resulting from the additional disturbed forest area available for wood supply, compared to the original base data, was determined, based on more recent country reports for the Global Forest Assessment 2010 (with data ranging from 2004 to 2008) [29]. This resulted in additional potential, particularly for Russia and Canada. The main differences were caused by the updated statistics on disturbed forest available for wood supply. The additional potential for the six countries was added to the ecological potential from [2], resulting in a total global potential for sustainable additional forest growth of ~27 EJ.

3.3.3.2 SUSTAINABLE SHARE OF TRADITIONAL BIOMASS USE

We also included the sustainable share of current use of biomass for traditional uses in the category 'Complementary fellings'. This biomass is currently primarily used for domestic heat production. We worked upon the assumption that the majority of this use is woody biomass, though other sources will contribute.

The demand side scenario, to which this work is related [1], postulates that any traditional use of biomass that is considered unsustainable today will be gradually phased out and replaced with more sustainable approaches, such as solar thermal heating.

No literature data was available on the sustainable share of current traditional biomass use. Therefore, as the scenario gradually phases out traditional biomass use, we have estimated that 30% of the phased out biomass can be harvested sustainably. This amounts to approximately 11 EJ of worldwide potential for this category.

3.3.4 SUSTAINABILITY OF BIOENERGY: USE OF RESIDUES AND WASTE

We performed a literature study on residues and waste potentials in 2050 for five main categories:

- Oil and fat residues and waste
- Forestry residues and wood waste
- Agricultural residues
- Wet waste and residues
- Dry waste

In our study we analysed the potential for the most important subcategories of each main category. After obtaining the literature values for the potential of each residue and waste (sub)category, we performed three additional analyses to arrive at the final residue and waste potential figures:

1. We adapted literature projections for 2050 for manure and waste animal fat potential to reflect the meat consumption level described above.
2. We altered the dry waste potential from municipal solid waste (MSW) to reflect the fact that not all MSW is renewable and that some MSW is wet and some is dry.
3. We updated the recoverable fraction, the share of residues and waste that is available for bioenergy production, for some of the categories because they were inconsistent with other developments such as improved future economic feasibility of residue collection or our study's

framework of closed nutrient loops. The recoverable fractions used always take into account other uses, e.g. the use of wood residues for production of fibre board, and sustainability considerations.

The results of our analysis are displayed in Table 4. Table 4 also includes data on the used recoverable fractions per subcategory. In some cases, the recoverable fraction was implicitly included in the analysis of the literature sources we consulted; in those cases the value cannot be reported. It should be noted that the reported recoverable fractions do not include the residue-to-product ratio, which is a measure of how much residue is produced per quantity of main product. This ratio was implicitly included in the analyses of the consulted literature sources.

3.3.5 SUSTAINABLE ALGAE

Our study uses algae oil to supply remaining demands in oil routes after the use of residues, waste and bioenergy crops. Because commercial scale algae growing and harvesting is currently still in development, we only include significant algae use from 2030 onwards. The approach to using algae in our work is based on a recent Ecofys study [30] on the worldwide potential of aquatic biomass. This study identified a number of different long-term feasible potentials for aquatic biomass. The most conservative scenario only contains algae oil from microalgae grown in open ponds on non-arable land filled with salt water. The total potential for algae oil from this technology was estimated at 90 EJ of oil. We envisage an algae cultivation system where the non-oil algae biomass are used to provide a nutrient loop where possible and the required energy for the cultivation and processing of the algae.

The maximum amount of algae oil actually used in the demand side scenario [1] is 21 EJ of oil in 2050. Based on the yields calculated in the Ecofys study, this amounts to approximately 300,000 km^2 (30 million hectares) use of non-arable land. The 21 EJ oil use is about 25% of the 90 EJ algae oil potential identified in the most conservative scenario containing only algae oil from microalgae grown in open ponds on non-arable land filled with salt water. Therefore, the algae oil use in the demand side scenario fits comfortably within the potential identified in the Ecofys study, especially as further potential from algae cultivation in open water may be tapped due to future technological progress.

TABLE 4: Results of our literature study on residue and waste potential for bioenergy in 2050.

(Sub)category	Recoverable fraction ranges across regions (%)	2050 potential (EJ)	References used
Oils and fats		1	[39], [40] and [41]
Animal fat	45%	<1	
Used cooking oil	–	<1	
Forestry residues and wood waste		25	[2]
Logging residues	–	~5	
Wood processing residues	–	~10	
Wood waste	–	~10	
Agricultural residues		25	[42], [43], [44], [45], [46], [47], [48], [49], [50], [51], [52], [53], [54], [55], [56], [57], [58], [59], [60], [61], [62], [63] and [64]
Cereals	30–40%	~22	
Rapeseed	30–40%	<1	
Coffee	75%	<1	
Soy	30–40%	~3	
Wet waste and residues		38	[42], [44], [47], [51], [55], [60], [62], [65] and [66]
Sugar beet processing residues	50%	<1	
Potato processing residues	50–55%	<1	
Manure	–	~30	
Oil palm empty fruit bunches	70%	~2	
Palm oil mill effluent	100%	<1	
Sugar cane	19%	~1	
Cassava	50%	<1	
Wet municipal solid waste	–	~4	
Dry waste		11	[67], [68], [69] and [70]
Dry municipal solid waste	–	~11	
Total		101	

3.3.6 SUMMARY OF RESULTS ON SUSTAINABLE BIOMASS AVAILABILITY

Sections 3.1 through 3.5 describe how we obtained the results for sustainable bioenergy availability. We summarise our results on availability here and compare them to the use found in the demand side scenario [1].

Our sustainable biomass availability only includes bioenergy supply that meets the sustainability criteria in Section 2.2. Fig. 5 shows that the demand side scenario [1] is capable of meeting demand with bioenergy using the available sustainable biomass and rain-fed land for energy cropping identified in this study.

3.3.7 GREENHOUSE GAS EMISSION SAVINGS

We have performed a life cycle analysis of the greenhouse gas (GHG) emissions associated with bioenergy use in the Ecofys Energy Scenario. We have included GHG emissions from six different contributors in the bioenergy life cycle as displayed in Table 5.

TABLE 5: Types of greenhouse gas emissions included in bioenergy life cycle analysis in the Ecofys Energy Scenario.

Type of emissions	Reference
Emissions from land use change when land is converted to bioenergy cropland.	[31]
Emissions from the production and application of nitrogen fertiliser for bioenergy crops and algae.	[30], [31] and [32]
Emissions from agricultural fuel inputs for cultivation of bioenergy crops.	[33]
Emissions from transport of biomass to the processing site.	[34]
Emissions from energy inputs during bioenergy conversion.	[35]
Emissions from transport of the bioenergy carrier to the end use location.	[34]

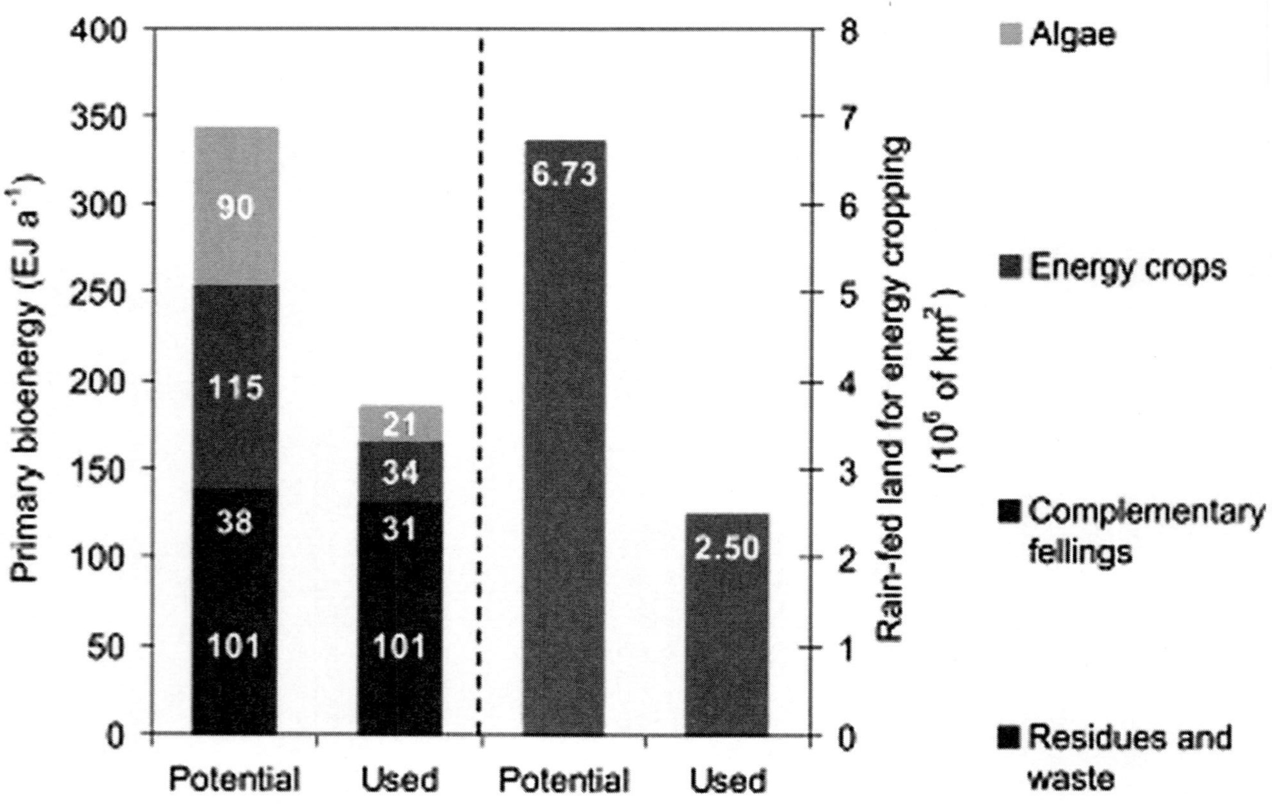

FIGURE 5: Bioenergy potential versus use. Overview of the Ecofys Energy Scenario's sustainable bioenergy use and rain-fed land use for energy cropping versus sustainable potential found in this study.

Most of these contributors include emissions associated with energy use. As the Ecofys Energy Scenario drastically increases the share of renewable energy technologies that have low or no GHG emissions, we have made two separate calculations: in one the emission factors for the energy inputs were extracted from the IPCC fossil fuel references [36] and in the other, they were taken from Ecofys Energy Scenario data. Therefore, we present results as a range.

Fig. 6 shows the results of the life cycle analysis. For 2050, we have calculated that the GHG emissions in CO_2 equivalent associated with bioenergy are 12–18 g MJ^{-1} final energy use. The values for the corresponding fossil references in CO_2 equivalent are 70–80 g MJ^{-1}. This means that even for the most conservative calculation (with fossil reference for energy inputs) average GHG emission savings are ~75%. When the corresponding Ecofys Energy Scenario values are used, average GHG emission savings are ~85%. This level of emission saving is consistent with an overall emission reduction from the energy system which would result in global average temperature increase of less than 2 °C above pre-industrial levels [1], [37] and [38].

3.4 DISCUSSION

In our work we have shown that the Ecofys Energy Scenario is capable of meeting demand with bioenergy within the sustainable potential we identified in this study. In this section we discuss this result in two ways: first, we compare our results on bioenergy use and the associated land use to energy and land potentials found in other studies. Then we show the sensitivity of our results to developments in the agricultural and food sectors such as diet and yield changes.

3.4.1 COMPARISON OF OUR BIOENERGY POTENTIAL TO OTHER STUDIES

We compare the land used for energy crops and the primary bioenergy use of bioenergy crops and algae in the Ecofys Energy Scenario with literature values on potentials ([2], [3], [6], [10] and [11]).

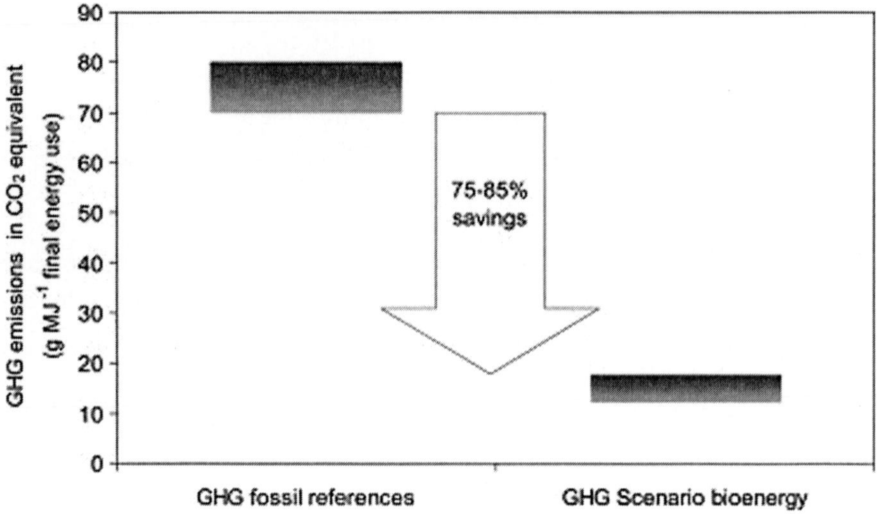

FIGURE 6: Overview of the Ecofys Energy Scenario's bioenergy greenhouse gas (GHG) emissions compared to fossil references.

In this comparison we differentiate between studies that applied no or few sustainability criteria (generally on food security and biodiversity) and those that applied a set of sustainability criteria in the same range of types as in this study (including e.g. criteria on water use, soil protection, degradation of land and deforestation and forest carbon stocks).

Fig. 7 shows that the land used for bioenergy cropping in the Ecofys Energy Scenario is at the lower end of the range of potentials found in literature. It is important to note that the given land use for energy crops in the Ecofys Energy Scenario is the maximum amount used during the 2005–2050 timeframe.

Fig. 8 demonstrates that the primary bioenergy use from energy crops and algae in the Ecofys Energy Scenario is at the lower end of the range of potentials for bioenergy from energy crops found in literature. It is important to note that the primary bioenergy use from energy crops in Fig. 8 is the maximum amount used during the 2005–2050 timeframe. This maximum use occurs in 2035 and use is lower in all other years.

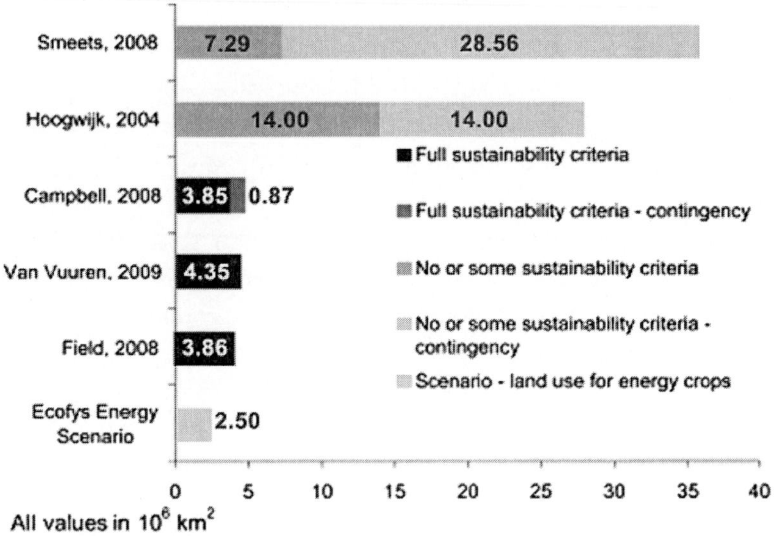

FIGURE 7: Area potential comparison. Comparison of global land use in the Ecofys Energy Scenario with bioenergy area potentials from literature ([2], [3], [6], [10] and [11]). Contingency indicates the author gave the potential as a range rather than a definitive number.

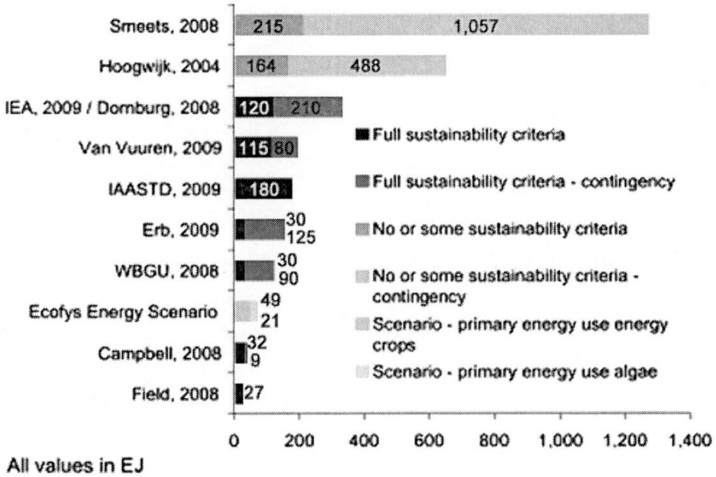

FIGURE 8: Biocrop energy comparison. Comparison of global primary energy use from energy crops and algae in the Ecofys Energy Scenario with primary bioenergy potentials for energy crops from literature ([2], [3], [4], [5], [6], [7], [8], [9], [10] and [11]). Contingency indicates the author gave the potential as a range, e.g. due to uncertainty in future yields. As [4] and [5] give the same numbers and partially have the same authors, they are grouped together.

3.4.2 SENSITIVITY OF OUR RESULTS TO DEVELOPMENTS IN THE AGRICULTURAL AND FOOD SECTOR

The vast majority of all agricultural production is directed at feeding the world's population; either directly, or indirectly by feeding livestock that supplies animal products. Other agricultural products, such as fibres for clothing, biomass for energy generation and tobacco production, make up a very small share of the total current agricultural production.

This means that developments in food demand and supply and the balance between them will heavily influence all agricultural and land use analyses, including the bioenergy potential analysis in this work. Forecasting these developments is a complex task for the following reasons:

- Demand: future food demand depends on the size of the world's population and the composition of its diet. This diet, in turn, depends on parameters such as wealth and cultural choice. Particularly important is the intensity of animal products in that diet, as these require a large amount of animal feed.
- Supply: future food supply depends on the area available for food cultivation and the food yield per unit of area. The evolution of this yield is hard to predict over larger timescales because it depends on numerous factors such as R&D results, technology adoption, education and sustainability requirements.
- Balance of demand and supply: the balance between food supply and demand is poorly understood. It is often argued that the current food supply is adequate for the entire global demand, but that unequal distribution leads to food shortages in parts of the world.

The land available for bioenergy cropping in our work is strongly dependent on the assumptions made in the food analysis. Where possible, we have used conservative assumptions (i.e. assumptions close to business-as-usual scenarios), with one notable exception; a constraint on the consumption of meat, creating a more sustainable diet, see Section 3.1.3. We performed a sensitivity analysis by varying parameters on food demand and supply and the balance between them. The effects of these variations on land available for bioenergy cropping are presented in Table 6. They show that our results on sustainable bioenergy potential are indeed sensitive to developments in food demand and supply and the balance between them.

TABLE 6: Sensitivity analysis. Effect of varying assumptions on food supply and demand and their balance on the land available for bioenergy cropping in this study.

Parameter	Example change compared to value used in this work	Cropland available for bioenergy (km^2)
Ecofys Energy Scenario results	–	6,730,000
Supply: Annual yield increase	0.4–1.5 % instead of 1%	3,000,000–10,800,000
Demand: Meat consumption	−30 to +40% instead of ~10% change in per capita consumption of animal products from 2005 to 2050	3,500,000–12,700,000
Balance of demand and supply	Supply is 90–110 % of demand in 2005 instead of being equal	5,000,000–8,000,000

3.5 CONCLUSION

The Ecofys Energy Scenario requires a significant share of bioenergy supply to meet the remaining energy demand after using other renewable energy options. We only include bioenergy supply that meets strict sustainability criteria and leads to high greenhouse gas emission savings when compared to fossil references. We safeguard this by applying a set of sustainability criteria to assess the sustainable bioenergy potentials from residues, wastes, complementary fellings, energy crops and algae.

We conclude that the potential identified in this study is capable of meeting the required demand of the Ecofys Energy Scenario [1] with bioenergy that meets these sustainability criteria and simultaneously accomplishes high greenhouse gas emission savings. The amount of sustainable bioenergy used and the amount of land used for sustainable bioenergy cropping in the Ecofys Energy Scenario are both near the low end of the range of potential values found in literature. In our sensitivity analysis we find, as expected, that the potential area available for bioenergy cropping is sensitive to developments in food demand and supply and the balance between them.

REFERENCES

1. Deng YY, Blok K, Van der Leun K. Transition to a fully sustainable, global energy system. 2012 [in preparation].
2. Smeets EMW. Possibilities and limitations for sustainable bioenergy production systems [dissertation]. Utrecht University; 2008.
3. Hoogwijk MM. On the global and regional potential of renewable energy sources [dissertation]. Utrecht University; 2004.
4. Bauen A (E4tech), Berndes G (Chalmers University of Technology), Junginger M (Copernicus Institute of the University of Utrecht), Londo M (ECN), Vuille F (E4tech). Bioenergy – a Sustainable and Reliable Energy Source. Main Report. Rotorua (New Zealand): IEA Bioenergy, Executive Committee; 2009 Jun.
5. V. Dornburg, Universiteit Utrecht, A. Faaij, Universiteit Utrecht, P. Verweij, Universiteit Utrecht, H. Langeveld, Wageningen UR, G. Van de Ven, Wageningen UR, F. Wester, Wageningen UR et al. E. Lysen, S. Van Egmond (Eds.), Assessment of global biomass potentials and their links to food, water, biodiversity, energy demand and economy. Main report, Netherlands Environmental Assessment Agency (WAB) MNP – WAB Secretariat, Bilthoven (Netherlands) (2008 Jan) Report 500102012
6. D.P. Van Vuuren, J. Van Vliet, E. Stehfest Future bio-energy potential under various natural constraints Energy Policy, 73 (11) (2009), pp. 4220–4230
7. B.D. McIntyre, H.R. Herren, J. Wakhungu, R.T. Watson (Eds.), Agriculture at a crossroads. Global report, IAASTD, Washington, DC (2009)
8. K.-H. Erb, H. Haberl, F. Krausmann, C. Lauk, C. Plutzar, J.K. Steinberger et al. Eating the Planet: feeding and fuelling the world sustainably, fairly and humanely – a scoping study Klagenfurt University, Faculty for Interdisciplinary Sciences, Institute of Social Ecology, Vienna (Austria) (2009 Nov) [Commissioned by Compassion in World Farming (UK) and Friends of the Earth (UK)]
9. R. Schubert, H.J. Schellnhuber, N. Buchmann, A. Epiney, R. Griesshammer, M. Kulessa et al. World in transition - future bioenergy and sustainable land use. Full report German Advisory Council on Global Change (WBGU), Berlin (Germany) (2009)
10. J.E. Campbell, D.B. Lobell, R.C. Genova, C.B. Field The global potential of bioenergy on abandoned agricultural lands Environ Sci Technol, 42 (15) (2008), pp. 5791–5794
11. C.B. Field, J.E. Campbell, D.B. Lobell Biomass energy: the scale of the potential resource Trend Ecol Evol, 23 (2) (2008), pp. 65–72
12. S. Teske, Greenpeace International, A. Zervos, C. Lins, J. Muth, EREC Energy [r] evolution – a sustainable world energy outlook. Version 3 Greenpeace International and EREC (2010 Jun.)
13. Shell International Shell energy scenarios to 2050. 2008 version. Energy scenarios: scramble and blueprints Shell International BV, The Hague (the Netherlands) (2008)
14. R. Priddle (Ed.), World energy outlook 2009, OECD/IEA, International Energy Agency, Paris (France) (2009)
15. G. Fischer, E. Hizsnyik, S. Prieler, M. Shah, H. Van Velthuizen Biofuels and food security Land Use Change and Agriculture Program International Institute for Ap-

plied Systems Analysis (IIASA), Laxenburg (Austria) (2009) [Commissioned by The OPEC Fund for International Development (OFID)]
16. FAOSTAT [Internet], Rome (Italy): Food and Agriculture Organization of the United Nations; 2009 [cited 2010 Mar]. Resources, ResourceSTAT, Land, Permanent meadows and pastures. Available from: http://faostat.fao.org.
17. H.J. Schellnuber, J. Kokott, F.O. Beese, K. Fraedrich, P. Klemmer, L. Kruse-Graumann et al. Conversation and sustainable use of the biosphere German Advisory Council on Global Change (WBGU), Bremerhaven (Germany) (1999)
18. World Database on Protected Areas [Internet]. Cambridge (UK): United Nations Environment Programme – World Conservation Monitoring Centre. [updated 2009 Jan, Cited 2010 Mar] Available from: http://www.unep-wcmc.org/wdpa/mdgs/WDPAPAstats_Jan09_download.xls.
19. R. Clarke, R. Lamb, D.R. Ward (Eds.), Global environmental outlook 3, United Nations Environment Programma, Nairobi (Kenya) (2002)
20. FAOSTAT [Internet] Food supply, crops primary equivalent [cited 2010 Mar] Food and Agriculture Organization of the United Nations, Rome (Italy) (2009) Available from: http://faostat.fao.org/
21. FAOSTAT [Internet] Food supply, livestock and fish primary equivalent [cited 2010 Mar] Food and Agriculture Organization of the United Nations, Rome (Italy) (2009) Available from: http://faostat.fao.org/
22. N. Alexandratos, J. Bruinsma, G. Bödeker, J. Schmidhuber, S. Broca, P. Shetty et al. World agriculture: towards 2030/2050-prospects for food, nutrition, agriculture and major commodity groups Food and Agriculture Organization of the United Nations, Rome (Italy) (2006 Jun) [Global Perspectives Studies Unit]
23. H. Blonk, A. Kool, B. Luske Milieueffecten van Nederlandse consumptie van eiwitrijke producten Blonk Milieu Advies BV, Gouda (The Netherlands) (2008 Oct) [Commissioned by Ministerie van VROM. Dutch]
24. D. Pimentel, M. Pimentel Sustainability of meat-based and plant-based diets and the environment Am J Clin Nutr, 78 (3) (2003), pp. 660S–663S
25. J. Linn, J. Salfer Feed efficiency Proceedings and Presentations of University of Minnesota Dairy Days (2006 Jan 6-20) multiple locations in Minnesota (US). Available from: http://www.ansci.umn.edu/dairy/dairydays/2006/proceedings/linn-salfer.pdf [cited 2010 Mar]
26. P. Vavra (Ed.), OECD-FAO agricultural outlook 2009 – 2018, OECD Publishing, Paris (France) (2009)
27. D. Van Vuuren, A. Faber Growing within limits. A report to the global assembly 2009 of the Club of Rome Netherlands Environmental Assessment Agency, Bilthoven (the Netherlands) (2009) [Publication NO.: 500201001]
28. G. Bull (Ed.), Global fibre supply model, FAO Forestry Department, Rome (Italy) (1998)
29. Global Forest Resources Assessment 2010. Country Reports. [Internet]. Rome (Italy): FAO Global Forest Resources Assessment. 2010. [cited 2010 Mar] Available from: http://www.fao.org/forestry/fra/67090/en/.
30. A. Florentinus, C. Hamelinck, S. De Lint, S. Van Iersel Worldwide potential of aquatic biomass Ecofys the Netherlands BV, Bio Energy Group, Utrecht (the Netherlands) (2008 May) [Commissioned by the Dutch Ministry of the Environment]

31. ,in: S. Eggleston, L. Buendia, K. Miwa, T. Ngara, K. Tanabe (Eds.), IPCC guidelines for national greenhouse gas inventories, Agriculture, forestry and other land use, vol. 4, Institute for Global Environmental Strategies, Kanagawa (Japan) (2006)
32. H. Sirviö (Ed.), Fertilizers, climate change and enhancing agricultural productivity sustainably (1st ed.), International Fertilizer Industry Association, Paris (France) (2009 Jul) [Task Force on Climate Change]
33. JEC Well-to-Wheel results [Internet]. JEC - Joint Research Centre-EUCAR-CONCAWE collaboration; Well-to-Wheel results. Version 3.0 [update 2008 Nov, cited 2010 Mar]. Available from: http://iet.jrc.ec.europa.eu/sites/about-jec/files/documents/WTT_App_2_v30_181108.pdf.
34. EC Well-to-Wheel results [Internet]. JEC - Joint Research Centre-EUCAR-CONCAWE collaboration; Well-to-Wheel results. Version 2C. Well-to-tank report Appendix 1. [update 2007 Mar, cited 2010 Mar]. Available from: http://iet.jrc.ec.europa.eu/sites/about-jec/files/documents/WTT_App_1_010307.pdf.
35. C. Hamelinck, K. Koop, M. Koper, Ecofys, H. Croezen, B. Kampman et al. Technical specification: greenhouse gas calculator for biofuels. Version 1.5 Ecofys BV, Utrecht (the Netherlands) (2007 Dec) [Commissioned by SenterNovem. Project No.: PBIONL062632]
36. ,in: S. Eggleston, L. Buendia, K. Miwa, T. Ngara, K. Tanabe (Eds.), IPCC guidelines for national greenhouse gas inventories, Energy, vol. 2, Institute for Global Environmental Strategies, Kanagawa (Japan) (2006)
37. N. Höhne, C. Ellermann, R. De Vos Emission pathways towards 2°C Ecofys, Utrecht (the Netherlands) (2009 Sep 14) [Commissioned by the Nordic COP 15 Group]
38. R.H. Moss, J.A. Edmonds, K.A. Hibbard, M.R. Manning, S.K. Rose, D.P. Van Vuuren et al. The next generation of scenarios for climate change research and assessment Nature, 463 (2010), pp. 747–756
39. F.D. Gunstone, J.L. Harwood, A.J. Dijkstra The lipid handbook with CD-rom Taylor & Francis Group, CRC Press (2007)
40. Clearpower Ltd A resource study on recovered vegetable oil and animal fats. Final report Sustainable Energy Ireland, Dublin (Ireland) (2003 Dec)
41. T. Caparella European renderers shape their future Render Mag (August 2009)
42. FAOSTAT [Internet], Rome (Italy): Food and Agriculture Organization of the United Nations. Statistics on crop and forestry production [cited 2010 Mar]. Available from: http://faostat.fao.org/site/377/default.aspx FAOSTAT [Internet], Rome (Italy): Food and Agriculture Organization of the United Nations; 2009 [cited 2010 Mar]. Production, Crops. Available from: http://faostat.fao.org/.
43. R. Siemons, M. Vis, D. van den Berg, BTG biomass technology group BV, Enschede, The Netherlands, I. Mc Chesney, M. Whiteley, ESD Ltd, Wiltshire, UK, N. Nikolaou, CRES, Pikermi Attiki, Greece Bio-energy's role in the EU energy market, a view of developments until 2020. Final report European Commission, Enschede (The Netherlands) (2004 Apr)
44. J. Cai, R. Liu, C. Deng An assessment of biomass resources availability in Shanghai: 2005 analysis Renew Sust Energy Rev, 12 (7) (2008), pp. 1997–2004
45. T. Chungsangunsit, S.H. Gheewala, S. Patumsawad Environmental assessment of electricity production from rice husk: a case study in Thailand Int Energy J, 6 (1) (2005), pp. 347–356

46. Damen K. Future prospects for biofuel production in Brazil. [thesis]. Utrecht University; 2002.
47. Dehue B. Palm Oil and its By-Products as a Renewable Energy Source: Potential, Sustainability and Governance. [thesis] Wageningen University; 2006.
48. European Environment Agency How much bioenergy can Europe produce without harming the environment? Final report European Environment Agency, Copenhagen (Denmark) (2006)
49. K. Ericsson, L.J. Nilsson Assessment of the potential biomass supply in Europe using a resource-focused approach Biomass Bioenergy, 30 (1) (2006), pp. 1–15
50. Gebrehiwot L, Mohammed J. The potential of crop residues, particularly wheat straw, as livestock feed in Ethiopia. Proceedings of the fourth annual workshop held at the Institute of Animal Research; 1987 Oct 20-27; Bamenda, Cameroon. [cited 2010 Mar]. Available from: http://www.fao.org/wairdocs/ilri/x5490e/x5490e0b.htm.
51. G. Fischer, H. van Velthuizen, M. Shah, F. Nachtergaele Global agro-ecological assessment for agriculture in the 21st century: methodology and results International Institute for Systems Analysis (IIASA), Laxenburg (Austria) (2002 Jan) [Includes CD-ROM. Report No.: RR-02–02]
52. G. Fischer, S. Prieler, H. Van Velthuizen, G. Berndes, A. Faaij, M. Londo et al. Biofuel production potentials in Europe: sustainable use of cultivated land and pastures, Part II: land use scenarios Biomass Bioenergy, 34 (2) (2010), pp. 173–187
53. P.K. Gupta, S. Sahai, N. Singh, C. Dixit, D.P. Singh, C. Sharma et al. Residue burning in rice–wheat cropping system: causes and implications Curr Sci India, 87 (12) (2004), pp. 1713–1717
54. JEC - Joint Research Centre-EUCAR-CONCAWE collaboration Well-to-Wheel results. Version 2C. Well-to-wheels report (2007 Mar) Available from: http://ies.jrc.ec.europa.eu/uploads/media/WTW_Report_010307.pdf
55. S. Kim, B.E. Dale Global potential bioethanol production from wasted crops and crop residues Biomass Bioenergy, 26 (4) (2004), pp. 361–375
56. I. Lewandowski, J. Weger, A. Van Hooijdonk, K. Havlickova, J. Van Dam, A. Faaij The potential biomass for energy production in the Czech Republic Biomass Bioenergy, 30 (5) (2006), pp. 405–421
57. I.C. Macedo, J.E.A. Seabra, J.E.A.R. Da Silva Greenhouse gas emissions in the production and use of ethanol from sugarcane in Brazil: the 2005/2006 averages and a prediction for 2020 Biomass Bioenergy, 32 (7) (2008), pp. 582–595
58. R. Lal World crop residues production and implications of its use as a biofuel Environ Int, 31 (4) (2005), pp. 575–584
59. S. Sokhansanj, S. Mani, M. Stumborg, R. Samson, J. Fenton Production and distribution of cereal straw on the Canadian prairies Can Biosystems Eng, 48 (2006), pp. 3.39–3.46
60. V. García Cidad, E. Mathijs, F. Nevens, D. Reheul Energiegewassen in de Vlaamse landbouwsector Publicatie 1 Steunpunt Duurzame Landbouw, Gontrode (Belgium) (2003)
61. Scarlat N, Dallemand JF, Martinov M, editors. Cereals straw and agricultural residues for bioenergy in New Member States and Candidate Countries. Workshop Proceedings; 2007 October 2–3; Novi Sad, Serbia. Luxembourg: Office for Official Publications of the European Communities; 2008.

62. M. De Wolf, J.J. De Haan Gewasresten afvoeren: Utopie of optie? Final report Praktijkonderzoek Plant & Omgeving, Wageningen University & Research Centre, Lelystad (2005) [Dutch]
63. X. Zeng, Y. Ma, L. Ma Utilization of straw in biomass energy in China Renew Sust Energy Rev, 11 (5) (2007), pp. 976–987
64. S. Yusoff Renewable energy from palm oil - innovation on effective utilization of waste J Clean Prod, 14 (1) (2006), pp. 87–93
65. Lysen EH (Universiteit Utrecht), coordinator. GRAIN – global restrictions on biomass availability for import to the Netherlands. Utrecht (NL): Novem Publication Centre. [Report No.: 2EWAB00.27. Project No.: 356598/1010].
66. Van Haandel A, van Lier JB. Vinasse treatment for energy production and environmental protection at alcohol distillery plants. In: Proceedings of DIVER 2006, International Conference on Production and Uses of Ethanol; 2006 June 19–23; Havana, Cuba. Havana, Cuba; 2006.
67. EPA Office of Solid Waste Municipal solid waste in the United states: 2007 facts & figures United States Environmental Protection Agency. Office of Solid Waste, Washington (2008 Nov) [Report No.:EPA530-R-08–010]
68. OECD. OECD Environmental Data - Compendium 2006-2008-Waste. OECD Working Group on Environmental Information and Outlooks, Environmental Performance and Information Division, OECD Environment Directorate; 2007.
69. Uitvoering Afvalbeheer Nederlands afval in cijfers: gegevens 2000–2006 Uitvoering Afvalbeheer, SenterNovem, Utrecht (NL) (2008) [Dutch]
70. CBS Statline [Internet]. Den Haag: Statistical data on municipal solid waste in the Netherlands for 2000–2005. [cited 2010 Mar]. Available from: http://statline.cbs.nl/StatWeb/publication/?VW=T&DM=SLNL&PA=7467&D1=84-87,89-93,109&D2=0,5-16&D3=a&HD=111129-1550&HDR=G2&STB=T.

Cornelissen, S., Koper, M., and Deng, Y. Y. The Role of Bioenergy in a Fully Sustainable Global Energy System. Biomass and Bioenergy Volume 41, June 2012, Pages 21–33. Copyright © 2012. Elsevier Ltd. All rights reserved. Printed with permission.

CHAPTER 4

BIOFUEL FROM ALGAE: IS IT A VIABLE ALTERNATIVE?

FIROZ ALAM, ABHIJIT DATE, ROESFIANSJAH RASJIDIN, SALEH MOBIN, HAZIM MORIA, and ABDUL BAQUI

4.1 INTRODUCTION

The global climate change, rising crude oil price, rapid depletion of fossil fuel reserves, and concern about energy security, land and water degradation have forced governments, policymakers, scientists and researchers to find alternative energy sources including wind, solar and biofuels. The biofuel production from renewable sources can reduce fossil fuel dependency and assist to maintain the healthy environment and economic sustainability. The biomass of currently produced biofuel is human food stock which is believed to cause the shortage of food and worldwide dissatisfaction especially in the developing nations. Therefore, microalgae can provide an alternative biofuel feedstock thanks to their rapid growth rate, greenhouse gas fixation ability (net zero emission balance) and high production capacity of lipids as microalgae do not compete with human and animal food crops. Moreover, they can be grown on non-arable land and saline water. Biofuels are generally referred to solid, liquid or gaseous fuels derived from organic matter [1]. The classification of biofuels is shown in Fig. 1. These classifications are: a) Natural biofuels, b) Primary biofuels, and c) Secondary biofuels. Natural biofuels are generally derived from organic sources and include vegetable, animal waste and landfill gas. On the other hand, primary biofuels are fuel-woods used mainly for cooking, heating, brick kiln or electricity production. The secondary biofuels are bioethanol and biodiesel produced by processing biomass and are used in transport

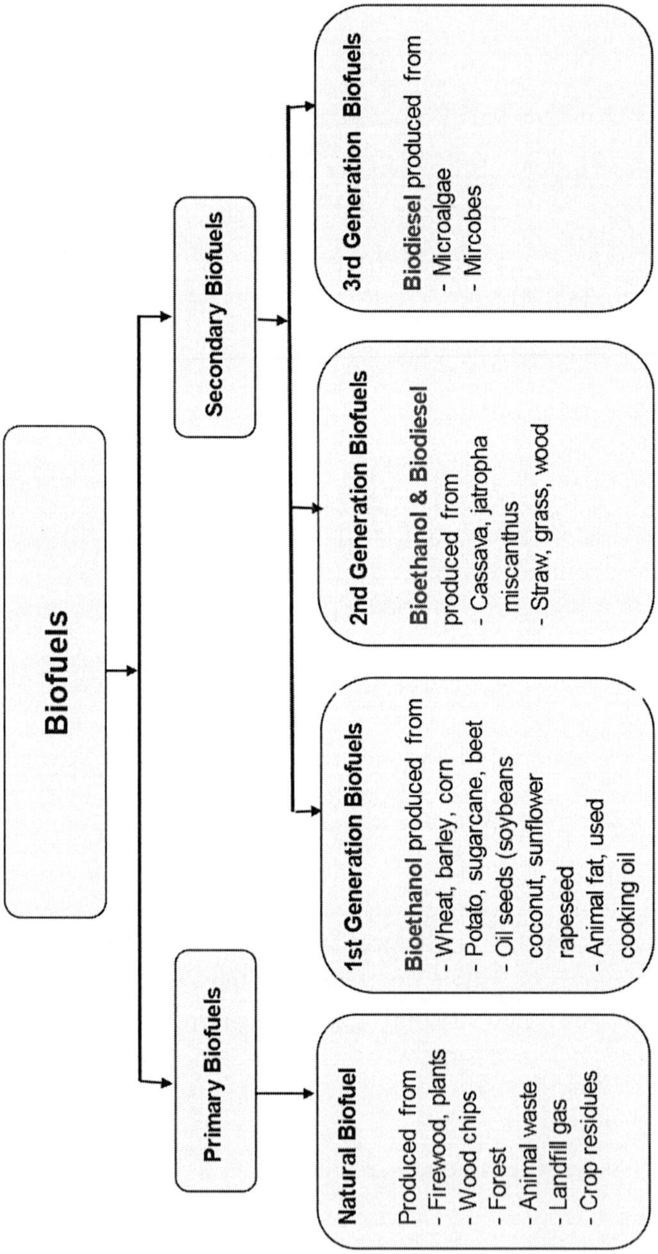

FIGURE 1: Biofuel production sources (biomasses) (adapted from [2])

sectors [1]. The secondary biofuels are sub classified into three so called generations, namely, a) First generation biofuels, b) Second generation biofuels, and c) Third generation biofuels based on their different features such types of processing technology, feedstock and or their development levels [2].

Despite having potential in producing carbon neutral biofuels, the first generation biofuels possess notable economic, environmental and political concerns. The most alarming issue associated with first generation biofuels is that with the increase of production capacity, more arable agricultural lands are needed for the production of first generation biofuel feedstock resulting in reduced lands for human and animal food production.

The increased pressure on arable land currently used for food production leads to severe food shortages, especially in developing countries of Africa, Asia and South America where over 800 million people have been suffering from hunger and malnutrition from severe shortages of food. With the growing world's population, the demand for food is increasing-while the arable land is decreasing. The intensive use of fertilizer, pesticides and fresh water on limited farming lands can reduce not only the food production capacity of lands but also cause significant environmental damage [15]. Therefore, enthusiasms about first generation biofuels have been demised. Increasing use of first generation biofuels will inevitably lead to increasing the price of food beyond the reach of the under privileged. The political consequences of this could be difficult to contain.

As first generation biofuels are not viable and receive lukewarm reception, researchers focused on second generation biofuels. The primary intention here is to produce biofuels using lignocellulosic biomass, the woody part of plants which do not compete with human food chain directly [2]. As shown in Fig.1, main sources for second generation biofuels are predominantly agricultural residues, waste (e.g., trimmed branches, leaves, straws, wood chips, etc.) forest harvesting residues, wood processing residues (e.g. saw dust) and non-edible components of corn, sugarcane, beet, etc. However, converting the woody biomass into fermentable sugars requires sophisticated and expensive technologies for the pretreatment with special enzymes making second generation biofuels economically not profitable for commercial production [2, 4].

Hence, the focus of research is drawn to third generation biofuels. The main component of third generation biofuels is microalgae as shown in

Fig. 1. It is currently considered to be a feasible alternative renewable energy resource for biofuel production overcoming the disadvantages of first and second generation biofuels [1- 2, 5, 16]. The potential for biodiesel production from microalgae is 15 to 300 times more than traditional crops on an area basis [2]. Furthermore compared with conventional crop plants which are usually harvested once or twice a year, microalgae possess a very short harvesting cycle (1 to 10 days depending on the process), allowing multiple or continuous harvesting with significantly increased yields [2, 15]. Additionally, the microalgae generally have higher productivity than land based plants as some species have doubling times of a few hours and accumulate very large amounts of triacylglycerides (TAGs). Most importantly, the high quality agricultural land is not required for microalgae biomass production [3].

4.2 BIOFUEL PRODUCTION POTENTIAL FROM MICROALGAE

Microalgae are single-cell microscopic organisms which are naturally found in fresh water and marine environment. Their position is at the bottom of food chains. Microalgae are considered to be one of the oldest living organisms in our planet. There are more than 300,000 species of micro algae, diversity of which is much greater than plants [3]. They are thallophytes - plants lacking roots, stems, and leaves that have chlorophyll as their primary photosynthetic pigment and lack a sterile covering of cells around the reproductive cells [4]. While the mechanism of photosynthesis in these microorganisms is similar to that of higher plants, microalgae are generally more efficient converters of solar energy thanks to their simple cellular structure. In addition, because the cells grow in aqueous suspension, they have more efficient access to water, CO_2, and other nutrients [2, 5]. Generally, microalgae are classified in accordance with their colours. The current systems of classification of microalgae are based on a) kinds of pigments, b) chemical nature of storage products, and c) cell wall constituents [2]. Some additional criteria are also taken into consideration including cytological and morphological characters: occurrence of flagellate cells, structure of the flagella, scheme and path of nuclear and cell division, presence of an envelope of endoplasmic reticulum around the chloroplast, and possible connection between the endoplasmic reticulum and the nuclear membrane [6]. Some major groups of microalgae are shown Table 1.

The oil contents of various microalgae in relation to their dry weight are shown in Table 2. It is clear that several species of microalgae can have oil contents up to 80% of their dry body weight. As mentioned earlier, some microalgae can double their biomasses within 24 hours and the shortest doubling time during their growth is around 3.5 hours which makes microalgae an ideal renewable source for biofuel production [7]. The oil content and types of microalgae available at fresh water and marine water are shown separately in Tables 3 & 4.

TABLE 1: Major microalgae groups based on their colours

	Colour	**Group**
1	Yellow-green algae	Xanthophyceae
2	Red algae	Rhodophyceae
3	Golden algae	Chrysophyceae
4	Green algae	Chlorophyceae
5	Brown algae	Phaeophyceae
6	Cyanobacteria	Cyanophyceae

TABLE 2: Oil contents of microalgae [7]

	Name of microalgae	**(% dry weight)**
1	*Botryococcus braunii*	25–75
2	*Chlorella sp.*	28–32
3	*Crypthecodinium cohnii*	20
4	*Cylindrotheca sp.*	16–37
5	*Dunaliella primolecta*	23
6	*Isochrysis sp.*	25–33
7	*Monallanthus salina*	20
8	*Nannochloris sp.*	20–35
9	*Nannochloropsis sp.*	31–68
10	*Neochloris oleoabundans*	35–54
11	*Nitzschia sp.*	45–47
12	*Phaeodactylum tricornutum*	20–30
13	*Schizochytrium sp.*	50–77
14	*Tetraselmis sueica*	15–23

TABLE 3: Oil contents of microalgae grown in fresh water [adapted from 2, 7-9, 14-16]

Where Grown		Name of microalgae species	(% dry weight)
Fresh Water Algae	1	*Botryococcus sp.*	25–75
	2	*Chaetoceros muelleri*	34
	3	*Chaetoceros calcitrans*	15–40
	4	*Chlorella emersonii*	25–63
	5	*Chlorella protothecoides*	15–58
	6	*Chlorella sorokiniana*	19–22
	7	*Chlorella vulgaris*	5–58
	8	*Chlorella sp.*	10 –48
	9	*Chlorella pyrenoidosa*	2
	10	*Chlorella sp.*	18–57
	11	*Chlorococcum sp.*	20
	12	*Ellipsoidion sp.*	28
	13	*Haematococcus pluvialis*	25
	14	*Scenedesmus obliquus*	11 –55
	15	*Scenedesmus quadricauda*	2–19
	16	*Scenedesmus sp.*	20–21

TABLE 4: Oil contents of microalgae grown in marine (salt) water [adapted from 2, 7-9, 14-16]

Where grown		Name of microalgae species	(% dry wt)
Marine Water Algae	1	*Dunaliella salina*	6 –25
	2	*Dunaliella primolecta*	23
	3	*Dunaliella tertiolecta*	18–71
	4	*Dunaliella sp.*	18 – 67
	5	*Isochrysis galbana*	7– 40
	6	*Isochrysis sp.*	7– 33
	7	*Nannochloris sp.*	20 –56
	8	*Nannochloropsis oculata*	23–30
	9	*Nannochloropsis sp.*	12–53
	10	*Neochloris oleoabundans*	29–65
	11	*Pavlova salina*	31
	12	*Pavlova lutheri*	36
	13	*Phaeodactylum tricornutum*	18–57
	14	Spirulina platensis	4 17

4.3 BIOFUELS PRODUCTION PROCESSES FROM MICROALGAE

The production of microalgae biomass for extraction of biofuels is generally more expensive and technologically challenging than growing crops. Photosynthetic growth of microalgae requires light, CO_2, water and inorganic salts. The temperature regime needs to be controlled strictly. For most microalgae growth, the temperature generally remains within 20°C to 30°C. In order to reduce the cost, the biodiesel production must rely on freely available sunlight, despite daily and seasonal variations in natural light levels [7, 17-20]. A number of ways the microalgae biomass can be converted into energy sources which includes: a) biochemical conversion, b) chemical reaction, c) direct combustion, and d) thermochemical conversion. Fig. 2 illustrates a schematic of biodiesel and bioethanol production processes using microalgae feedstock [10]. As mentioned previously, microalgae provide significant advantages over plants and seeds as they: i) synthesize and accumulate large quantities of neutral lipids (20 50 % dry weight of biomass) and grow at high rates; ii) are capable of all year round production, therefore, oil yield per area of microalgae cultures could greatly exceed the yield of best oilseed crops; iii) need less water than terrestrial crops therefore reducing the load on freshwater sources; iv) cultivation does not require herbicides or pesticides application; v) sequester CO_2 from flue gases emitted from fossil fuel-fired power plants and other sources, thereby reducing emission of greenhouse gas (1 kg of dry algal biomass utilise about 1.83 kg of CO_2). In addition, microalgae offer wastewater bioremediation by removing of NH_4, NO_3, PO_4 from wastewater sources (e.g. agricultural run-off, concentrated animal feed operations, and industrial and municipal wastewaters). Their ability to grow under harsher conditions and reduced needs for nutrients, microalgae can be cultivated in saline/brackish water/coastal seawater on non-arable land, and do not compete for resources with conventional agriculture. Depending on the microalgae species other compounds may also be extracted, with valuable applications in different industrial sectors, including a large range of fine chemicals and bulk products, such as polyunsaturated fatty acids, natural dyes, polysaccharides, pigments, antioxidants, high-value bioactive compounds, and proteins [2, 8, 10, 21-28].

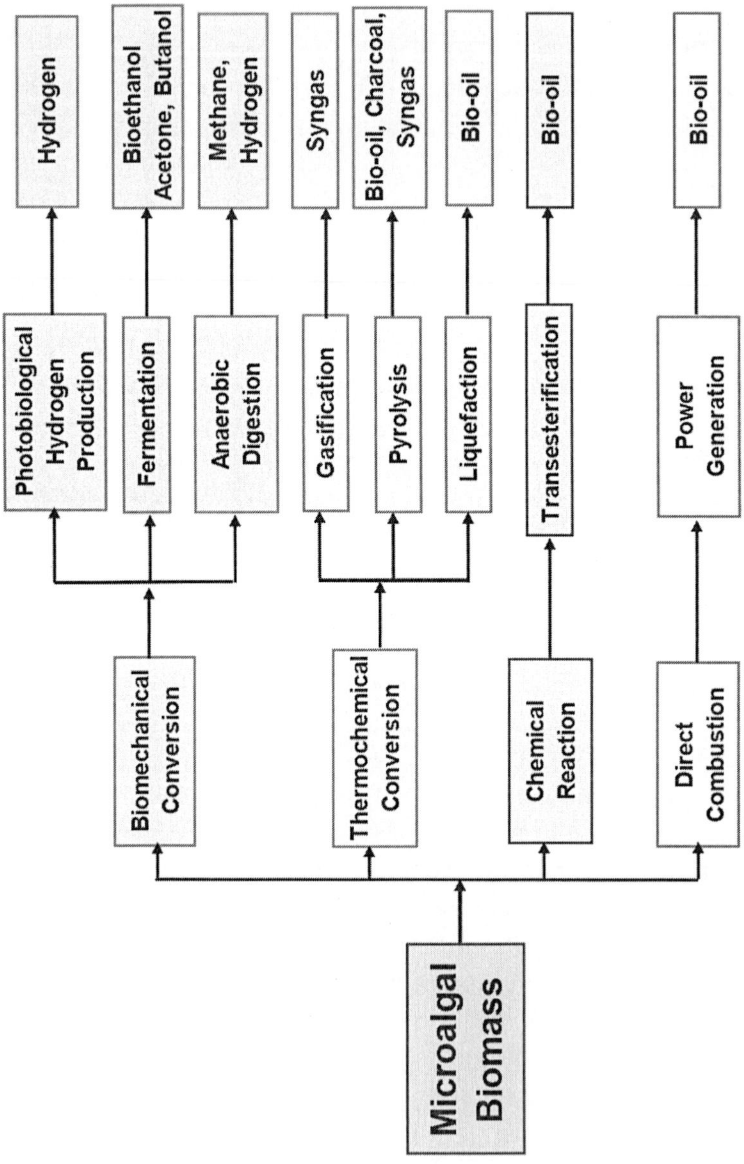

FIGURE 2: Biofuel production processes from microalgae biomass, adapted from [2, 11]

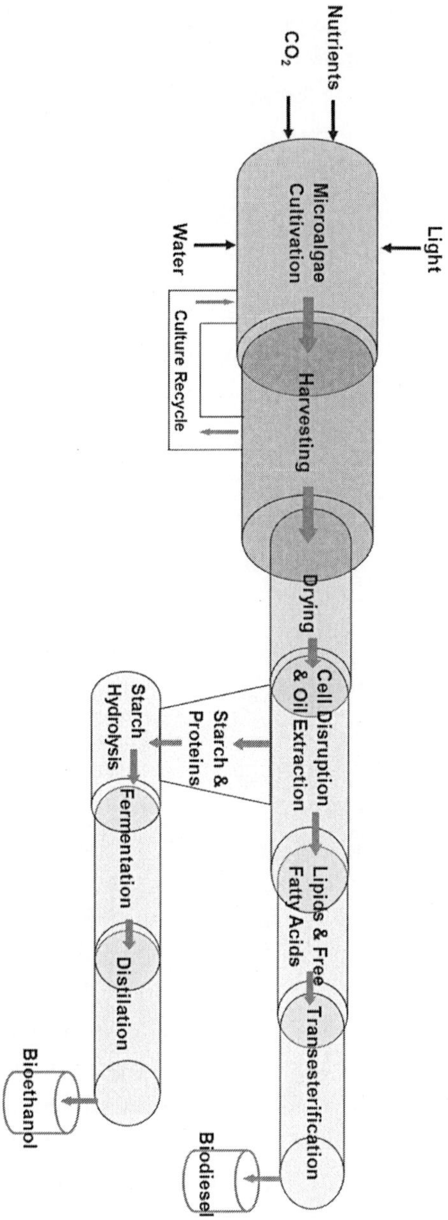

FIGURE 3: Biodiesel and Bioethanol production processes from microalgae, adapted from [2]

There are different ways microalgae can be cultivated. However, two widely used cultivation systems are the open air system and photobioreactor system. The photoreactor system can be sub-classified as a) tabular photoreactor, b) flat photoreactor, and c) column photoreactor. Each system has relative advantages and disadvantages. More details about these cultivation systems can be found in [2-3, 7].

The production of biofuel is a complex process. A schematic of biofuel production processes from microalgae is shown in Figure 3. The process consists of following stages: a) stage 1 - microalgae cultivation, b) stage 2 harvesting, drying & cell disruption (cells separation from the growth medium), c) stage 3 - lipid extraction for biodiesel production through trans-esterification and d) stage 4 starch hydrolysis, fermentation & distillation for bioethanol production. However, these processes are complex, technologically challenges and economically expensive. A significant challenge lies ahead for devising a viable biofuel production process [2, 28-30].

4.4 DISCUSSION AND CONCLUSION

Many countries including the European Union (EU) have adopted policies on certain percentage of renewable energy use for transport and other relevant sectors. In December 2008, the EU signed a directive that requires 10% of member to come from renewable sources (biofuels, hydrogen and green el policy towards mitigation of climate change effect and global warming. The EU directive also obliges the bloc to ensure that biofuels offer at least 35% carbon emission savings compared to fossil fuels and the figure should rise to 50% in 2017 and debates among governments, policymakers, scientists and environmentalists as currently most commercially produced biofuels are derived from sources that compete with or belong to feedstock for human and animal consumption.

In terms of greenhouse gas emission, the biofuels produced from microalgae is generally carbon neutral. The CO_2 emitted from burning biofuel is assumed to be neutral as the carbon was taken out of the atmosphere when the algae biomass grew. Therefore, biofuels from microalgae do not add new carbon to the atmosphere. Biofuels can be a viable alternative to fossil fuels on short and medium terms. Additionally, advanced biofuels made from residues or waste have the potential to reduce CO_2 emissions with 90% compared to petrol/diesel.

While many years of research and development still lie ahead, if successful, algae-based fuels can help meet the world's growing demand for transportation fuel while reducing greenhouse gas emissions. However, a number of challenges remain before algae can be used for mainstream commercial applications as the uncertainty of cost constitutes the biggest obstacle. There is no doubt that research work on microalgae is still in primary stages. Currently, it is not clear that what kinds or families of algae would be most appropriate in order to produce commercially viable biofuels. Researchers are currently working on appropriate commercial cultivation processes of algae biomasses. At this point in time, there is no definitive answer to an open question if it is better to grow algae in photobioreactor system or open air (pond) system. As algae are micro-organisms of a size ten times smaller than human hair, it is a great challenge to harvest them. At present, microalgae harvestings are based on either centrifugation or chemical flocculation, which push all the microalgae together, but these processes associated with high cost [16-23].

Biodiesel or bioethanol production from algae biomass cannot be commercially viable unless by-products are optimally utilised. As mentioned earlier, the lipid or the oil part is around 30% of the total algae biomass and the remaining 70% is currently wasted which can be used as nutrients, pharmaceuticals, animal feed or bio-based products. The use of lipid as well as all by-products will allow exploring the full potential of microalgae towards sustainable environment and economy. At present, 70 -90% of the energy put into harvesting microalgae for fuel usually gets used into extracting the lipids (oil) they produce under current factory designs. It is obvious that new technologies are needed for reducing huge energy losses [13, 23-27].

Microalgae have immense potentials for biofuels production. However, these potentials largely depend on utilisation of technology, input feedstock (CO_2, wastewater, saltwater, natural light), barren lands and marine environment. Based on energy content, available technology, land, it is hard to overemphasize that biofuels are a realistic short-term, but definitely not a long-term and large scale solution to energy needs and environmental challenges. Microalgae can be temporary sources of energy, and with the appropriate growth protocols they may address some of the concerns raised by the use of first and second generation biofuels.

REFERENCES

1. Nigam, P.S. and Singh, A. (2011), Production of liquid biofuels from renewable resources, Progress in Energy and Combustion Science, 37(1): 52 68
2. Dragone, G., Fernandes, B., Vicente, A.A. and Teixeira, J.A. (2010), Third generation biofuels from microalgae in Current Research, Technology and Education Topics in Applied Microbiology and Microbial Biotechnology, Mendez-Vilas A (ed.), Formatex, 1355-1366
3. Scott, S.A., Davey, M.P., Dennis, J.S., Horst, I., Howe, C.J., Lea-Smith, D.J. and Smith, A.G. (2010), Biodiesel from algae: challenges and prospects, Current Opinion in Biotechnology, 21:277-286.
4. Brennan L, Owende P. (2010), Biofuels from microalgae--A review of technologies for production, processing, and extractions of biofuels and coproducts. Renewable and Sustainable Energy Reviews, 14:557-577.
5. Chisti Y. (2007), Biodiesel from microalgae. Biotechnology Advances., 25:294-306.
6. Tomaselli, L. (2004), The microalgal cell. In: Richmond A, eds. Handbook of Microalgal Culture: Biotechnology and Applied Psychology. Oxford: Blackwell Publishing Ltd; 2004: 3-19.
7. Chisti, Y (2007), Biodiesel from microalgae, Biotechnology Advances, 25:294-306
8. Um B-H, Kim Y-S. (2008), Review: A chance for Korea to advance algal-biodiesel technology, Journal of Industrial and Engineering Chemistry, 15:1-7.
9. Sydney, E.B, Sturm, W., de Carvalho, J.C., Thomaz-Soccol, V., Larroche, C., Pandey, A., Soccol, C.R. (2010), Potential carbon dioxide fixation by industrially important microalgae, Bioresource Technology, 101:5892-5896.
10. Mata TM, Martins AA, Caetano NS. (2010), Microalgae for biodiesel production and other applications: A review, Renewable and Sustainable Energy Reviews, 14:217-232.
11. Wang, B., Li, Y., Wu, N. and Lan, C. (2008), CO2 bio-mitigation using microalgae. Applied Microbiology and Biotechnology, 79:707-718.
12. EU news & policy debates, retrieved on 22 August, 2012 from www.euractiv.com/transport/eu-agrees-10-green-fuel-target-r-news-220953
13. Lardon, L., Hélias, A., Sialve, B., Steyer, J. P., & Bernard, O. (2009), Life-Cycle Assessment of Biodiesel Production from Microalgae, Environmental, Science & Technology, 43(17):6475-6481
14. Koh, L.P., Ghazoul, J. (2008), Biofuels, biodiversity, and people: understanding the conflicts and finding opportunities, Biological Conservation, 141:2450-2460.
15. Schenk, P., Thomas-Hall, S., Stephens, E., Marx, U., Mussgnug, J., Posten, C., Kruse, O., and Hankamer, B. (2008), Second generation biofuels: high efficiency microalgae for biodiesel production, BioEnergy Research, 1:20-43
16. Li, Y., Horsman, M., Wu, N., Lan, C.Q. and Dubois-Calero, N. (2008), Biofuels from microalgae, Biotechnology Progress, 24:815-820.
17. Chaumont, D. (1993), Biotechnology of algal biomass production: a review of systems for outdoor mass culture, Journal of Applied Phycology, 5:593-604.
18. Borowitzka, M.A. (1999), Commercial production of microalgae: ponds, tanks, tubes and fermenters, Journal of Biotechnology, 70:313-321.

19. Borowitzka, M.A. (2005), Culturing microalgae in outdoor ponds In: Andersen RA, eds. Algal Culturing Techniques. Burlington, MA: Elsevier Academic Press, 205-218.
20. Pulz, O. (2001), Photobioreactors: production systems for phototrophic microorganisms, Applied Microbiology and Biotechnology, 57:287-293.
21. Spolaore, P., Joannis-Cassan, C., Duran, E., and Isambert, A. (2006), Commercial applications of microalgae, Journal of Bioscience and Bioengineering, 101:87-96.
22. Carvalho, A.P., Meireles, L.A., Malcata, F.X. (2006), Microalgal reactors: A review of enclosed system designs and performances, Biotechnology Progress, 22:1490-1506.
23. Benemann, J.R., Tillett, D.M. and Weissman, J.C. (1987), Microalgae biotechnology, Trends in Biotechnology, 5:47-53.
24. Eriksen, N., Poulsen, B., Lønsmann, I.J. (1998), Dual sparging laboratory-scale photobioreactor for continuous production of microalgae, Journal of Applied Phycology, 10:377-382.
25. Tredici, M.R. (1999), Bioreactors, photo. In: Flickinger MC, Drew SW, eds. Encyclopedia of Bioprocess Technology: Fermentation, Biocatalysis, and Bioseparation. New York, NY: Wiley, 395-419.
26. Molina, G.E., Belarbi, E.H., Acién, F.G., Robles, M.A. and Chisti, Y. (2003), Recovery of microalgal biomass and metabolites: process options and economics, Biotechnology Advances, 20:491-515.
27. Harun, R., Singh, M., Forde, G.M., Danquah, M.K. (2010), Bioprocess engineering of microalgae to produce a variety of consumer products, Renewable and Sustainable Energy Reviews, 14:1037-1047.
28. Mendes-Pinto, M.M., Raposo, M.F.J., Bowen, J., Young, A.J., Morais, R. (2001), Evaluation of different cell disruption processes on encysted cells of Haematococcus pluvialis: effects on astaxanthin recovery and implications for bio-availability, Journal of Applied Phycology, 13:19-24.
29. Vasudevan, P. and Briggs, M. (2008), Biodiesel production - current state of the art and challenges, Journal of Industrial Microbiology and Biotechnology, 35:421-430.
30. Harun, R., Danquah, M.K., Forde, G.M. (2010), Microalgal biomass as a fermentation feedstock for bioethanol production, Journal of Chemical Technology & Biotechnology, 85:199-203.

Alam, F., Date, A., Rasjidin, R., Mobin, S., Moria, H., and Baqui, A. Biofuel from Algae: Is It a Viable Alternative? Procedia Engineering 2012; 49: 221–227. Copyright © 2012. Published by Elsevier Ltd. Printed with permission.

CHAPTER 5

COMPREHENSIVE EVALUATION OF ALGAL BIOFUEL PRODUCTION: EXPERIMENTAL AND TARGET RESULTS

COLIN M. BEAL, ROBERT E. HEBNER, MICHAEL E. WEBBER, RODNEY S. RUOFF, A. FRANK SEIBERT, and CAREY W. KING

5.1 INTRODUCTION

The aspiration for producing algal biofuel is motivated by the desire to: (1) displace conventional petroleum-based fuels, which are exhaustible, (2) produce fuels domestically to reduce energy imports, and (3) reduce greenhouse gas emissions by cultivating algae that re-use carbon dioxide emitted from industrial facilities. In theory, algae have the potential to produce a large amount of petroleum fuel substitutes, while avoiding the need for large amounts of fresh water and arable land [1–3]. These attributes have created widespread interest in algal biofuels. In practice, however, profitable algal biofuel production faces several important challenges. The goal of the research presented in this paper is to examine and quantify the extent of some of those challenges with an eye towards identifying critical areas for advances in the development of algal biofuels.

For algae to be a viable feedstock for fuel production: a significant quantity of fuel must be produced, the energy return on investment (EROI) of the life cycle must be greater than 1 (and practically greater than 3 [4]), the financial return on investment (FROI) should be greater than 1, the water intensity of transportation using algal biofuels should be sustainable, and nutrient requirements should be manageable. This study examines these criteria for two cases using second-order analysis methods described by Mulder and Hagens [5], which include direct and indirect operating expenses, but neglect all capital expense. Process-specific terminology is based on the reporting framework established by Beal et al. [6].

There are several energy carriers and co-products that can be produced from algae, such as renewable diesel, electricity, hydrogen, ethanol, pharmaceutics, cosmetics, and fertilizers [7–9]. While non-energy co-products might enable economic viability of algal biofuel products in the short term, large scale production would quickly saturate co-product markets. Thus, in the long term, production of domestic, renewable, low-carbon fuels as an alternative to conventional fuel sources remains the main motivation for researching large-scale algae production. Consequently, this research focuses on the energy products. While bioelectricity from algal feedstocks is one possible pathway for energy production, this work considers only the co-production of bio-oil (a petroleum fuel substitute) and bio-gas (i.e., methane, which is a natural gas substitute) because those two fuels are produced from the experimental process at UT and align more directly with displacing petroleum [10–12]. Further, both bio-oil and bio-gas are feedstocks that can be combusted within additional technologies to produce electricity.

Because the intent of this research is to analyze and anticipate a mature algal fuels industry that does not yet exist, researchers have two options for conducting a process analysis as in this paper: (1) use data derived from experimental processes followed by scaling analyses (recognizing that lab-scale experiments are inherently sub-optimal) or (2) use estimated data from models of future commercial-scale systems. Both of these approaches are used in this study. Firstly, an Experimental Case is described, which is based on unique direct end-to-end measurements (from growth through biocrude separations) performed in a controlled indoor/outdoor laboratory setting at The University of Texas at Austin. Secondly, a Highly Productive Case is described, which is based an optimistic analytical model that incorporates the technology and pathways of the Experimental Case.

We encourage other researchers to present (life cycle) metrics of alternative algal technology pathways in the step-by-step manner we demonstrate. The reasons for presenting life cycle metrics at multiple stages are threefold: (1) easier facilitation of future life cycle assessment (LCA) harmonization and meta-analyses that can effectively compare many independent studies, (2) better tracking of technological progress over time, and (3) better comparison of competing technologies (e.g., capital intensive versus resource intensive). The benefits of LCA harmonization were demonstrated by Farrell et al. [13] in comparing net energy for corn ethanol. The National Renewable Energy Laboratory of the US Department of Energy tests and tracks photovoltaic cell efficiencies over time such that specialists and the general public can easily track the rate of progress, which is beneficial for the community as a whole. By doing so, one is able to observe the improvements that were made to photovoltaic cell designs over the course of research and development, providing a vantage point for researchers and investors alike to gauge the progress in that energy production technology. The authors believe algal energy processes would benefit from similar indicators and analyses, and this manuscript presents its results in that spirit of tracking technological metrics starting at the experimental batch scale. Additionally, the calculation of multiple life cycle indicators (e.g., EROI, FROI, water use, resource consumption, land use, air emissions, etc.) from the same experimental or modeled processes provides congruent indicators that emphasize the real design tradeoffs (e.g., water versus electricity inputs).

The work presented adds to research in the authors' prior publications, which presented the second-order energy return on investment (2nd O EROI) analyses for an Experimental Case and a modeled Highly Productive Case. In the previous work, the 2nd O EROI, which is a ratio of the energy output of a system to the energy input for that system, for these two cases was determined to be 9.2×10^{-4} and 0.22, respectively [14]. That study illustrated the energetic challenges associated with producing algal biofuel. The present study extends the previous work with five new analytical thrusts to determine (1) the partial FROI, (2) the second-order water intensity of transportation using the algal biofuels produced, (3) the nitrogen constraints, (4) the carbon constraints, and (5) the electricity resource constraints for the Experimental Case and the Highly Productive Case, respectively. The cost, water, and resource results from this new work are presented in conjunction with the previously determined energy results. Thus, for our two cases (one experimentally measured and one analytically derived), this present research serves as a comprehensive and coherent evaluation of the algal biofuel process. It is important that LCAs demonstrate relationships among multiple metrics that are calculated. By reporting multiple metrics for the same algal energy processes, this paper presents an understanding of how one metric (e.g., water consumption) is linked to another (e.g., energy production). Although the Experimental Case is not representative of commercial biofuel production due to significant artifacts that are inherent to lab-scale (vs. industrial scale) production, it represents the first known end-to-end experimental characterization of algal biocrude production at relatively large scale (thousands of liters). While other experiments have been performed at similar scale, they did not conduct the comprehensive mass and material balances that are presented here. Conversely, the data used for the Highly Productive Case are based on optimistic assumptions for operating within the specific production pathway in this study. To place the Highly Productive Case in context with other analyses that have been published, each assumption is compared with those from other studies in the literature.

Many prior studies have been performed, each with a slightly different focus: some have emphasized algal biomass productivity, estimated algal oil productivity per acre of land, or evaluated only a few constraints on algal biofuel production (e.g., energy requirements, cost, etc.) [15–18]. This paper takes the approach of considering many constraints simultaneously (energy, cost, water, and resources) to give a more complete assessment. To this end, quantitative targets are presented in the "Conclusions" that, if achieved, would enable algal biofuel production at large scale.

5.2 METHODS AND MATERIALS

The production pathway and experimental methods used in this analysis has been described in detail in previous publications [6,14,19,20]. Furthermore, the materials and energy consumption data used in the Experimental Case and the Highly Productive Case are taken from Beal et al. that calculated the second-order energy return on (energy) investment (2nd O EROI) [14]. The term "second-order" refers to the inclusion of direct energy inputs (e.g., electricity consumed for pumping) and indirect energy inputs for consumed materials (e.g., the energy embedded in nitrogen fertilizer that is consumed). Details regarding data collection and uncertainty analysis in the Experimental Case and modeling calculations in the

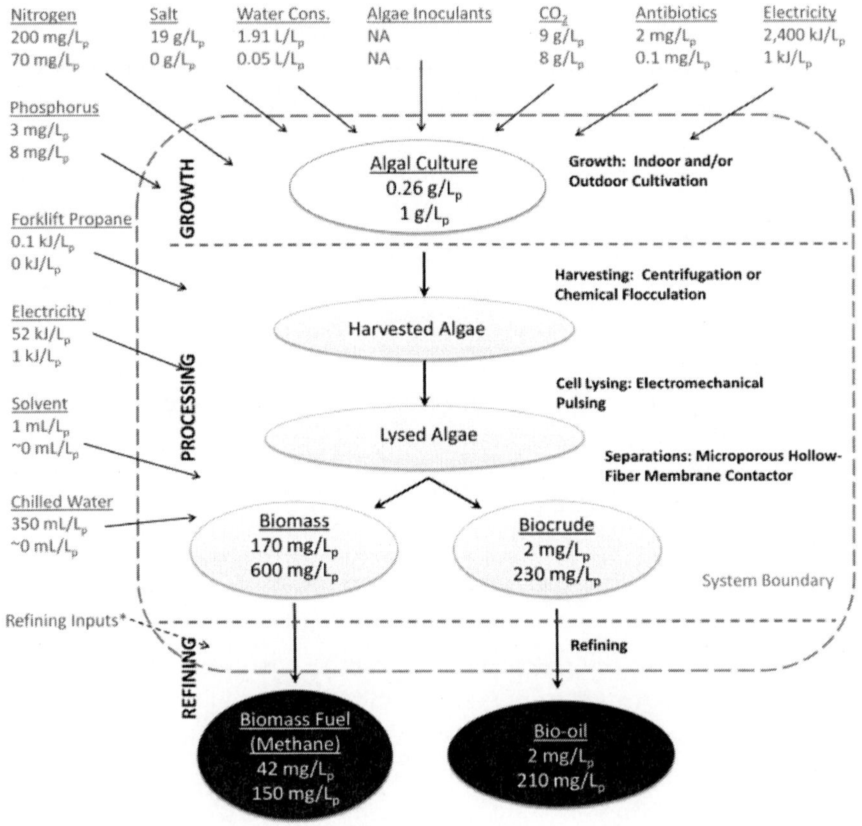

FIGURE 1: The production pathway is represented as three phases: growth, processing, and refining [6]. The data used for each input and output are shown for the Experimental Case (top) and Highly Productive Case (bottom). For the Experimental Case, the material and energy inputs crossing the system boundary were measured for five relatively large scale batches (970–2000 L, each), grown and processed at The University of Texas at Austin, except for the refining inputs (which were modeled from literature data, and are noted in the figure with an asterisk (*) [14]. The Highly Productive Case is an analytical model of a similar production pathway operated more efficiently.

Highly Productive Case can be found in the previous publication [14] and at greater length in a publically available doctoral dissertation (cf. Chapter 4, Appendix 4A, and Appendix 4B of [19]). The work presented herein expands those datasets to incorporate the new analyses mentioned above (water use, nutrients use, FROI, etc.).

Figure 1 shows that the biocrude production process for both of our analyzed cases consists of algal cultivation, harvesting (i.e., concentration) with centrifugation or chemical flocculation, cell lysing via electromechanical pulsing, and neutral lipid recovery using a microporous hollow-fiber membrane contactor. The Experimental Case is comprised of growth and processing data from five relatively large batches (970–2000 L each), with a total processed volume of about 7600 L. The energy and material inputs that were measured are shown in Figure 1 and the energy outputs are modeled to include bio-oil and biomass fuel (methane) (refining was not conducted during the experiments). The Highly Productive Case models energy-efficient growth and processing methods with higher biomass and lipid productivities.

5.2.1 EXPERIMENTAL CASE

The Experimental Case is comprised of five batches, ranging in volume from 970 L to 2000 L each. A marine species of *Chlorella* (KAS 603, provided by Kuehnle AgroSystems, Inc.) was used for all batches and was grown in four different growth stages: flasks, airlift photobioreactors, greenhouse tanks, and covered raceway ponds.

This growth process provided a stable method for scaling up cultivation volumes, although, the inherent inefficiencies of operating at lab-scale required high energy and material inputs (an artifact described in detail by Beal et al. [14]) and yielded relatively low biomass and lipid productivities, as listed below. Energy and material consumption were measured throughout the entire cultivation process and these data have been reported previously [14]. The amounts of resources consumed in the smaller growth volumes (e.g., energy required for bioreactor lighting) were allocated to the larger growth volumes as the algae were transferred through the system during scale-up (cf. Appendix 4A of [19] for details).

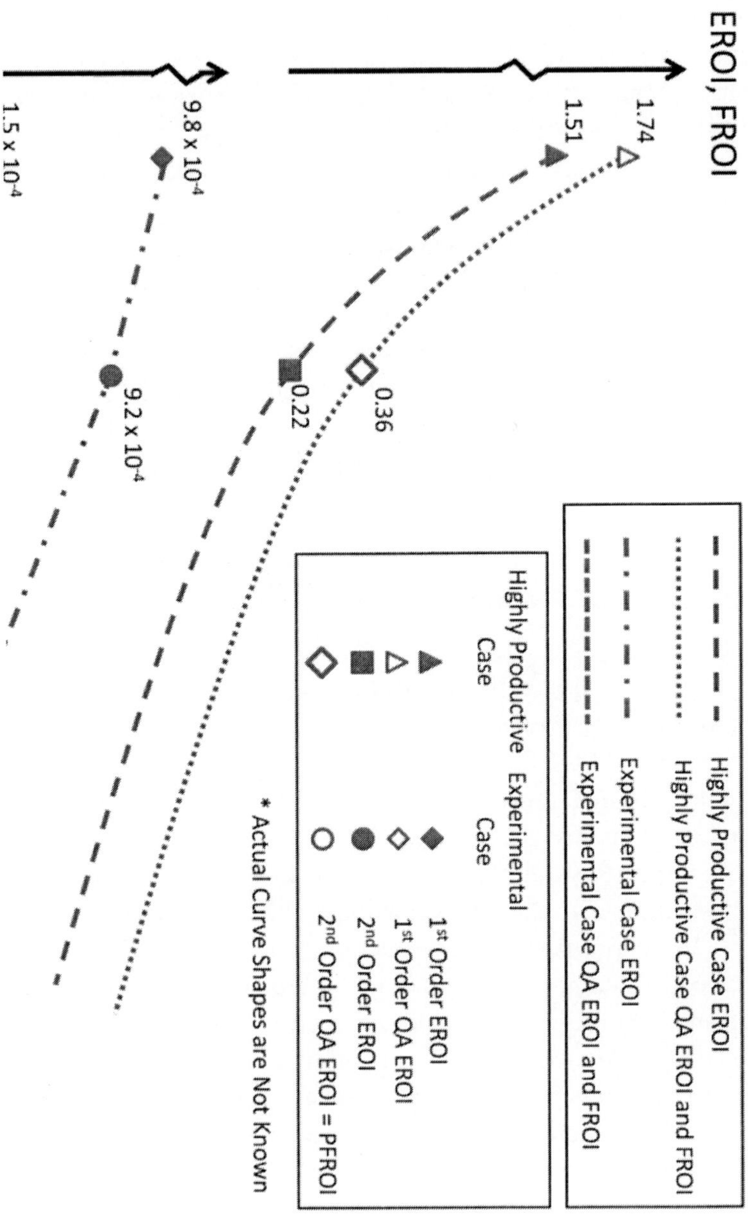

FIGURE 2: Frequency of matching first and last names.

The algal biomass was tracked during each batch by measuring the dry cell weight of multiple samples collected throughout the production pathway. These samples were centrifuged and the pellet was rinsed three times to remove salts. Then, the samples were maintained at 70 °C until a constant weight was obtained. High performance liquid chromatography (HPLC) was used to calculate the lipid content and lipid composition for each batch according to methods developed at The University of Texas at Austin [21], which are refinements of standard methods [22–24].

All five of the experimental batches were processed using a centrifuge for harvesting, electromechanical cell lysing, and a microporous hollow-fiber membrane contactor for separations. While the energy and materials consumed during each of these steps, and the associated uncertainty, has been described in detail by Beal et al. [14,19], this study, combines these data with monetary costs, water impacts, and resource constraints for each input.

2.2.2 HIGHLY PRODUCTIVE CASE

The Highly Productive Case is an analytical model that was constructed to represent a system with greater biomass productivity (80 mg/L-d) and a higher neutral lipid fraction (30%) than the Experimental Case (which had productivity of 2 mg/L-d and lipid fraction of 2%). The Highly Productive Case assumes the same basic production pathway as the Experimental Case, but it substitutes bioreactors for growth and an advanced flocculation technique in place of centrifugation. In addition, several modifications are modeled to improve energy efficiency in the Highly Productive Case. In addition, it is assumed that there is no water loss from evaporation and 95% of the water used for cultivation was recycled. In this sense, the Highly Productive Case is an optimistic, but not wholly unreasonable, scenario for achieving low operating expense in commercial-scale algal biofuel production based on current technologies. The less optimistic assumption is the requirement of "full-price" inputs (such as nitrogen fertilizer and carbon dioxide from ammonia production plants). The ability to achieve each of the specified conditions in the Highly Productive Case in practice is assumed to be possible and the capital required to do so is not considered in this study. Each assumption used in the Highly Productive Case is compared with those from several literature sources in the "Discussion" section, below.

The ability to utilize discounted inputs, such as waste forms of carbon, nitrogen, and phosphorus and cheap energy inputs, would further improve the return on investment for producing algal biofuels with respect to the Highly Productive Case [11,14,25–28]. The Highly Productive Case is not intended to represent the optimum scenario for algal biofuels nor is it presented as the final arbiter of the fuel's prospects for success; rather, it is intended to serve as a useful benchmark. The optimum scenario might utilize discounted inputs, high productivity algal strains (e.g., genetically modified organisms), and improved growth, processing, and harvesting methods that might be developed in the future. Instead, the Highly Productive Case models a similar production pathway as the Experimental Case, but with significantly higher fuel productivity and significantly more efficient growth and processing methods.

2.2.3 BIOMASS AND LIPID PRODUCTIVITY FORMULAE

The bio-oil and biomass fuel (methane) productivities of this system can be reported as:

$$P_{BO} = P_{GM} \cdot \varphi_{harv} \cdot \varphi_{cellys} \cdot \varphi_{sep_{BC}} \cdot \varphi_{ref_{BO}} \left[\frac{g}{L-d}\right] \tag{1}$$

and:

$$P_{BMF} = P_{GM} \cdot \varphi_{harv} \cdot \varphi_{cellys} \cdot \varphi_{sep_{BS}} \cdot \varphi_{ref_{BMF}} \left[\frac{g}{L-d}\right] \tag{2}$$

where is the productivity (of bio-oil (BO), biomass fuel (BMF), and grown mass (GM)) and φ represents the efficiency of harvesting (harv), cell lysing (cellys), separations (sep) (of biocrude (BC) and biomass in the post-extraction slurry (BS)), and refining (ref). Each efficiency is defined as the mass of the output divided by the mass of the input for that step (cf. [6]). For example, the biocrude separations efficiency, φ_{sepBC}, is defined as the mass of biocrude recovered divided by the lysed algal biomass, and the neutral lipid fraction is embedded in this efficiency [6].

5.2.4 PHOTOSYNTHETIC EFFICIENCY

The photosynthetic efficiency can be calculated as the energy content of the glucose produced during photosynthesis divided by the incident radiation. This value is different than the overall energy efficiency of growth, which includes the cost of living (i.e., respiration to enable cell functions), conversion of glucose to biomass, and required energy inputs (e.g., mixing, nutrient supply, etc.) [29]. The energy content of the glucose produced (per liter processed) can be calculated as:

$$\acute{E}_{CH_2O} = \left(\acute{I} \cdot PAR \cdot PTE \cdot PUE \cdot \alpha\right)\left[\frac{kJ}{L_p}\right] \tag{3}$$

where t_c is the volumetric irradiance (in joules per liter of processed volume), PAR is the photosynthetically active radiation fraction (0.46), PTE is the photon transmission efficiency, and PUE is the photon utilization efficiency [30]. The volumetric irradiance can be converted to an areal irradiance, I, according to:

$$\acute{I}\left[\frac{MJ}{L_p}\right] \cdot \frac{1}{t_c}\left[\frac{1}{yr}\right] \cdot \frac{d}{1}\left[\frac{m}{1}\right] \cdot \frac{1000}{1}\left[\frac{L}{m^3}\right] = I\left[\frac{MJ}{m^2 - yr}\right] \tag{4}$$

where t_c is the cultivation time (123 days for the Experimental Case and 12.5 days for the Highly Productive Case) and d is the pond depth (0.2 m for both cases). In Equation 3, the variable α characterizes the efficiency by which photons used for photosynthesis are converted to glucose through Z-scheme photosynthesis. With a quantum requirement of 8 mol photons per mol of glucose (the energy content of glucose is 467.5 kJ/mol) and an average photon energy content of 225 kJ/mol [30], α = (467.5/8 ·

225) = 0.26 [29]. The photosynthetic efficiency, PE, is the ratio of $\acute{E}D_{CH2)}$ to Í, which becomes:

$$PE = \frac{\acute{E}D_{CH_2O}}{\acute{I}} = PAR \cdot PTE \cdot PUE \cdot \alpha \, [-] \tag{5}$$

The amount of energy contained in the growth volume (as algal biomass), $\acute{E}D_{GV}$, can be calculated from:

$$\acute{E}D_{GV} = P_{GM} \cdot HHV_{GM} \cdot t_c \left[\frac{kJ}{L}\right] \tag{6}$$

where HHVGM is the higher heating value of the grown algal biomass. The amount of energy contained in the growth volume can also be calculated from:

$$\acute{E}D_{GV} = \acute{I} \cdot PAR \cdot PTE \cdot PUE \cdot \alpha \cdot (1 - CoL) \cdot \tau \left[\frac{kJ}{L}\right] \tag{7}$$

where CoL is the cost of living, which is defined as the fraction of glucose consumed for cellular operations [30]. The energy conversion of glucose to biomass energy can be grossly simplified as a single-step process, represented by τ. Assuming algae have the Redfield stoichiometry defined by Clarens et al. [15] ($C_{106}H_{181}O_{45}N_{15}P$), the conversion of glucose to algal biomass can be approximated as:

$$106CH_2O + 15NaNO_3 + 0.5P_2O_5 + C_{106}H_{181}O_{45}N_{15}P + 8H_2O + 42.75O_2 + 15NaOH \tag{8}$$

The higher heating value (HHV_{GM}) for algae can be estimated as stated by Clarens et al.;

$$HHV_{GM} = 35160 \cdot x_c + 116225 \cdot x_H - 11090 \cdot x_0 + 6280 \cdot x_n \text{ [kJ/kg]} \tag{9}$$

where x is the mass fraction of each element (carbon, hydrogen, oxygen, and nitrogen) [15]. The HHV_{GM} for the algae considered here is 24.49 MJ/kg (59.12 MJ/mol). The energy conversion of glucose (106 mol with a HHV of 467.5 kJ/mol) to biomass energy is represented by τ = 1.19. Combining several of these relations, the PE can be calculated as:

$$PE = PAR \cdot PTE \cdot PUE \cdot \alpha = \frac{\acute{E}D_{CH_2O}}{\acute{I}} = \frac{\acute{E}D_{GV}}{(1 - CoL) \cdot \tau \cdot \acute{I}} = \frac{P_{GM} \cdot HHV_{GM} \cdot t_c}{(1 - CoL) \cdot \tau \cdot \acute{I}} \; [-] \tag{10}$$

5.2.5 ENERGY RETURN ON INVESTMENT FORMULAE

The second-order energy return on investment, 2nd O EROI is calculated as:

$$2^{nd}O\ EROI = \frac{\acute{E}D_{out}}{\hat{E}_G + \hat{E}_p + \hat{E}_R} = \frac{\acute{E}D_{BO} + \acute{E}D_{BMF}}{\hat{E}_G + \hat{E}_p + \hat{E}_R} \tag{11}$$

As a second-order analysis, direct and indirect energy inputs are included, while a first-order analysis would include only direct energy inputs [5]. An apostrophe accent is used to denote units that are reported with respect to the growth volume processed, such as the energy inputs in units of J per L processed (J/Lp).

To account for differences in energy quality among the inputs and outputs, the quality-adjusted second-order energy return on investment (QA 2nd EROI) was calculated by multiplying each input and output by priced-based quality factors. The quality factors (QF) were calculated for energy flows based on the energy price (EP), which is the price of each energy source per joule, and correlates the relative value of each fuel [31]. Using coal as the arbitrary standard with a quality factor equal to 1 ($1.5/MMBtu, $1.4/GJ), the quality factors used in this study were: electricity 19.5 ($27.8/GJ, ¢10/kWh), petroleum 14.5 ($20.6/GJ, $0.66/L), and natural gas 2.7 ($3.8/GJ, $4/MMBtu) [32]. The bio-oil was assigned the QF of petroleum and methane was assigned the QF of natural gas. For materials, the quality-factor was determined as:

$$QF = \frac{MP}{EE \cdot EP_{coal}} = \frac{[\frac{\$}{kg}]}{\left[\frac{MJ}{kg}\right] \cdot [\frac{\$}{MJ}]_{coal}} \tag{12}$$

where MP is the price (in \$/kg), EE is the energy equivalent (in MJ/kg), and EP_{coal} is the energy price for coal (\$1.4/GJ).

5.2.6 PARTIAL FINANCIAL RETURN ON INVESTMENT ANALYSIS FORMULAE

The overall financial return on investment, FROI, can be calculated as:

$$FROI = \frac{R_{BO} + R_{BMF} + R_S}{(CC + OC + L)_G + (CC + OC + L)_p(CC + OC + L)_R(CC + OC + L)_D} [-] \tag{13}$$

where R is revenue (from bio-oil (BO), biomass fuel (BMF), and subsidies (S)), and the total investment is the sum of the capital costs (CC), operating costs (OC), and labor costs (L) for growth (G), processing (P), refining (R), and distribution (D). To parallel the 2nd O EROI, the partial financial return on investment, PFROI, is defined as:

$$PFROI = \frac{\acute{R}_{BO} + \acute{R}_{BMF}}{(\acute{O}C)_G + (\acute{O}C)_p(\acute{O}C)_R} [-] \tag{14}$$

and is equivalent to the QA 2nd O EROI.

In Equation (14), $\acute{R}_{BO}$ is revenue from bio-oil, $\acute{R}_{BMF}$ is revenue from biomass fuel (methane), and ÓC is the operating cost for growth (G), processing (P), and refining (R). Capital, labor, fuel distribution, discounting, and potential subsidy revenue would need to be included to determine an overall FROI.

5.2.7 WATER INTENSITY ANALYSIS FORMULAE

The water consumption and water withdrawal required for transportation via bio-oil and biomass fuel (methane) produced in this production pathway are calculated based on the methodology presented by King and Webber [33]. Consumption and withdrawal are defined as:

> "Water consumption describes water that is taken from surface water or a groundwater source and not directly returned. For example, a closed-loop cooling system for thermoelectric steam power generation where the withdrawn water is run through a cooling tower and evaporated instead of being returned to the source is consumption. Water withdrawal pertains to water that is taken from a surface water or groundwater source, used in a process, and (may be) given back from whence it came to be available again for the same or other purposes. To determine the water consumption or withdrawal for each input, the amount of each energy or material input is multiplied by the water equivalent for that input." [33].

The water consumption intensity, WCI, is defined as:

$$WCI = \frac{\acute{W}C}{\left(\acute{V}_{BO} \cdot FE_{BO} + V_{BMF} + FE_{BMF}\right)} \left[\frac{\text{L H}_2\text{O consumed}}{\text{km traveled}}\right] \tag{15}$$

and the water withdrawal intensity, WWI, is defined as:

$$WWI = \frac{\acute{W}W}{\left(\acute{V}_{BO} \cdot FE_{BO} + \acute{V}_{BMF} + FE_{BMF}\right)} \left[\frac{\text{L H}_2\text{O withdrawn}}{\text{km traveled}}\right] \tag{16}$$

where WC is the water consumed per liter of growth volume processed, WW is the water withdrawn per liter of growth volume processed, V_{BO} and V_{BMF} are the volumes of bio-oil and biomass fuel (methane) produced per liter of growth volume processed, FEBO and and FEBMF are the fuel economy values for transportation via bio-oil (28 miles/gallon, 11.8 km/L) and methane fuels (0.2 miles/standard cubic foot, 0.01 km/L). Thus, these metrics are calculated as the water required (consumed or withdrawn) for operating the production pathway shown in Figure 1 divided by the total distance that could be traveled using the bio-oil and the biomass fuel produced (assuming typical conversion efficiencies). The water consumption and water withdrawal include direct water inputs (e.g., water supplied to the growth volumes) and indirect water inputs (e.g., water used during nitrogen fertilizer production and electricity generation), thereby yielding a second-order water analysis. The energy return on water investment (EROWI) is a similar metric for evaluating water intensity [9,34] and can be calculated from the data in this study that are reported in Tables 3A and 4A. However, this metric does not consider the energy quality of the fuels produced, and therefore the WCI and WWI were used as the main metrics for evaluating water intensity in this study.

5.3 RESULTS

5.3.1 BIOMASS, LIPID, AND BIOFUEL PRODUCTIVITIES

The average algal concentration of the growth volume in the Experimental Case was 0.26 g/L, the neutral lipid fraction was estimated to be 0.02, and cultivation required 123 days. The neutral lipid content was determined by HPLC and the lipid composition included hydrocarbons, triglycerides, diglycerides, and monoglycerides. On average, 2 mg of biocrude and 165 mg of post-extraction biomass were recovered per liter of processed volume. These are not high productivity values as the research focus was on processing rather than growth. It was assumed that the bio-oil refining efficiency was 1 and the biomass fuel refining efficiency (converting post-extraction biomass to methane) was 0.25 [14,35].

The grown mass productivity (P_{GM}), estimated bio-oil productivity (P_{BO}), and estimated methane productivity (P_{BMF}) were calculated by combining these values, yielding 2.1 g of bio-oil per thousand liters of processed growth volume (0.0026 L/kLp, 0.00069 gal/kLp where Lp is the liters of processed growth volume) and 41.6 g of methane per kLp (cf. Table 2).

For the Highly Productive Case, the algal concentration was modeled to be 1 g/L (a factor of four improvement over the Experimental Case), requiring 12.5 days of cultivation with a grown mass productivity of 0.08 g/L-d. The neutral lipid fraction was assumed to be 0.3 and the production efficiencies were specified as: $\varphi_{harv} = 0.9$, $\varphi_{cellsys} = 0.95$, $\varphi_{sepBC} = 0.9$, $\varphi_{refBO} = 0.9$, $\varphi_{sepBS} = 1$, $\varphi_{refBMF} = 0.25$ [14]. The grown mass productivity (P_{GM}), bio-oil productivity (P_{BO}), and methane productivity (P_{BMF}) were calculated from these values and are listed in Table 1. As listed, 210 g (0.26 L, 0.069 gal) of bio-oil and 150 g of methane are produced for each kLp. The Highly Productive Case yields an energy output that is 7 times greater than that for the Experimental Case, and the energy inputs are described below.

TABLE 1: Grown mass, bio-oil, and biomass fuel (methane) productivities for the Experimental Case and the Highly Productive Case are listed in terms of volume and surface area. Culture depth is assumed to be 0.2 m.

	Photosynthetic Efficiency (%)	Grown Mass Productivity (P_{GM}) mg/L-d, (g/m²-d)	Bio-oil Productivity (P_{BO}) mg/L-d (g/m²-d)	Biomass Fuel Productivity (P_{BMF}) mg/L-d (g/m²-d)
Experimental Case	NA	2.17 (0.43)	0.02 (0.004)	0.34 (0.07)
Highly Productive Case	3.7	80.0 (16.0)	16.6 (3.32)	12.1 (2.42)
Theoretical Optimum Case	11.9	921 (184)	NA	NA

5.3.2 PHOTOSYNTHETIC EFFICIENCY

The photosynthetic efficiency (PE) cannot be determined for the Experimental Case because the incident radiation was not measured. For the Highly Productive Case, the PE can be determined from Equation (10), using the grown mass productivity and the cultivation time as specified inputs ($P_{GM} = 0.08$ g/L-d and $t_c = 12.5$ d). Therefore, inserting values for the HHV (24.49 MJ/kg) and τ (1.19) into Equation (10) and specifying values for CoL (0.5, based on the results of Weyer et al. [30]) and I (1100 kJ/Lp, which converts to 6500 MJ/m²-yr [30]), the PE for the Highly Productive Case can be calculated as:

$$PE = PAR \cdot PTE \cdot PUE \cdot \alpha = \frac{ED'_{CH2O}}{\acute{I}} = \frac{ED_{GV}}{(1 - CoL) \cdot \tau \cdot \acute{I}} = \frac{P_{GM}' \cdot HHV \cdot t_c}{(1 - CoL) \cdot \tau \cdot \acute{I}} = 0.037\ [-] \tag{17}$$

Similarly, the PE for an idealized Idealized Case can be calculated from Equation (10) by setting PAR = 0.46 , PTE = 1 , PUE = 1 , and α = 0.26 , yielding a PE = 0.119 . Based on

5.3.3 ENERGY RETURN ON INVESTMENT FOR ALGAL BIOFUEL

The 2nd O EROI for the Experimental Case and the Highly Productive Case, which have been reported previously by Beal et al. [14], are 9.2 × 10−4 ± 3.3 × 10−4 (cf. [19] for uncertainty analysis) and 0.22, respectively.

For algal biofuels to be produced commercially, the EROI must be competitive with that of conventional fuels (e.g., over the last few decades the EROI for oil and gas, including industrial capital, has typically been 10–20 [36] with delivered gasoline between 5 and 10 [37]). Several other studies have presented hypothetical energy analyses of algal biofuel production, and although the scope and systems evaluated vary, each of these studies has also found that without discounted inputs, the EROI is not competitive with conventional fuels [11,15,17,27,38]. The 2nd O EROI results from this study are plotted in Figure 2, along with the first-order EROI, which only includes direct energy inputs (and thereby neglects energy embedded in material inputs).

For the Experimental Case, 90% (2308 kJ/Lp) of the total energy input (2572 kJ/Lp) was associated with bioreactor lighting, air compression (for supplying CO_2), and pond mixing; all of which are considered to be artifacts of inefficient research-scale growth methods. Conversely, in the Highly Productive Case, which modeled efficient growth equipment, embedded energy in nutrients accounted for 85% (63 kJ/Lp) of the total energy input (75 kJ/Lp). The Highly Productive Case assumes 8 kg of CO_2, 70 g of nitrogen, and 8 g of phosphorus consumed per kg of algae produced.

Based on conservation of mass, the minimum possible CO_2, nitrogen, and phosphorus consumption can be approximated as 1.8 kg, 70 g, and 8 g per kg of generic algal biomass, respectively [2,15,17,44–46]. Using these minimum data, and the associated energy equivalents (with values of 7.3 MJ/kg CO_2, 59 MJ/kg N, and 44 MJ/kg P [15,19,46–51]), the minimum possible energy embedded in the (full-price) nutrients alone requires more energy (17.7 kJ/Lp) than the total energy produced (16.6 kJ/Lp), which prevents a positive net energy yield, and illustrates the need to use waste forms of nutrients. The energy embedded in carbon, nitrogen, and phosphorus is dependent on the stoichiometric requirement and energy intensity of production for each element. However, the embedded energy in these elements is independent of growth rate [14], demonstrating the limited ability for growth optimization to alter the overall EROI for algal biofuels.

The EROI was adjusted using quality factors reported by Beal et al. [14] that were calculated according to the price of each input, yielding a QA 2nd O EROI that directly parallels the PFROI analysis. For the Experimental Case and the Highly Productive Case, the QA 2nd O EROI was 9.2 × 10−5 and 0.36, respectively [14].

5.3.4 FINANCIAL RETURN ON INVESTMENT OF ALGAL BIOFUEL

The PFROI is equivalent to the QA 2nd O EROI and is calculated using Equation (14). This relation serves as a standard way to compare energy and cost analyses at a systems level. By doing so, the energetic profitability of an energy system (which is the most important metric for researchers interested in global energy production and consumption or thermodynamics of energy systems) can be compared with the financial profitability of an energy system (which is most important to businesses and investors).

The cost of growing algae was calculated for the Experimental Case by applying electricity and material prices, yielding a total cost of growth of \$105.2/kLp. With 2.1 g of bio-oil produced from each kL of processed

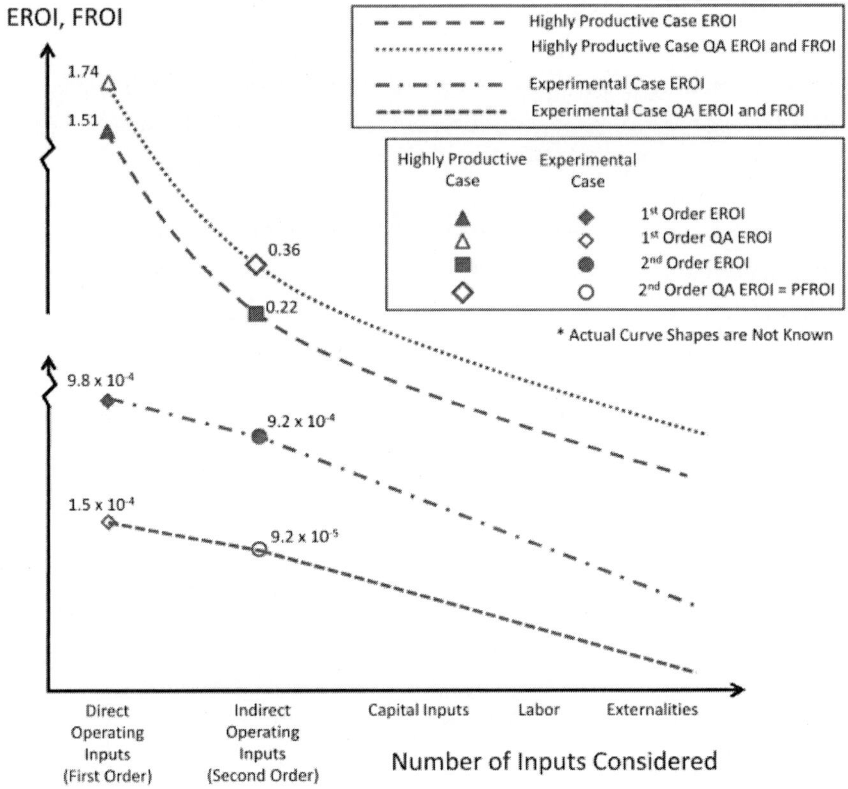

FIGURE 2: The EROI and PFROI for the Experimental Case and the Highly Productive Case decline as more inputs are considered. The curves are presented for illustration only, as the curve shapes are unknown.

volume, these cultivation costs are \$40,000/L of bio-oil (\$150,000/gal). The Highly Productive Case data results in a total cultivation cost of \$0.42/kLp, which is equivalent to \$0.42/kg of algae or \$1.6/L of bio-oil (\$6.1/gal) based on the bio-oil productivity calculated above (210 g bio-oil/kLp). The combined cost of processing and refining was calculated to be \$7.71/kLp and \$0.13/kLp for the Experimental and Highly Productive Cases, respectively (cf. Tables 1A and 2A). Based on the resulting bio-oil productivities, these values correspond to \$2900/L of bio-oil (\$11,000/gal) and \$0.5/L of bio-oil (\$1.9/gal) for these cases, respectively. Davis et al. present a comprehensive techno-economic analysis of a similar production system (including capital costs) and determined that operating costs for both open-pond and enclosed bioreactor settings would be near \$1.3/L of bio-oil (\$5/gal) [16]. This result is similar to the total operating cost of the Highly Productive Case (\$2.1/L of bio-oil, \$7.99/gal). In the Experimental Case, 2.1 g of bio-oil were produced per kLp (0.0026 L/kLp) and 41.6 g of methane were produced per kLp. Assuming market prices of \$0.66/L of bio-oil (\$0.66/L, \$0.83/kg) and \$4/MMBtu of methane (\$0.21/kg) yields revenues of \$0.0017/kLp for bio-oil and \$0.0087/kLp for methane in the Experimental Case (yielding \$0.010/kLp of total revenue). In the Highly Productive Case, 210 g (0.26 L) of bio-oil and 150 g of methane are produced for each kLp, resulting in \$0.17/kLp of bio-oil revenue and \$0.03/kLp of methane revenue. Until 2012, a production subsidy of \$0.13/L was provided for corn ethanol in the United States, and if an equal subsidy was provided for algal fuels, the production plant would gain incremental income of \$0.0004/kLp for the Experimental Case and \$0.035/kLp for the Highly Productive Case.

The partial financial returns on investment (PFROI) are calculated from Equation (14) for the Experimental Case and the Highly Productive Case to be 9.2×10^{-5} and 0.37, respectively. The challenge in obtaining a PFROI greater than 1 is growing, processing, and refining high-yield biomass cheaply, especially since many of the costs scale directly with biomass productivity (e.g., nutrient costs increase as biomass productivity increases). The overall FROI would be lower than the PFROI as capital, labor, and distribution costs will be significant expenses, which are not included in the PFROI. For example, Lundquist et al. and Davis et al. provide analyses for capital costs of similar production systems and demonstrate that capital costs might contribute roughly 50% of the total cost for open-pond systems (this fraction increases substantially for bioreactors) [16,52]. Figure 2 illustrates the relationships between the EROI, QA EROI, and PFROI with respect to the number of inputs that are considered in the analysis, and is based on the work of Henshaw, King, and Zarnikau in relating EROI to full business costs, or cash flows [53]. For a given biofuel output, as more inputs are included in the calculations, the return on investment values decrease.

5.3.5 WATER INTENSITY OF ALGAL BIOFUEL

The WCI is calculated using Equation (15). Figure 3 plots the second-order water intensity of transportation (consumption and withdrawal) using algal biofuels produced in this system (bio-oil and methane) for the two cases considered. These data are shown alongside equivalent results for a variety of transportation fuels, including fossil fuels, electricity for electric vehicles, and biofuels reported previously by King and Webber [33] (note the logarithmic scales).

As shown, the Experimental Case water intensity (which includes significant research-scale artifacts, no recycling, evaporation from the ponds, and relatively low biofuel yields) far exceeds any of the other transportation fuels. Meanwhile, the Highly Productive Case water consumption intensity is lower than that of biofuels from irrigated crops, while its water withdrawal intensity is similar to, or slightly greater than, that of biofuels from irrigated crops. Still, the Highly Productive Case, which assumes very efficient water use (no evaporation and 95% recycling), is much more water intensive than traditional fossil fuels or non-irrigated biofuels from conventional feedstocks. While the WCI and WWI metrics are useful to evaluate the magnitude of water required for fuel production, they do not consider water quality (that is, algae can be grown in degraded, brackish, or saline sources, for which the concerns about water quantity are muted as compared with freshwater). The relationship between water requirements (considering magnitude and quality) and water availability (including precipitation, which is not considered here for the algal biofuel cases) is more important than the water intensity, alone. However, this relationship is dependent on location and must be evaluated on a case-by-case, site-specific basis for all of the fuels shown.

Several other studies have been conducted to determine the water intensity of algal biofuel production and the system boundaries used in each study vary [9,11,15,17,54,55]. Analogous to energy inputs, the water inputs for a production pathway include direct and indirect parts. Additionally, the water consumption required to produce capital equipment can be included (e.g., water required for producing glass bioreactors [54]). Finally, the water intensity is dependent on co-product allocation, as the

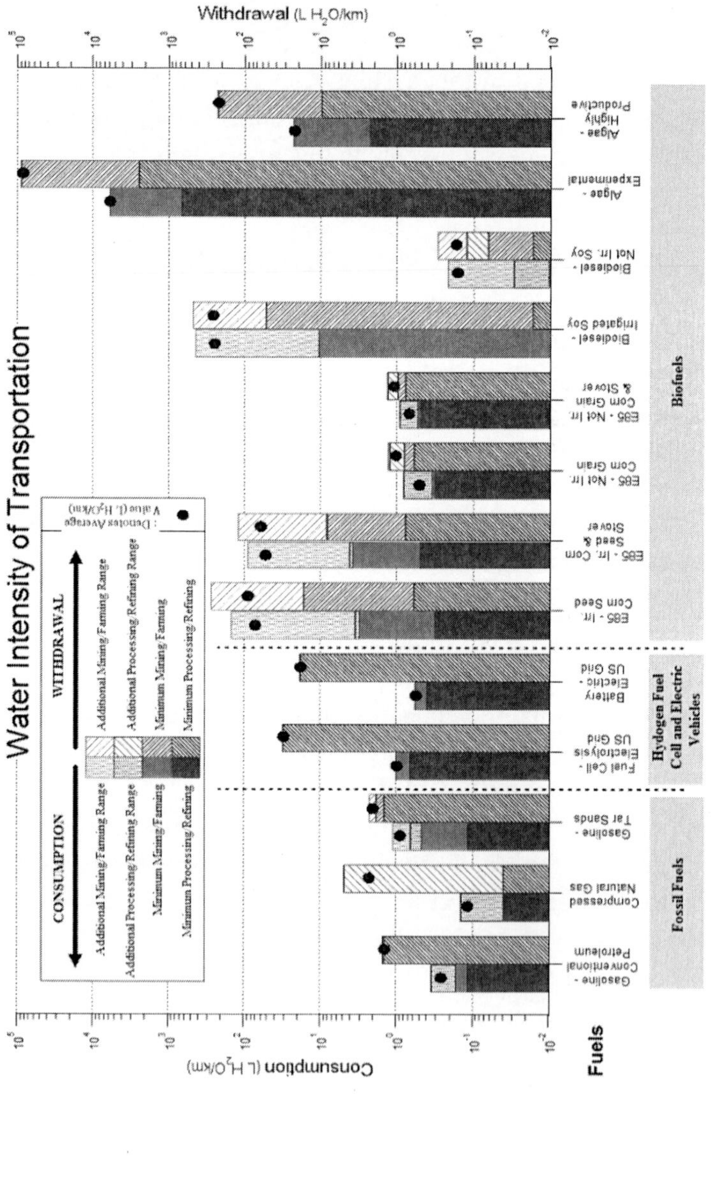

FIGURE 3: Second-order water intensity of transportation for several fuels [33], including the bio-oil and methane co-products from the two algal biofuel cases: the Experimental Case and the Highly Productive Case. *Note the logarithmic scale. To evaluate sustainability, the water intensity and required water quality must be considered in conjunction with water availability.

total water consumed to operate the production pathway should be allocated between the bio-oil and co-products (e.g., methane).

The WWI is calculated according to Equation (16). The WWI for the Experimental Case and the Highly Productive Case are 87,000 L/km and 220 L/km, respectively. Like the nutrient analysis presented below, the water analysis underscores the advantages of using nutrient-rich low-quality water, like waste water or agricultural runoff [28]. In these cases, the incremental water usage is minimized and the discharge water can be of higher quality (e.g., higher purity) than the water input.

5.3.6 RESOURCE REQUIREMENTS FOR 5 BGAL/YR OF ALGAL BIO-OIL

While re-use of CO_2 would be desirable and some water requirements could be met with wastewater or saline water, the increased demand for fertilizer and electricity could have negative economic impacts. Depending on the scale of production, this electricity input requirement could impact electricity prices and yield a significant, unintended increase in carbon emissions.

5.3.6.1 CARBON DIOXIDE

Under ideal conditions, algae require roughly 2 kg of CO_2 for each kg of algal biomass produced [2,15,17]. However, in the experiments, most of the CO_2 delivered to the growth volumes was not retained in biomass (and released as outgas). As a result, 9.35 g of CO_2 were consumed per liter of pond water processed, which only contained 0.26 g of algae, on average. Based on this consumption, 3.7 Mg of CO_2 were consumed per L of bio-oil. For the Highly Productive Case (with 1 kg algal biomass/kL of processed volume, 8 kg of CO_2 per kL of processed volume, and 0.26 L of bio-oil per kL of processed volume) 31 kg of CO_2 would be required for each L of bio-oil produced. For 19 GL/yr of bio-oil (5 Bgal/yr), this equates to 5.8 × 1011 kg of CO_2 consumed per year, which is ~11% of the total CO_2 emissions from the U.S. [42].

5.3.6.2 NITROGEN FERTILIZER

For ideal conditions, roughly 70 g of nitrogen are required for each kg of algal biomass [15,17,44]. In the experiments, 0.20 kg of nitrogen was consumed per kLp. This amount translates to 77 kg of nitrogen per L of bio-oil produced (which is 769 g of N per kg of algal biomass). In the Highly Productive scenario (with 1 kg of algal biomass/kL of processed volume, 70 g of N per kL of processed volume, and 0.26 L of bio-oil per kLp) 0.27 kg of N are required per L of bio-oil produced, or 5.1 × 109 kg of N would be required for 19 GL of bio-oil (5 Bgal), which is 45% of the total amount of nitrogen fertilizer consumed in the U.S. annually [43].

5.3.6.3 ELECTRICITY

In the Experimental Case, 2.4 GJ of electricity were consumed per kLp, resulting in ~0.92 × 1012 J of electricity consumption per L of bio-oil. In the Highly Productive Case, 2.59 MJ of electricity is consumed per kLp, which yields 0.26 L of bio-oil. Thus, 9.9 MJ of electricity would be consumed per L of bio-oil, or 0.19 EJ per year for 19 GL of bio-oil per year (5 Bgal/yr). This amount is 1.3% of the annual U.S. electricity generation in 2009 [41].

5.3.6.4 METHANE CO-PRODUCT

Based on the methane productivity presented above, 15.9 kg of methane would be produced per L of bio-oil in the Experimental Case, which yields 16.6 EJ/yr of methane energy produced for 19 GL/yr of bio-oil (5 Bgal/yr). This methane yield would displace 60% of the total U.S. natural gas consumption (~28.1 EJ/yr in 2009 [39]) (although this result is not a realistic expectation, as the EROI for this scenario is several orders of magnitude less than 1). In the Highly Productive Case a smaller portion of the biomass is used to produce methane (70% rather than ~95%–99% in the Experimental Case) because of the much higher lipid fraction. As a result, only 0.6 kg of methane would be co-produced for each L of bio-oil, yielding 0.60 EJ/yr of methane co-product for 19 GL/yr of algal bio-oil (5 Bgal/yr). This methane production could replace ~2.2% of the total U.S. natural gas consumption [39].

5.4 CONCLUSIONS

5.4.1 CURRENT FEASIBILITY

As shown above, the 2nd O EROI and PFROI are less than 1 for algal biofuels produced in this production system, even for the Highly Productive Case, which assumes efficient growth and processing methods. Including additional expenses that were omitted by this analysis (i.e., capital, labor, externalities, etc.) would further reduce profitability. Additionally, transportation using algal biofuels produced in these cases is more water intensive and resource intensive than conventional fuels. The challenge for achieving energy-positive, profitable biofuel production from algae is rooted in the thermodynamic challenges associated with converting materials with low energy density (such as dispersed photons, CO_2, and nutrients) to energy-dense fuels [29]. This conversion requires a significant reduction in specific entropy, which thereby requires a significant amount of work input (i.e., energy expense). This body of work demonstrates that producing petroleum fuel substitutes from algae without using discounted electricity, nutrients, and/or CO_2 is not energetically favorable with the existing technologies considered. Although improving algal biomass productivity by optimizing physical conditions, biochemical conditions, and by genetic engineering might improve the overall biofuel yield, the required amounts of carbon, nitrogen, and phosphorus are dictated by stoichiometry. While some variation in algal stoichiometry exists (evidenced, for example, by the change in chemical composition that can occur under different growth conditions, such as nitrogen starvation), any algal species will inevitably be constrained by stoichiometric conditions. Thus, producing more algae (e.g., by increasing photosynthetic efficiency) also increases the nutrient requirement and the associated energy embedded in the nutrients. As shown in this study, there can be more energy and cost embedded in the nutrients consumed than produced in the resulting algal biomass. As a result, low-energy and low-cost sources of carbon, nitrogen, phosphorus, and water (for example, from waste streams) would likely be needed.

Researchers have two options for conducting analyses for non-commercial algal energy production processes: (1) use data from experimental processes (which are devoid of the efficiencies that accompany large-scale production) or (2) use data from models of future commercial-scale systems. The Experimental Case and the Highly Productive Case represent these two approaches, respectively. The presentation of results for both measured lab and up-scaled estimates is important because it enables more informed modeling of the innovation process from lab to production. It is unclear how well a lab-scale experiment needs to perform before moving to the next stage of development. We see this simultaneous presentation of multiple metrics (EROI, FROI, water intensity, nutrient constraints, and CO_2 constraints) as part of a critical due diligence process for inventors and investors. Ongoing, it can be possible to have standard experimental test conditions that enable consistent comparison and tracking progress as new technologies are incorporated into the process chain. This tracking of progress can mimic that of the photovoltaic cell industry.

Since the Experimental Case contains many lab-scale artifacts, the constraints on the Highly Productive Case are more representative of the challenges that will be faced by the algal biofuels industry. Most of the conclusions in this study are based on the Highly Productive Case and the targets provided in the following section for achieving profitable algal biofuel production rely on the Highly Productive Case for comparison. 2, 5 1967

5.4.2 TARGETS TO ACHIEVE SUSTAINABLE PRODUCTION

Based on the results of this study, targets can be set for producing algal biofuel that will enable a 2nd O EROI and PFROI equal to 1 (i.e., break-even) without exceeding water availability constraints or drastically increasing national fertilizer consumption. Since these targets are devised only with consideration of operating inputs, productivity would need to be increased and/or expenses would need to be decreased significantly to achieve an overall EROI and overall FROI (including capital costs) greater than 1 for the delivered energy carriers (a requirement for fuels to make a net energetic contribution), or greater than 3 (for practical purposes). The guiding targets for research stakeholders for comparison to the Experimental Case and the Highly Productive Case are:

1. Algal concentration of 3 g/L with a lipid fraction of 0.3, which would yield approximately 25 kJ of bio-oil and 25 kJ of methane per liter of processed volume (which is about 800 L BO/MLp and 450 kg methane/MLp, estimated to be roughly \$600 of revenue per million liters of growth volume (assuming \$0.66/L BO (\$2.50/gal BO) and \$3.8/GJ (\$4/MMBtu) of methane));
2. In conjunction with item 1, an energy input for growth, processing, and refining that is less than 50 kJ per liter of processed volume enables a 2nd O EROI > 1 and requires using discounted inputs;
3. The FROI is dependent upon market prices, and therefore can vary substantially depending on market conditions (e.g., oil price). However, based on the price assumptions used in this study, if the targets listed above can be achieved, the PFROI would be greater than 1 if the cost of growth, processing, and refining is less than \$600 per million liters of growth volume processed (which is equivalent to \$0.20/kg of grown mass). Achieving a total cost less than \$600 per million liters of growth volume processed would yield an overall FROI greater than 1 for this scenario (assuming no subsidy revenue);
4. A fresh water consumption intensity on the order of 2.4 L/km (1 gal/mi), achieved by consuming roughly 25 liters of fresh water per thousand liters of processed volume (which corresponds to no evaporation during growth, minimal processing water use, and greater than 97.5% recycling for fresh water cultivation). This consumption corresponds to about 33 liters of fresh water per liter of bio-oil produced (with a methane co-product of about 0.58 kg/L BO). Using saline water or waste water could also enable a low fresh water consumption intensity;

5. A net nutrient consumption that would enable large-scale production while only marginally increasing the national fertilizer consumption. For example, to produce 5 Bgal of fuel per year (19 GL/yr), the net nitrogen consumption for each liter of fuel produced should be less than about 26 g to prevent a national increase in nitrogen fertilizer consumption of more than 5% (which is about 6 × 108 kg N/yr [43]). In this scenario, one liter of bio-oil is produced from about 4 kg of algae, and therefore the nitrogen consumption should be less than about 7 g per kg of algae, which is roughly 10% of the minimum possible nitrogen requirement for algae (~70 g of nitrogen per kg of algae). Therefore, nitrogen recycling or utilization of waste nitrogen of 90% or more is required.

5.4.3 INNOVATION PATHWAYS

Based on this analysis and consistent with some earlier research, there are a few approaches and areas of opportunity where innovations would make the biggest impact in terms of improving the energy balance, economic profitability, and water intensity of algal biofuel production. These improvements include:

1. using waste and recycled nutrients (e.g., waste water and animal waste) [11,15,25–28,65,71,72];
2. using waste heat and flue-gas from industrial plants [44,59], carbon in wastewater [28], or developing energy-efficient means of using atmospheric CO_2;
3. developing ultra-productive algal strains (e.g., genetically modified organisms) [73–75];
4. minimizing pumping [58,76,77];
5. establishing energy-efficient water treatment and recycling methods [55];
6. employing energy-efficient harvesting methods, such as chemical flocculation [66,78,79], and
7. avoiding separation via distillation.

The development of genetically modified organisms that secrete oils might provide parallel reductions in energy expense, as the oil might be more easily collected. Policies (e.g., carbon legislation) and externalities could change algal biofuel economics, but not energy accounting. Additionally, algae can produce nutraceutical and pharmaceutical co-products, which could significantly improve the overall process economics. For comparison, co-products account for approximately 20% of the energy value for corn ethanol [13]; because co-products from algae find markets in higher value industries, algal fuels will likely have higher co-product allocation than from corn seed. The most favorable scenario for algal biofuel production is one that can use each of the improvements listed above. Implementing growth and processing technology advancements, in conjunction with co-locating facilities with discounted energy and materials (i.e., electricity plants, waste water treatment plants, livestock feed lots, etc.) offers the potential for profitable algal biofuel production, and this concept has been proposed by several researchers [11,15,25,26,28,44]. However, relying on waste materials as feedstock relegates algal biofuel production to relatively low volumes [11,28,71].

Overall, it is most important that the EROI for the energy sector is greater than unity, including contributions from all energy resources. Although the results of this study suggest that the EROI for algal fuels will remain less than one without significant biotechnology innovations, algae represent one of the few alternative feedstocks capable of producing petroleum fuel substitutes directly (without expensive gasification or Fischer-Tropsch processes) for applications that require high energy-density, such as aviation. Thus, even though algal biofuels face significant hurdles before becoming large-scale substitutes for petroleum, they have the potential to satisfy niche markets in the short-term, while implementation of "game-changing" biotechnology advances are needed for sustainable large-scale algal biofuel production.

When looking forward towards those potential advances, it is the authors' hope that the analytical approach presented in this manuscript will provide a useful framework with which progress can be tracked. Specifically, we think this framework will be useful for tracking energy, cost, water and other resource inputs and outputs of cultivation.

REFERENCES

1. Wijffels, R.H.; Barbosa, M.J. An outlook on microalgal biofuels. Science 2010, 329, 796–799.
2. Schenk, P.; Thomas-Hall, S.; Stephens, E.; Marx, U.; Mussgnug, J.; Posten, C.; Kruse, O.; Hankamer, B. Second generation biofuels: high-efficiency microalgae for biodiesel production. BioEnergy Res. 2008, 1, 20–43.
3. Sheehan, J.; Dunahay, T.; Benemann, J.; Roessler, P. A Look Back at the U.S. Department of Energy's Aquatic Species Program, Biodiesel from Algae; NREL and U.S. Department of Energy's Office of Fuels Development: Golden, CO, USA, 1998.
4. Hall, C.A.S.; Balogh, S.; Murphy, D.J.R. What is the minimum EROI that a sustainable society must have? Energies 2009, 2, 25–47.
5. Mulder, K.; Hagens, N.J. Energy return on investment: toward a consistent framework. AMBIO J. Hum. Environ. 2009, 37, 74–79.
6. Beal, C.M.; Smith, C.H.; Webber, M.E.; Ruoff, R.S.; Hebner, R.E. A framework to report the production of renewable diesel from algae. BioEnergy Res. 2011, 4, 36–60. Energies 2012, 5 1977
7. Amin, S. Review on biofuel oil and gas production processes from microalgae. Energy Convers. Manag. 2009, 50, 1834–1840.
8. Brennan, L.; Owende, P. Biofuels from microalgae—A review of technologies for production, processing, and extractions of biofuels and co-products. Renew. Sustain. Energy Rev. 2010, 14, 557–577.
9. Subhadra, B.G.; Edwards, M. Coproduct market analysis and water footprint of simulated commercial algal biorefineries. Appl. Energy 2011, 88, 3515–3523.
10. Collet, P.; Helias, A.; Lardon, L.; Ras, M.; Goy, R.; Steyer, J. Life-cycle assessment of microalgae culture coupled to biogas production. Bioresour. Technol. 2011, 102, 207–214.
11. Lundquist, T.J.; Woertz, I.C.; Quinn, N.W.T.; Benemann, J.R. A Realistic Technology and Engineering Assessment of Algae Biofuel Production; Energy Biosciences Institute: Berkeley, CA, USA, 2010.
12. Sialve, B.; Bernet, N.; Bernard, O. Anaerobic digestion of microalgae as a necessary step to make microalgal biodiesel sustainable. Biotechnol. Adv. 2009, 27, 409–416.
13. Farrell, A.E.; Plevin, R.J.; Turner, B.T.; Jones, A.D.; O'Hare, M.; Kammen, D.M. Ethanol can contribute to energy and environmental goals. Science 2006, 311, 506–508.
14. Beal, C.M.; Hebner, R.E.; Webber, M.E.; Ruoff, R.S.; Seibert, A.F. The energy return on investment for algal biocrude: Results for a research production facility. BioEnergy Res. 2012, 5, 341–262.
15. Clarens, A.F.; Resurreccion, E.P.; White, M.A.; Colosi, L.M. Environmental life cycle comparison of algae to other bioenergy feedstocks. Environ. Sci. Technol. 2010, 44, 1813–1819.
16. Davis, R.; Aden, A.; Pienkos, P.T. Techno-economic analysis of autotrophic microalgae for fuel production. Appl. Energy 2011, 88, 3524–3531.
17. Lardon, L.; Helias, A.; Sialve, B.; Steyer, J.; Bernard, O. Life-cycle assessment of biodiesel production from microalgae. Environ. Sci. Technol. 2009, 43, 6475–6481.

18. Pate, R.; Klise, G.; Wu, B. Resource demand implications for US algae biofuels production scale-up. Appl. Energy 2011, 88, 3377–3388.
19. Beal, C.M. Constraints on Algal Biofuel Production. Ph.D. Thesis, University of Texas at Austin, Austin, TX, USA, May 2011.
20. Beal, C.M.; Hebner, R.E.; Romanovicz, D.; Mayer, C.C.; Connelly, R. Progression of lipid profile and cell structure in a research production pathway for algal biocrude. Renew. Energy 2012, in press.
21. Jones, J.; Manning, S.; Montoya, M.; Keller, K.; Poenie, M. Analysis of Algal Lipids by HPLC and Mass Spectroscopy. J. Am. Oil Chem Soc. 2012, doi:10.1007/s11746-012-2044-8,.
22. Nordbäck, J.; Lundberg, E.; Christie, W.W. Separation of lipid classes from marine particulate material by HPLC on a polyvinyl alcohol-bonded stationary phase using dual-channel evaporative light-scattering detection. Mar. Chem. 1998, 60, 165–175.
23. Gillan, F.T.; Johns, R.B. Normal-phase hplc analysis of microbial carotenoids and neutral lipids. J. Chromatogr. Sci. 1983, 21, 34–38.
24. Graeve, M.; Janssen, D. Improved separation and quantification of neutral and polar lipid classes by HPLC-ELSD using a monolithic silica phase: Application to exceptional marine lipids. J. Chromatogr. B 2009, 877, 1815–1819. Energies 2012, 5 1978
25. Pittman, J.K.; Dean, A.P.; Osundeko, O. The potential of sustainable algal biofuel production using wastewater resources. Bioresour. Technol. 2011, 102, 17–25.
26. Sturm, B.S.M.; Lamer, S.L. An energy evaluation of coupling nutrient removal from wastewater with algal biomass production. Appl. Energy 2011, 88, 3499–3506.
27. Batan, L.; Quinn, J.; Willson, B.; Bradley, T. Net energy and greenhouse gas emission evaluation of biodiesel derived from microalgae. Environ. Sci. Technol. 2010, 44, 7975–7980.
28. Beal, C.M.; Stillwell, A.S.; King, C.W.; Cohen, S.M.; Berberoglu, H.; Bhattarai, R.; Connelly, R.; Webber, M.E.; Hebner, R.E. Energy return on investment for algal biofuel production coupled with wastewater treatment. Water Environ. Res. 2012, in press.
29. Beal, C.M.; Hebner, R.E.; Webber, M.E. Thermodynamic analysis of algal biofuel production. Available online: http://dx.doi.org.ezproxy.lib.utexas.edu/10.1016/j.energy.2012.05.003 (accessed on 8 June 2012).
30. Weyer, K.; Bush, D.; Darzins, A.; Willson, B. Theoretical maximum algal oil production. BioEnergy Res. 2010, 3, 204–213.
31. Cleveland, C.J.; Kaufmann, R.K.; Stern, D.I. Aggregation and the role of energy in the economy. Ecol. Econ. 2000, 32, 301–317.
32. Energy Information Agency. Annual Energy Outlook 2010: DOE/EIA-0383(2010); Energy Information Agency: Washington, DC, USA, 2010.
33. King, C.W.; Webber, M.E. Water intensity of transportation. Environ. Sci. Technol. 2008, 42, 7866–7872.
34. Mulder, K.; Hagens, N.; Fisher, B. Burning water: A comparative analysis of the energy return on water invested. AMBIO J. Hum. Environ. 2010, 39, 30–39.
35. Oyler, J. Catalytic Hydrothermal Gasification. Genifuel, Halethorpe, MD, USA. Personal Communication, 2010.
36. Cleveland, C.J. Net energy from the extraction of oil and gas in the United States. Energy 2005, 30, 769–782.

37. King, C.W. Energy intensity ratios as net energy measures of United States energy production and expenditures. Environ. Res. Lett. 2010, 5, doi:10.1088/1748-9326/5/4/044006.
38. Xu, L.; Brilman, D.W.F.; Withag, J.A.M.; Brem, G.; Kersten, S. Assessment of a dry and a wet route for the production of biofuels from microalgae: Energy balance analysis. Bioresour. Technol. 2011, 102, 5113–5122.
39. Natural Gas Overview: Table 6.1, EIA. Available online: http://www.eia.gov/totalenergy/data/annual/txt/ptb0601.html (accessed on 13 June 2012).
40. Austin Water, Austin City Connection. Available online: http://www.austintexas.gov/department/austin-water-utility-statistics (accessed on 14 April 2011).
41. Summary Statistics for the United States: Electric Power Annual data for 2009, EIA; 2011. Available online: http://205.254.135.7/electricity/annual/archive/03482009.pdf (accessed on 8 February 2011).
42. Hockstad, L.; Cook, B. 2011 Draft. U.S. Greenhouse Gas Inventory Report. 2011. Available online: http://www.epa.gov/climatechange/emissions/usinventoryreport.html (accessed on 13 June 2012). Energies 2012, 5 1979
43. Huang, W. Impact of Rising Natural Gas Prices on U.S. Ammonia Supply. Available online: http://www.ers.usda.gov/publications/wrs0702/ (accessed on 31 October 2010).
44. Campbell, P.K.; Beer, T.; Batten, D. Life cycle assessment of biodiesel production from microalgae in ponds. Bioresour. Technol. 2010, 102, 50–56.
45. Chisti, Y. Biodiesel from microalgae. Biotechnol. Adv. 2007, 25, 294–306.
46. Murphy, C. Analysis of Innovative Feedstock Sources and Production Technologies for Renewable Fuels: Chapter 6. Algal Oil Biodiesel, EPA: XA-83379501-0; University of Texas: Austin, TX, USA, 2010.
47. Ramírez, C.A.; Worrell, E. Feeding fossil fuels to the soil: An analysis of energy embedded and technological learning in the fertilizer industry. Resour. Conserv. Recycl. 2006, 46, 75–93.
48. Sheehan, J.; Camobreco, V.; Duffield, J.; Shapouri, H.; Graboski, M.; Tyson, K.S. An Overview of Biodiesel and Petroleum Diesel Life Cycles; Technical Report. 2000; NREL/TP-580-24772.
49. Wu, H.; Fu, Q.; Giles, R.; Bartle, J. Production of mallee biomass in Western Australia: Energy balance analysis. Energy Fuels 2007, 22, 190–198.
50. U.S. Life-Cycle Inventory Database; NREL: Golden, CO, USA, 2008; Volume 1.6.0.
51. Worrell, E.; Phylipsen, D.; Einstein, D.; Martin, N. Energy Use and Energy Intensity of the U.S. Chemical Industry; 2000; p. 40.
52. Lundquist, T.J.; Woertz, I.C.; Quinn, N.W.T.; Benemann, J.R. A Realistic Technology and Engineering Assessment of Algae Biofuel Production; EBI: Berkeley, CA, USA, 2010.
53. Henshaw, P.; King, C.W.; Zarnikau, J. System Energy Assessment (SEA), Defining a Standard Measure of EROI for Energy Businesses as Whole Systems. Sustainability 2011, 3, 1908–1943.
54. Harto, C.; Meyers, R.; Williams, E. Life cycle water use of low-carbon transport fuels. Energy Policy 2010, 38, 4933–4944.

55. Yang, J.; Xu, M.; Zhang, X.; Hu, Q.; Sommerfeld, M.; Chen, Y. Life-cycle analysis on biodiesel production from microalgae: Water footprint and nutrients balance. Bioresour. Technol. 2010, 102, 159–165.
56. Energy Independence and Security Act of 2007, One Hundred Tenth Congress of the United States of America. Available online: http://www.govtrack.us/congress/bills/110/hr6 (accessed on June 2012).
57. Benemann, J.; Oswald, W. Systems and Economic Analysis of Microalgae Ponds for Conversion of CO2 to Biomass; U.S. Department of Energy: Washington, DC, USA, 1996.
58. Jorquera, O.; Kiperstok, A.; Sales, E.A.; Embiruçu, M.; Ghirardi, M.L. Comparative energy life-cycle analyses of microalgal biomass production in open ponds and photobioreactors. Bioresour. Technol. 2010, 101, 1406–1413.
59. Kadam, K.L. Environmental implications of power generation via coal-microalgae cofiring. Energy 2002, 27, 905–922.
60. Pruvost, J.; van Vooren, G.; Cogne, G.; Legrand, J. Investigation of biomass and lipids production with Neochloris oleoabundans in photobioreactor. Bioresour. Technol. 2009, 100, 5988–5995.
61. Li, X.; Xu, H.; Wu, Q. Large-scale biodiesel production from microalga Chlorella protothecoides through heterotrophic cultivation in bioreactors. Biotechnol. Bioeng. 2007, 98, 764–771. Energies 2012, 5 1980
62. Hu, Q.; Sommerfeld, M.; Jarvis, E.; Ghirardi, M.; Posewitz, M.; Seibert, M.; Darzins, A. Microalgal triacylglycerols as feedstocks for biofuel production: Perspectives and advances. Plant J. 2008, 54, 621–639.
63. Miao, X.; Wu, Q. Biodiesel production from heterotrophic microalgal oil. Bioresour. Technol. 2006, 97, 841–846.
64. Meng, X.; Yang, J.; Xu, X.; Zhang, L.; Nie, Q.; Xian, M. Biodiesel production from oleaginous microorganisms. Renew. Energy 2009, 34, 1–5.
65. Wang, B.; Lan, C.Q. Biomass production and nitrogen and phosphorus removal by the green alga Neochloris oleoabundans in simulated wastewater and secondary municipal wastewater effluent. Bioresour. Technol. 2011, 102, 5639–5644.
66. Molina Grima, E.; Belarbi, E.H.; Acién Fernández, F.G.; Robles Medina, A.; Chisti, Y. Recovery of microalgal biomass and metabolites: Process options and economics. Biotechnol. Adv. 2003, 20, 491–515.
67. Choi, J. Pilot Scale Evaluation of Algae Harvesting Technologies for Biofuel Production. M.S. Thesis, University of Texas at Austin: Austin, TX, USA, 2009.
68. Umdu, E.S.; Tuncer, M.; Seker, E. Transesterification of Nannochloropsis oculata microalga's lipid to biodiesel on Al2O3 supported CaO and MgO catalysts. Bioresour. Technol. 2009, 100, 2828–2831.
69. Demirbas, A. Progress and recent trends in biodiesel fuels. Energy Convers. Manag. 2009, 50, 14–34.
70. Meher, L.C.; Vidya Sagar, D.; Naik, S.N. Technical aspects of biodiesel production by transesterification—A review. Renew. Sustain. Energy Rev. 2006, 10, 248–268.
71. Christenson, L.; Sims, R. Production and harvesting of microalgae for wastewater treatment, biofuels, and bioproducts. Biotechnol. Adv. 2011, 29, 686–702.

72. Woertz, I.; Feffer, A.; Lundquist, T.; Nelson, Y. Algae grown on dairy and municipal wastewater for simultaneous nutrient removal and lipid production for biofuel feedstock. J. Environ. Eng. 2009, 135, 1115–1122.
73. Radakovits, R.; Jinkerson, R.E.; Darzins, A.; Posewitz, M.C. Genetic engineering of algae for enhanced biofuel production. Eukaryot. Cell 2010, 9, 486–501.
74. Robertson, D.; Jacobson, S.; Morgan, F.; Berry, D.; Church, G.; Afeyan, N. A new dawn for industrial photosynthesis. Photosynth. Res. 2011, 107, 269–277.
75. Rosenberg, J.N.; Oyler, G.A.; Wilkinson, L.; Betenbaugh, M.J. A green light for engineered algae: Redirecting metabolism to fuel a biotechnology revolution. Curr. Opin. Biotechnol. 2008, 19, 430–436.
76. Bolhouse, A.; Ozkan, A.; Berberoglu, H., Rheological Study of Algae Slurries for Minimizing Pumping Power. In Proceedings of the ASME-IMECE 2010, Vancouver, Canada, 13–18 November 2010.
77. Murphy, C.; Allen, D. The energy water nexus of mass cultivation of algae. Environ. Sci. Technol. 2011, 45, 5861–5868.
78. Henderson, R.; Parsons, S.; Jefferson, B. The impact of algal properties and preoxidation on solid-liquid separation of algae. Water Res. 2008, 42, 1827–1845. Energies 2012, 5 1981
79. Poelman, E.; De Pauw, N.; Jeurissen, B. Potential of electrolytic flocculation for recovery of micro-algae. Resour. Conserv. Recycl. 1997, 19, 1–10.
80. Jiménez-González, C.; Kim, S.; Overcash, M. Methodology for developing gate-to-gate Life cycle inventory information. Int. J. Life Cycle Assess. 2000, 5, 153–159.

The article as it appears here has been abridged. To view additional tables and an appendix, please view the original article. This chapter was originally published under the Creative Commons Attribution License. From: Beal, C. M., Hebner, R. E., Webber, M. E., Ruoff, R. S., Seibert, A. F., and King W. C. Comprehensive Evaluation of Algal Biofuel Production: Experimental and Target Results. Energies 2012, 5(6), 1943-1981.

CHAPTER 6

ELECTROMAGNETIC BIOSTIMULATION OF LIVING CULTURES FOR BIOTECHNOLOGY, BIOFUEL AND BIOENERGY APPLICATIONS

RYAN W. HUNT, ANDREY ZAVALIN, ASHISH BHATNAGAR, SENTHIL CHINNASAMY, and KESHAV C. DAS

6.1 INTRODUCTION

Electromagnetic fields are capable of eliciting in vivo and in vitro effects in many biological systems [1]. Increasing attention is being directed towards bioelectromagnetic stimulation of living cultures for biotechnology and bioenergy applications using the low frequency electromagnetic fields (EMF). A number of bioprocesses could be successfully integrated with electromagnetic or electrochemical stimulation if the cultivation conditions are properly engineered using specialized reactors viz. electrolytic bioreactors, electro-bioreactors and bioelectro-reactors [2]. Most recently, a strong initiative in bioenergy research has been taken up to investigate methods for enhancing productivity and metabolic processes for biomass production and biorefining of biomass for production of biofuels, energy and other added value products. Currently, microalgae are considered to be the most promising candidates for biomass production because of their ability to grow fast, produce large quantities of lipids, carbohydrates and proteins, thrive in poor quality waters, sequester and recycle carbon dioxide from industrial flue gases and remove pollutants from industrial, agricultural and municipal wastewaters. Microalgae are novel feedstocks for renewable biomass production that is capable of meeting the global demand for transportation fuels because the oil productivity of many strains of microalgae greatly exceeds that of the most productive oil crops such as oil palms and soybean [3]. Although biomass production may be most effectively performed by large-scale algae cultivation, yeast and bacteria are the most common groups of organisms used in bioprocessing and conversion technologies like fermentation, composting, anaerobic digestion and bioremediation. Considering the current importance of waste management and recycling in conserving natural resources, bioenergetic stimulation technologies may be used as a potential tool for bioremediation by stimulating the uptake rates of various polluting components found in the waste streams by microbes.

Extensive studies have been conducted over both eukaryotic (algae, yeasts and molds) and prokaryotic microorganisms using various electromagnetic regimes. The biological effects have been found to depend on field strength, frequency, pulse shape, type of modulation, magnetic intensity, and length of exposure [4]. Some results have been difficult to replicate due to various hidden parameters typically not monitored, such as local intensity and orientation of Earth's geomagnetic field, cosmic radiations, solar winds and sunspot events.

Electromagnetism may affect organisms in both negative and positive manner which includes acceleration of growth and metabolism. This paper however focuses on the facilitative effects of electromagnetism on various microorganisms. The research attempts in this area can be divided into several groups based on implemented EMF parameters. Simplest initial classification can be based on time behavior of EMF and relative representation of the electric and magnetic components of the field. As it follows from the recent research results, a spatial configuration and topology of the EMF may also have significant impact on processes in living cultures. This paper also summarizes our own data regarding the effects of multipolar electromagnetic influences on biological systems and the future potential biostimulation techniques for improving microalgae biomass and lipid productivity for producing biofuels.

6.2 ELECTROMAGNETIC EXPERIMENTS

Three primary classes of experiments of electromagnetic influence (Figure 1) can be distinguished viz.:

1. Predominantly magnetic fields: Near-field regime (Permanent, slowly changing, and pulsed fields from magnetic coils)
2. Predominantly electric fields: Near-field regime (Permanent or slowly changing)
3. Fields with both electric and magnetic components, with ratios between 0.1 and 10: Far-field regime (typical EMF oscillation frequency is 100 kHz or more)
4. Fields from (I, II, or III) with unique spatial and/or temporal topology

Group I is represented relatively larger, mostly because of simplicity of experimental setup and extended penetration depth of magnetic field inside the water containing systems (Figure 4). The generated fields are either static magnetic fields or oscillating magnetic fields created by either permanent magnets or electromagnets, like Helmholtz and Solenoid coils. The biological experiments generally use a standard bipolar configuration with a N/S magnetic or +/- electric field for stimulation.

Group II is most often used in electroporation where strong pulsed electric fields (or PEF's) are used for reversible membrane permeabilization to induce the uptake or release of some cell ingredients or foreign molecules. Group III is electromagnetic energy that propagates as a wave at higher frequencies and is considered as the far-field regime via an antenna, magnetron, or klystron. This classification encompasses non-ionizing radiowaves and microwaves, as well as optical and ionizing radiations such as IR, visible, UV, X-ray and Gamma radiation.

The following section on the effects of electromagnetic fields has been organized by the type of the EM treatment and further categorized on the basis of growth and physiological processes that have been studied within each treatment group.

6.3 BIOSTIMULATION BY ELECTROMAGNETIC FIELDS

6.3.1 GROUP I: TREATMENTS INVOLVING MAGNETIC FIELD PREDOMINANCE

Experiments involving a predominant magnetic field have been conducted on a range of microorganisms that represent both prokaryotes (eubacteria, archaea) and eukaryotes (algae, fungi, protozoa).

6.3.1.1 GROWTH

Growth is a physiological response of an organism and a positive effect on growth indicates that some of the biosynthetic pathways are being stimulated. Erygin et al. [15] grew a gram-positive bacterium *Bacillus mucilaginous* in a magnetic field of ~0.26 T under different media compositions and compared it with unexposed control cultures. The magnetically treated liquid medium consisting of ferromagnetic salts showed rapid growth of the bacterium over control in 3 h. Similarly magnetically treated dry whey medium yielded three times higher cell count than the untreated medium. However, there was an overall increased response from the exposed dry whey illustrating how the culture medium composition may influence the effect of magnetic field.

Moore [17] studied five strains of bacteria and a yeast under a magnetic flux of 5–90 mT and reported maximum stimulation of growth at 15 mT (at 0.3 Hz) and maximum inhibition at 30 mT. Experiments with varying time especially using oscillating magnetic fields have uncovered new effects related to resonant phenomena in the living systems. The biostimulation of a denitrifying gramnegative bacterium *Pseudomonas stutzeri* by a magnetic field of 0.6–1.3 mT pulses via inductively coupled Helmholtz coils for 8–10 h resulted in a proliferation of biomass that was 10–30% more than the control [18].

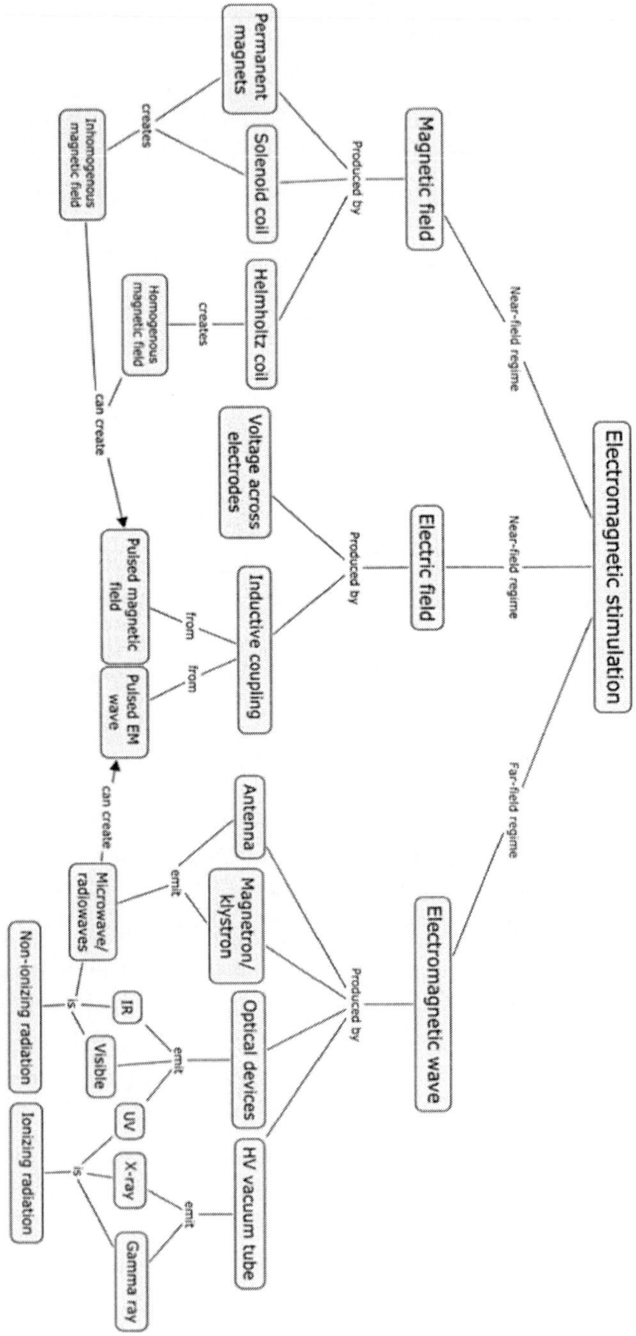

FIGURE 1: Overview of various electromagnetic stimulation modalities from fields and waves.

Other than the medium conditions, magnetic flux and type of magnetic field, the exposure time is another major factor that governs the intensity of response. Justo et al. [8] observed that the growth of *Escherichia coli* could be stimulated or inhibited by exposure to an oscillating 100 mT extremely low frequency (ELF) magnetic field for 6.5 h. Exposed cells had 100 times greater viability than unexposed cells, however the viability varied with duration of exposure. It was suggested that the effect was a result of magnetic field driven alteration of membrane permeability and availability of ions in the culture medium.

Research groups in Japan and China have focused on investigating ways to improve the cultivation of the cyanobacterium *Spirulina platensis* for production of nutraceuticals using permanent magnetic fields. Hirano et al. [26] reported a significantly higher specific growth rate of 0.22 d-1 in S. platensis exposed to 10 mT magnetic field when compared to 0.14 d-1 for untreated culture. The growth of *S. platensis* was maximum when it was cultured phototrophically at lower light intensities; but did not show improvement under heterotrophic conditions.

Magnetic field induced growth stimulation in S. platensis has also been reported by Li et al. [27]. They observed a 47% increase in dry biomass on the sixth day of cultivation, and a 22% increase over control by day eight under the exposure of a 250 mT homogeneous magnetic field from a Helmholtz coil.

Chlorella vulgaris is another algal strain of interest for its nutraceutical value and is a promising producer of starch-glucose. This microalga can yield starch to the tune of 60 t ha^{-1} yr^{-1} which is 7.7 times more than that of traditional corn [50]. Takahashi et al. [31] used magnetic flux densities of 6–58 mT for cultivating *Chlorella sp.* and obtained facilitative growth at 40 mT. The specific growth rate of *Chlorella vulgaris* almost doubled from 0.07 to 0.12 d-1 under magnetic field generated using a dual-yoke electromagnet, which concentrates a magnetic field into a small cross-sectional area [30].

The static magnetic field strengths ranging from 0 to 230 mT on Dunaliella salina were used by Yamaoka et al. [32]. They observed an improvement in growth rate that peaked at 10 mT with the addition of 1 mg L^{-1} of Fe-EDTA. A ~0.26 T magnetic field exposure using different growth media for the yeast *Saccharomyces fragilis* showed that rapid growth (27–

36% over the control in 3 h) occurred on magnetic treatment when a dry whey nutrient medium was used, but it turned inhibitory on using a liquid nutrient medium [15]. On the other hand Fiedler et al. [36] used an oscillating magnetic field generated by a Helmholtz coil via inductive coupling to produce 0.28–12 mT magnetic field at 50 Hz for 9 h to treat S. cerevisiae. They observed a maximum growth of 8 g L-1 of biomass under 0.5 mT magnetic field exposure and 6.8 g L-1 of biomass for the cells untreated.

6.3.1.2 PHOTOSYNTHESIS AND CELL CONSTITUENTS

Hirano et al. [26] observed acceleration of the rates of O_2 evolution as well as synthesis of sugar during photosynthesis in *Spirulina platensis* when exposed to 10 mT geomagnetic field. They opined that the treatment using magnetic field increased the phycocyanin content in *S. platensis*, which plays an important role in the activation of photosystem II to help the activation of electron transfer reactions during photosynthesis. Their results also suggested that the magnetic fields accelerate the light excitation of chlorophyll radical pair.

Li et al. [27] subjected the same cyanobacterium *S. platensis*, to a range of static magnetic field intensities among which some stimulated its growth, uptake of carbon and light energy utilization. They observed that the levels of micro and trace elements (Ni, Sr, Cu, Mg, Fe, Mn, Ca, Co and V) and essential amino acids such as histidine improved at 250 mT magnetic field treatments. Also, chlorophyll a content of the magnetically treated sample was higher than the control, suggesting better light harvesting for photosynthesis. However there was slight decrease in lipid synthesis.

In *Dunaliella salina*, β-carotene content could be raised when treated with 10–23 mT of static magnetic field and the maximum was obtained at 10 mT with addition of 1 mg L-1 of Fe-EDTA. It also showed higher accumulation of the heavy metals viz. Co, Cd, Cu and Ni in the magnetically treated cultures, indicating its potential for bioremediation of heavy metals [32].

Singh et al. [29] investigated the use of permanent magnets and found that the physiological response of a cyanobacterium *Anabaena doliolum*, was dependent on exposure time and magnetic pole orientation. They reported that N, S and N+S poles from 0.3 T permanent magnets produced different effects depending on the exposure time from 1 to 6 h. The effect was significant on a two hour exposure with combined N+S poles, where one culture was exposed to only N pole, which was then mixed with another culture exposed to S pole only. Treated cultures recorded 150, 110, 38, 34 and 20% increase in phycocyanin, chlorophyll a, carbohydrates, carotenoid and protein content, respectively and 55% increase in optical density over the control.

5.3.1.3 OTHER PHYSIOLOGICAL PROCESSES

5.3.1.3.1 ETHANOL FERMENTATION

The biotechnology of fermentation using yeasts, like *Saccharomyces cerevisiae*, has a long history in many sectors of industry from alcoholic beverages to ethanol production. A vital focus of ongoing research is the study of the key enzymes responsible for the production of the metabolites of interest, namely ethanol. Increasing the activity of key enzymes, like alcohol-dehydrogenase, is a primary goal of metabolic and enzyme engineering. The glucose dehydrogenase and alcohol dehydrogenase were studied in S. cerevisiae under the influence of a non-uniform pulsed magnetic field of 30 mT for 60 minutes [34]. They found that in the presence of NAD the glucose dehydrogenase activity increased 18%, while no effect was observed in the absence of NAD or NADP. The activity of alcohol dehydrogenase in the absence of co-enzymes rose to 10.7% in the anaerobically cultivated cells and 19.9% in those cultivated aerobically. The activity of this enzyme increased by 20.5% when NAD was added to this enzyme in the aerobic culture, while an 8.5% decrease was observed in the anaerobic culture. Thus, the non-homogenous pulsed magnetic field of 30 mT stimulated the activity of the dehydrogenases, but behaved differently in the absence or presence of NAD and NADP.

The effects on ethanol fermentation by *S. cerevisiae* under the influence of two styles of oscillating magnetic fields were studied by Perez et al. [35]. The primary magnetic field generator was composed of several permanent magnets stacked in series, while the recirculating culture broth was directed through the intervening space of the magnetic fields where spatial orientation determined the desired intensity of 5–20 mT for each exposure. The recirculation velocity passing through the array of static magnets modulated the frequency. The secondary generator was a double layer solenoid coil that produced 8 mT. Two magnetic field generators were coupled to the bioreactor, which were operated conveniently in simple or combined ways. The overall volumetric ethanol productivity enhanced by 17% over control at an optimum magnetic field treatment of 0.9–1.2 m s^{-1} velocity and 20 mT plus 8 mT solenoid. These results made it possible to verify the effectiveness of the dynamic magnetic treatment since the fermentations with magnetic treatment reached their final stage, 2 h earlier than the control. Perez et al. [35] postulated that membrane permeability and the redox system that are affected by the electromagnetic field might have resulted in alterations of ion transport of the substrates. As a consequence, the cellular metabolism was stimulated for higher ethanol production.

6.3.1.3.2 ANTI-OXIDANT DEFENSE SYSTEM

Wang et al. [30] used a magnetic field concentrated to a small area and observed that it helped to regulate the anti-oxidant defense system of Chlorella vulgaris at a threshold magnetic flux intensity of 10–35 mT. The authors proposed that this is probably due to the free radicals altered by the magnetic field, which accelerated the relative biological reactions. The analysis of hydroxyl radical (^{-}OH) showed that it increased simultaneously with increasing magnetic flux density suggesting an oxidative stress induced by the exposure compared to the control.

6.3.1.3.3 BIODEGRADATION

A study using airlift reactors showed that the influence of magnetic fields enhanced the degradation of phenolic waste liquors by submersed microorganisms at a magnetic field intensity of 22 mT [23].

6.3.1.4.GENETIC MACHINERY AND MOLECULAR MECHANISMS

E. coli cells when placed under extremely low frequency (ELF) magnetic field sine wave of 30 μT at 9 Hz, exhibited a change in the conformational state of the genome, which was maximum at 4 × 108 cells mL^{-1} while there was no such response at lower cell densities of 3 × 105 cells mL^{-1}. Other than cell density, time of exposure also affected genomic conformation. The change in the conformational state of the genome is considered to be dependent on DNA parameters, i.e. molecular weight and the number of proteins bound to the DNA [9]. Thus the ELF field which is close to the ion cyclotron resonance parameters for a medium weight ion might be influencing these factors that ultimately elicit response on the conformation. It was also proposed that the possibility of a resonance fluorescence effect where recombination of fluorescing radicals may act as signals for intercellular communication and participate in the synchronization of gene expression. Weak, static magnetic fields (0–110 μT) are shown affecting DNA-protein conformations in *E. coli.* The analysis by Binhi et al. [51] represented a dose-response curve for the static magnetic field. The curve however is peculiar in having three prominent maxima unlike other dose-response curves in nature that usually follow rising or decaying exponential functions. They explained this peculiarity in the context of the ion interference mechanism. No alteration in the profile of stress proteins of *E. coli* was observed by Nakasono et al. [52] on exposure to AC fields (7.8–14 mT, 5–100 Hz). In *Saccharomyces cerevisiae* no changes were observed under AC magnetic fields (10–300 mT, 50 Hz) in differential gene expression and protein profile that were determined using microarray and 2-D protein profile analysis, respectively [53]. But, Gao et al. [54] reported that strong magnetic fields (14.1 T) could lead to transcriptional up-regulation of 21 genes and down-regulation of 44 genes in a gram-negative anaerobic bacterium Shewanella oneidensis that did not show any significant effect on growth. In the anoxygenic photosynthetic bacterium, Rhodobacter sphaeroides, AC magnetic fields of 0.13–0.3 T induced a 5-fold increase in porphyrin synthesis, and enhanced expression of the enzyme 5-aminolevulinic acid dehydratase, while very strong DC fields (0.13–0.3 T) also induced synthesis of this enzyme predominantly at the magnetic North Pole. The effects are attributed to elevated gene expression that ultimately resulted in increased porphyrin production [25].

Mitotic delay of 0.5 to 2 h was observed in a slime mold *Physarum polycephalum* in presence of ELF electromagnetic fields (45, 60 and 75 Hz) by Goodman et al. [44]. Removal of the mold from magnetic field recovered normal mitosis in 40 days.

6.3.2 GROUP II: TREATMENTS INVOLVING ELECTRIC FIELD PREDOMINANCE

Stimulation in the growth of immobilized *E. coli* cells by 140% over control, was reported by Chang et al. [10], which was attributed to the enhanced removal of inhibitory products from the gel through electro-osmosis and electrophoresis as well as an augmented glucose supply. Kerns et al. [19] reported growth stimulation in *Trichoderma reesei* by using pulsed EMF's for electroporation via inductively coupled electric currents from a Helmholtz coil. The use of electric fields has also been investigated with yeasts in either a static mode or an oscillating/pulsed mode. The survival rate of *Saccharomyces cerevisiae* was investigated under bipolar electric field pulses from 0–1.5 kV/cm by measuring plating efficiency. The maximum growth after plating appeared at 0.85 kV/cm which demonstrated a 100% increase over the control [55]. An electrostimulation in *S. cerevisiae* from electric field application at 10 mA DC and 100 mA AC resulted in an increase in growth rate by 60% in AC mode and 50% in DC along with an increase in the production of the acetic acid, lactic acid and acetaldehyde. The results suggest that the acceleration of growth rate from a DC exposure stimulated cell budding during the early stages of cultivation, which could be due to a 60% decrease in inhibitory concentration of dissolved

CO_2 and other chemical modifications of the culture medium [38]. Zrimec et al. [12] have shown that external AC electric fields of low intensity stimulated membrane bound ATP synthesis in starving *E. coli* cells with electric field amplitudes of 2.5–50 V/cm and a frequency optimum at 100 Hz. The model of electro conformational coupling was used to analyze the frequency and amplitude responses of ATP synthesis. Two relaxation frequencies of the system were obtained at 44 and 220 Hz, and an estimate of roughly 12 elementary charges was obtained as the effective charge displacement for the catalytic cycle of ATP synthesis.

An actinomycetous eubacterium *Streptomyces noursei* used for antibiotic production was electrostimulated via PEMF's using a pair of Helmholtz coils via inductive-coupling producing 5 ms bursts of 220 μs duration in intervals of 60 ms by Grosse [20]. The process resulted in a mean inductive electric field strength of approximately 1.5 mV cm^{-1}. An increase was observed in the formation of the product but only during the first 50 hours of the starting phase although the exposed culture exhibited an overall increase in O_2 consumption and glucose utilization.

Electric field stimulation may also be used to improve the substrate utilization efficiency in microbial processes. Cells when subjected to electric field pulses of 0.25 kV for 10 ms in the presence of the enzyme cellobiose showed enhanced utilization of cellobiose and conversion of substrate into ethanol by a thermotolerant yeast, Kluyveromyces marxianus. As a result, ethanol yield increased by nearly 40% over the control [43].

Kerns et al. [19] showed that pulsed EMF's at 1.5 mVcm-1 bursts for 115 hours used for electroporation lead to ~60% increase in cellulase activity and ~80% increase in cellulase secretion in *Trichoderma reesei*. They concluded that the effect occurred inside the cells on either the formation of the cellulase enzyme complex at the genetic level or the secretion into the medium via altered membrane permeability. A 62% increase in biosorption of uranium was observed using pulsed electric fields of 1.25 to 3.25 kV cm^{-1}, suggesting that the application of short and intense pulses might enhance the biosorption of toxic metals and radionuclides from wastewater streams [24].

6.3.3 GROUP III: TREATMENTS INVOLVING BOTH ELECTRIC AND MAGNETIC FIELDS IN FAR-FIELD REGIME

Some of the original pioneering work with the bioeffects from weak electromagnetic radiation in the form of microwaves was performed in Russia and extended into Europe in the 1970's. The work by Grundler et al. [39], investigated the use of very weak microwave irradiation of a few mW/cm^2 at a frequency around 42 GHz ±10 MHz on Saccharomyces cerevisiae. The experiments demonstrated multiple resonance dependent effect of coherent millimeter electromagnetic waves in the frequency region of 41.83 to 41.96 GHz that increased growth rates up to 15% or decreased the growth rate by 29% depending on frequency.

Banik et al. [5] investigated the use of electromagnetic irradiation at the microwave frequency from 13.5 to 36.5 GHz on *Methanosarcina barkeri* DSM-804, a methanogenic archaebacterium used in anaerobic digestion for biogas production. The bacteria were exposed for 2 h duration for three days before inoculation into the anaerobic digesters. Significant increases in methane (CH_4) concentration were observed that peaked at 76.3% CH_4 at 31.5 GHz, compared to 52.3% CH_4 in control. Furthermore, an increase in specific growth rate was observed for every frequency with a significant reduction in the lag phase. The irradiated cultures had higher cell numbers and the cell diameter was enlarged by 20%. It was concluded that the growth rate and biomethanation potential of *M. barkeri* DSM-804 could favorably induce catalytic abilities via a thermal microwave irradiation at 31.5 GHz.

Tambiev and co-workers (cited in [28]) observed that exposure of high frequency microwaves for 30 min at 2.2 mW cm^{-2} and 7.1 mm wavelength enhanced the growth of the cyanobacterium *Spirulina platensis* by 50%. Belyaev et al. [56] suggested that there was frequency-specific resonant interaction between low-intensity microwave and chromosomal DNA in *E. coli.*

6.3.4 GROUP IV: TREATMENTS WITH SPATIAL/TEMPORAL TOPOLOGY

6.3.4.1 SPATIAL SUPERPOSITION

The investigation of using multiple independent field sources has led to studies where the treatment area exhibits spatial topology from superposition. A magnetic therapeutic device that uses four nonuniform static magnets in four-pole symmetry demonstrates an increased rate of Myosin phosphorylation over control. The notion that the magnetic field amplitude is the only parameter involved to determine the outcome with magnetobiology experiments has been shown to be false and it is suggested the topological parameters in a spatial domain, such as field gradient and symmetry might also be of relevance [57].

Mazur investigated the use of multiple magnetic fields in superposition on biological samples. He exposed *S. cerevisiae* to a six-pole electromagnet with coils of alternating polarity at a magnetic field of 0.39–0.52 T, while saturating it with pure molecular oxygen. The magnetic field has an influence on the biosynthesis of yeast and changes their enzymatic activity when grown under aerobic conditions as opposed to anaerobically cultivated yeast. He found that in the presence of a magnetic field, the oxygen saturation increased from 5.37 to 39.9 mg L^{-1} and simultaneously stabilized the pH. The initiation of fermentation occurred immediately after mixing of the dough. It was found that there was an improvement in the physical qualitative property of rising strength, which was decreased from 76 minutes to 53 minutes in the presence of oxygen saturation and a magnetic field. It was also found that the increase in CO_2 production was 3.7 times greater in the magnetic treated culture than the control, which indicates a significant increase in maltase activity. The amount of dissolved oxygen in water increased and was sharply activated in the presence of a magnetic field [58].

6.3.4.2 SPATIAL AND TEMPORAL SUPERPOSITION

Aspects of EMF topology in the time domain have been studied by researchers looking at the influence on biological systems from combined AC and DC EMFs in superposition [58–61]. It has been shown that the cellular response to the orientation of the fields is distinct depending whether the AC and DC fields are perpendicular and parallel to each other. It was found that the perpendicular orientation is dominant in an intensity-dependent non-linear manner [61]. There is a fundamental difference in the spatial pattern of cellular response between DC and pulsed stimulation [62]. Several studies report that the relative orientation of AC and DC magnetic fields appears to be critical for a number of calcium-dependent cell processes. The data suggests that DC magnetic fields influence biological membranes in a somewhat different manner than low frequency AC magnetic fields [1].

6.3.4.3 MULTIPOLAR ELECTROMAGNETIC SYSTEMS

The advent of quantum theories on the molecular scale has inspired the development of electromagnetic exposure systems that mimic the complex interactions and symmetry found in nature from endogenous electromagnetic signals and their destructive interference between interdependent cells. The idea of using multiple interdependent electromagnetic emitters has led into a novel investigation of complex configurations using specific geometric orientations of multiple electrodes generating electromagnetic fields with precise phase orientation and relationships, which may lead to even more significant coupling with biological systems.

The interdependent Multipolar (MP) electromagnetic systems were devised and developed by Lensky [63], and Zavalin and his co-workers [13,14]. The MP system may contain a variety of number of poles, i.e., 2, 3, 5, 6, 9, 12, in the symmetrical electrode configuration (Cn, where n = 2, 3, 5, 6, 9, 12 correspondingly, in notation of the crystallographic groups of symmetry) and complex driving system of interdependent multidimensional transformers that is of most importance. For research with biostimulation of microorganisms, preliminary studies by Zavalin have found that six-pole systems are most effective for microorganisms compared to other configurations. The MP system used in their research consisted of six electrodes in a symmetric hexagonal geometric arrangement (group of symmetry C6), driven by a hexapole interdependent transformer system, powered by an amplified function generator. The frequencies of the EMF oscillations are lower than 100 kHz, providing the near-field regime of the MP EMF during the treatment. The MP EMF generated is fine tuned such that the superpositional field, composed of oscillating electric fields from each electrode in the near-field regime undergoes complete destructive interference with a resultant zero-vector electric and magnetic field within a certain area, located near the center of symmetry and called the "compensation zone". The compensation zone can produce a "breathing" mode where all coils are energized simultaneously to achieve the multipolar compensation zone. A scheme for the 6-polar EMF treatment for the test tube culture studies is shown in Figure 2. The multiple pole EMF configurations have a substantial effect on growth of microorganisms. Maximum achieved growth or gas production increases up to approximately 200% (see Figure 3) were observed in various bacteria, yeast, and protozoa under a 5 or 6-pole configuration at 1 kHz [13], 60 Hz, 0.35–2.1 kHz [14]. The AC voltages at the electrodes were applied 180 degrees out of phase for each opposing set of electrodes, resulting in rather pulsating than a rotating EMF pattern. Figure 3 shows maximum increase in growth of E. coli cultures in test tubes under treatment at different frequencies of 6-polar AC EMF. In the plot a maximal achieved ratio of concentration of stimulated E. coli culture to concentration of control *E. coli* culture at the same conditions is shown in the right vertical axis. A corresponding time, required to achieve such a maximal relative stimulated increase is shown in the left vertical axis. It should be noted that a depression in growth was observed in 2 and 4-pole system at the similar parameters of the EMF at each electrode. The stimulatory effect was greatest in the lag and log phases of the growth curve.

These studies show great promise considering the uniform frequency being emitted was chosen arbitrarily and are open for future research on

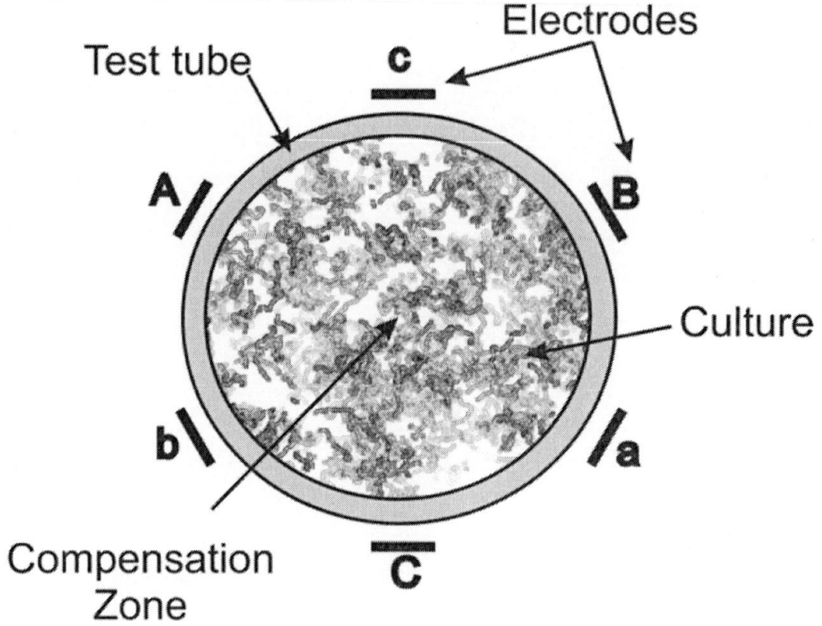

FIGURE 2: Cross-section of a test tube and a 6-polar electrode configuration for biostimulation of *E.coli.*

the optimization of output signal for growth stimulation. The results of studies conducted by Lensky and Zavalin indicate that higher topological EMF, having specific group of rotational symmetry is biologically active. This phenomenon has been previously observed using other types of self-cancelling coil windings [64–66] although the groups of symmetry have not been disclosed. Preliminary evidence indicates that these nonclassical designs may be more effective at delivering vibrational information by coupling with interdependent harmonic oscillating cells because these methods produce relatively large biological effects experimentally [13,14,66]. Thus, the multipolar configuration is a strong prospect for exhibiting unique and distinct biological effects.

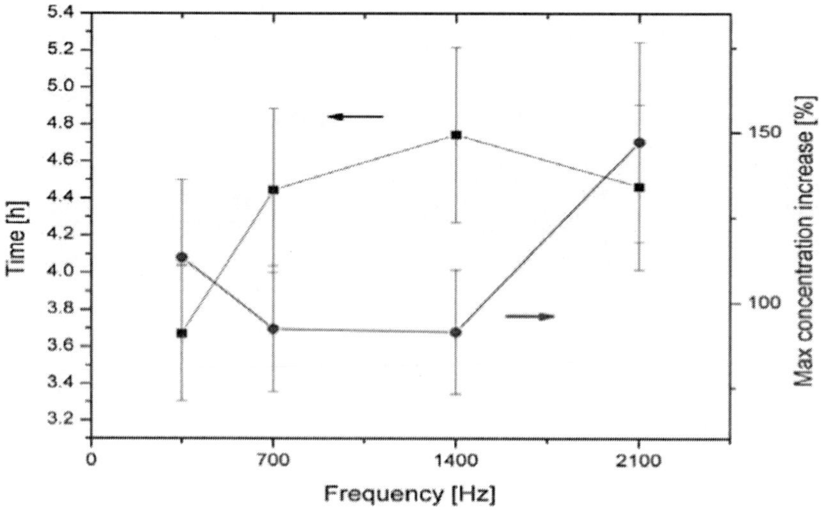

FIGURE 3: Maximum growth increase, achieved in *E. coli* cultures in test tubes versus frequency of the 6-polar AC EMF treatment (right vertical axis). The left vertical axis shows time to achieve the maximum, while the right axis shows concentration increase with respect to the control.

6.4 MECHANISM OF ELECTROMAGNETIC EFFECTS

Above observations show growth stimulation by magnetic treatment in a diverse array of organisms (from prokaryotic to eukaryotic) and a variety of stimulative responses by each organism under varied conditions of treatment and growth. While former indicates at some general mode of mechanism(s), the later gives an impression in contrast to it. Lack of adequate information eludes a consensus on the mechanism(s). Several factors appear to be affecting the stimulation process. The flux generating system, intensity of the flux, type of the flux (oscillatory or static), orientation of magnetic poles, duration of exposure, cell density and cell environment (for example type of medium and its ingredients) and other physicochemical conditions affect the process of biostimulation through electromagnetic forces. It has also been marked that the results sometimes do not show repeatability at other locations suggesting that local geomagnetic realities might also affect the process of stimulation.

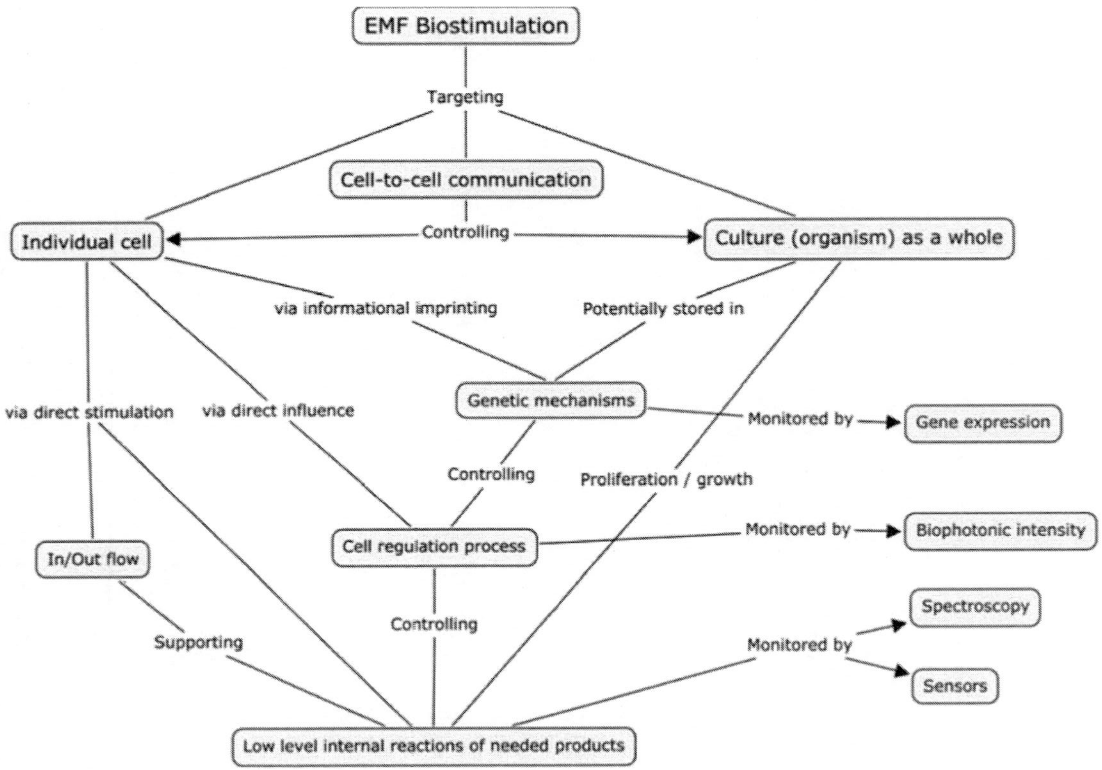

FIGURE 4: Concept map of an EMF biostimulation at different levels of living systems.

There are physiological effects other than growth that have been observed. These are processes such as carbon uptake, sugar synthesis and oxygen evolution in photosynthesis, synthesis of pigments (chlorophyll, carotenoids and phycocyanins), carbohydrates and proteins, accumulation of micro and trace metals and essential amino acids, fermentative activity and even genetic processes like transposition. They can be stimulated under specific conditions adopted in the experiments. Only one study [29] specifically referred to lipids reported a decline in lipid content under the particular set of treatment. It may be worth noting that an exposure to surprisingly low levels of exogenous electromagnetic fields can have a profound effect on a large variety of biological systems [1]. A number of mechanisms have been proposed for observable magnetobiological and bioelectromagnetic effects at different levels [51]. A concept map, demonstrating different levels of the EMF influence is shown in Figure 4

6.4.1 IONIZATION AND FREE RADICAL RELEASE

Magnetic fields cause oxidative stress in organisms by altering energy levels and spin orientation of electrons and concentration and lifetime of free radicals, which change the relative probabilities of recombination of other interactions with possible biological consequences [67]. Oxidative stress due to the radical pair mechanism becomes applicable around 1 mT which can be common in industrial or laboratory settings, while the geomagnetic field intensity stays below 0.07 mT. Studies with *Chlorella vulgaris* demonstrated that hydroxyl ions increase in magnetically treated medium suggesting alteration of free radical levels in the medium that might hyperactivate antioxidant defense system of the organism. This situation also affects the membrane permeability and ion transport process and might be responsible for the acceleration of chlorophyll excitation by the light [30].

6.4.2 ELECTROCHEMICAL MODELS

These models explain biological processes considering electromagnetic fields as modulators of molecular information transfer. It is considered that the EMF either itself acts as signal(s) and/or intercepts or modifies the processes of molecular interaction.

6.4.2.1 ION CYCLOTRON RESONANCE CONCEPT

Many authors have developed the idea of ion cyclotron resonance (ICR) of specific ions like Ca^{2+} and Na^{+} [68] which predicts ELF magnetic effects at the cyclotron frequencies and there harmonics. Later, it was modified to the ion parametric resonance (IPR) model, which includes the cyclotron sub harmonics. The IPR is composed of a number of theoretical models based on classical and quantum electrodynamics where biomagnetic effects are considered as magnetically modulated ion binding in ion-ligand interactions [69]. Free ions move with the cyclotron frequency in a static magnetic field and can be influenced by ELF magnetic fields or appropriate frequencies [70]. The main focus of these studies was the essential role of Ca^{2+} ions in magnetobiology experiments. It is proposed that ion behavior in channels like the acetylcholine receptor have constrictions in them, which cause thermal collisions. Under certain magnetic field parameters the wall collisions could be avoided at certain amplitudes and frequencies determined for the Lorentz force equation [71]. Under these conditions, the ions are predicted to "fly" through the channel unimpeded increasing the membrane permeability. ICR allows circulation of ions through selective enhancement, which affects the rate of biochemical reactions [72].

The fact that magnetic fields can modulate enzyme activities in vitro is a crucial observation, because it indicates that enzymes may function as magnetoreceptors [69]. EMF modulations could also initiate changes in the distribution of protein and lipid domains in the membrane bilayer, as well as conformational changes in lipid-protein associations [1]. The interface between cell membrane and extra and intercellular fluids can be electrified on the order of 106 to 1010 V cm^{-1} [73]. The impact of an electric field on a biological cell membrane and its change with time may constitute a relevant mechanism of information transmission influencing the membrane properties. The electric field, mainly generated by ions flowing to the membrane from the external environment, can change the molecular distribution of electronic charge inside each lipid molecule, producing perturbations of collective excitations in the mechanical and electrical properties of the lipid chain which can be treated as a mechanism for intermembrane communication, analogous to a damped harmonic oscillations [74].

Pilla et al. [73] presented a working model of electrochemical information transfer by which the injection of low-level current can provide functional selectivity in the kinetic modulation of cell regulation. His theory was based on ion/ligand binding being a possible transduction mechanism for the detection of exogenous EMF's at the cell membrane [75]. In order to derive the specifications for electromagnetic field signals having optimal biological effects, it is first necessary to develop a model for the underlying biological processes which are assumed to be complex physical systems that may be modeled mathematically as non-linear, time-varying, finite-dimensional dynamic systems. They developed a method for the systematic analysis of electrical impedance for each relevant electrochemical pathway of a cellular system [62]. The electrochemical transfer hypothesis postulated that the cell membrane would be the site of interaction of low level electromagnetic fields by altering the rate of binding of calcium ions to enzymes or receptor sites [1]. The Ca^{2+} pathway can be influenced by EMFs on the complex chain of transduction, amplification, and expression.

Experimental results have shown that specific ion/ligand binding pathways such as Ca^{2+} binding to calmodulin (CaM) and the ensuing steps of calcium-dependent signaling to intracellular enzymes may act as primary transduction mechanisms for EMF detection leading to an increase in the instantaneous reaction velocity and enzyme kinetics [75,76]. Calmodulin also plays a role in many other important biochemical processes such as cell proliferation, Ca^{2+} membrane transport and plant cell function [77]. An alteration of cell signaling events can lead to changes in cell proliferation and differentiation, which can be initiated, promoted or co-promoted [70]. The capability of the weak EMF to have a bioeffect appears to reside in the informational content of the waveform [1]. The waveform duration and the voltage dependence are the most important parameters to increase the activity of the specific adsorption of an enzyme [62]. The proposed interfacial membrane model reveals that it is entirely reasonable to expect specific electrochemical effects as a result of electrical stimulation with signals of relatively low frequency and amplitude [73].

The incorporation of quantum states into ion interference has also been involved in the explanation of the physical nature of magnetoreception [78]. Variations in magnetic field magnitude affect the phase of ion wave functions and the interference of these phase changes affect the physical observables in quantum mechanics. This theory predicts magnetobiological effects for magnitude/direction modulated magnetic fields, pulsed magnetic fields and weak AC electric fields among others [51]. In these cases, ions of calcium, magnesium, zinc, hydrogen, and potassium appear to be relevant. However, the most prominent example of a proven bioelectromagnetic mechanism is the radical pair recombination mechanism, which has been demonstrated biochemically in vitro. Radical pairs are formed as reaction intermediates in many biochemical reactions within complex reaction chains under the influence of exogenous electromagnetic influence [70]. The recent breakthrough regarding the radical pair mechanism in the blue light receptor protein, cryptochrome, by Schulten and his colleagues, supports the concept that radical pair recombination is involved in magnetoreception in avian navigation. Molecular modeling and calculations showed that the signaling of cryptochrome, which involves a photoreduction process, can be modulated in the presence of a magnetic field on the order of 1 mT inducing an increase in the signaling activity of the protein by ~10% [79,80]. This prediction appears to be consistent with the experimental results on the effect of magnetic fields on cryptochrome-dependent responses in *Arabidopsis thaliana* seedlings attained by Ahmad and coworkers [81]. It is then suggested that the magnetic navigation capability could be mediated by the presence of cryptochrome that is localized in the retinas of migratory birds which could alter how the bird perceives colors enabling something akin to an internal magnetic compass [82]. This radical pair mechanism is probably coupled with the alternative magnetite-based mechanism of magnetoreception and navigation, which poses that the Earth's magnetic field exerts a minute mechanical force on the magnetite particles found in the upper beaks of migrating birds providing positional information due to fluctuations in the geomagnetic strength in different locations [83].

6.4.2.2 STOCHASTIC RESONANCE AMPLIFICATION

Electromagnetic bioeffects from relatively weak signals are often due to a time-varying electric field, induced by a time-varying magnetic field [1]. However, the ability of weak oscillating EMF fields to interact with living cells has been a source of controversy since thermal and other noise poses restrictions to the detection of weak signals by a cell. Activation of signal pathways by external stimuli connects the physical interactions of the applied EMF to the biological response [70]. In nonlinear systems such as biological sensory apparatus, presence of noise can actually enhance the detection of weak signals, called stochastic resonance [84]. Very small changes in the underlying non-linear kinetics caused by very weak coherent signals and noise can lead to strong, but reversible alterations in the internal nonlinear processes and associated biological function such as ELF influences on G-protein activation dynamic, magnetic field influence on radical pair recombination reactions and weak signal amplification by stochastic resonance incorporated within the Ca^{2+} signal pathway models [70]. The mechanism of stochastic resonance has shown an amplification factor that may exceed a factor of 1,000. This is because in a nonlinear system, the reaction to an external signal may be much greater when acting as a whole than the response of the system's individual elements. This resonance manifests itself by the appearance of sharp peaks in the power spectrum of the system at the driving frequency and in some of the higher harmonics. Currently, the cell membrane is considered the most likely cellular site for interactions with EMF's and the possible role of ionic channels of the membrane in the amplification process. The potential well-like structure of an ionic channel makes it the ideal system for stochastic resonance amplification [85].

6.4.2.3 LONG RANGE MOLECULAR ORGANIZATION

The application of the nanosized voltmeter, used to measure the electric fields throughout the interior of cellular structures, has indicated that the theoretical calculation of electric field penetration into a cell's cytosol

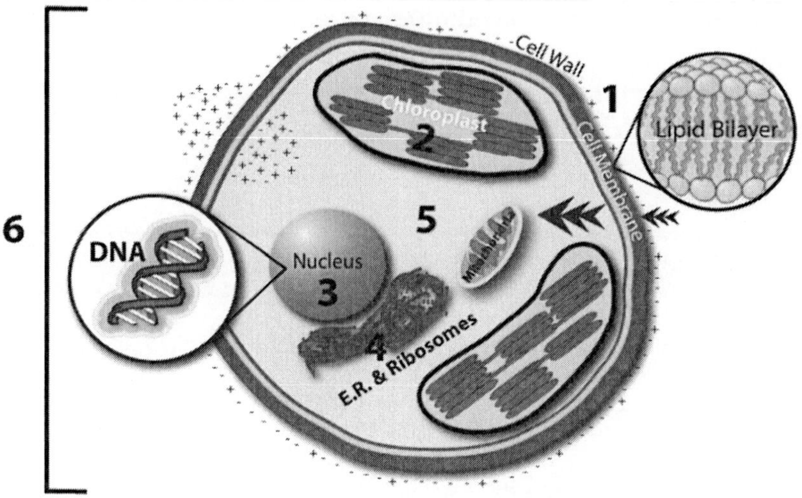

FIGURE 5: Molecular interaction sites of electromagnetic influences.

arising from the membrane and mitochondrial potential do not match the empirically measured values. It is proposed that this may be due to the traditional model using saline solution to simulate the physical properties of the cytoplasm, where alternatively the cytoplasmic structure has been described as having a complex gel-like composition [86,87]. One such possibility for a heterogeneous substance with distinct microdomains is liquid crystal. Liquid crystals are phases of matter that are exhibited by anisotropic organic materials as they undergo cascades of transitions between solid and the liquid states [88]. These mesophases possess symmetry and mechanical properties of long-range orientational order intermediate between those of liquids and of solid crystals. Liquid crystals can undergo rapid changes in orientation of phase transition upon electric or magnetic exposure, or changes in temperature, pH, pressure, hydration, and concentrations of inorganic ions. These properties are ideal for organisms, and it has been found that lipids of membranes, DNA in chromosomes, all proteins, especially cytoskeletal proteins are liquid crystalline in nature [89]. Ho's group observed that electrodynamic activities might be acting on endogenous non-equilibrium electrodynamic processes involved in phase ordering and patterning domains of liquid crystals [65]. Their findings support that organisms are polyphasic liquid crystals where different mesophases may have important implications for biological organization and function [90].

1. Cell Membrane

- Magnetic field oscillations may increase membrane permeability under ion cyclotron resonance
- Increased circulation and selective enhancement of ion flow may affect the rate of biochemical reactions
- Alter the rate of binding of calcium ions to enzymes or receptor sites
- Change distribution of protein and lipid domains, and conformational changes in lipid-protein associations
- Change internal molecular distribution of electronic charge inside lipid molecule in the membrane bilayer
- May play the primary role in the stochastic resonance amplification process

2. Chloroplast

- May modulate the quantity of pigments, such as chlorophyll, phycocyanin, and beta-carotene

3. Nucleus/DNA

- Magnetic field affects specific gene expression
- Individual DNA sequences may function as antennae
- Leads to changes in DNA conformation
- May activate different DNA sequences depending on field intensity
- Can affect enzyme activity

4. Proteins:

- Breathing motions are the source and receiver of multipole EMF
- Potential coupling mechanism for external multipolar influences

5. Protoplasm

- Static magnetic fields influence the speed of protoplasm movement, miotic activity, and quantity of organic acids in plants

6. Whole Cell

- Biophotonic emission and interaction with nearby cells
- Endogenous electric field modulation may alter natural processes

6.4.2.4 JOSEPHSON SEMICONDUCTOR MODEL

From a geometric perspective, it is possible to compare two dividing cells in living systems with a Josephson junction of superconductivity [91]. The

Josephson junction may represent a gap junction between two nearby cells coupled via electromagnetic interactions, which provides a mechanism forthe transfer of correlated charged particles, electrons, and ions. The gap junctions serve to transmit electrical signals between adjacent cells without the need for mediation by a neurotransmitter or messenger substance [92]. Positive experimental results were attained in yeast cells by examining their current-voltage characteristics and radiofrequency oscillation spectra during cell division [91].

6.4.2.5 PROTEIN SYMMETRY

The macroscopic ordering displayed in living systems is an "emergent" property arising from a collective behavior of the elementary microscopic components [93]. The low-frequency internal motions in protein molecules play a key role in biological functions where it is suggested that there is a direct relationship between low-frequency motions and enzymatic activity [94]. The symmetry of protein molecules is also a very important factor in understanding its structure and function, which depends on stability, number of subunits, and folding efficiencies that limits the functionality of the protein. The functionality requirements of symmetry and asymmetry can drive the evolution of proteins to have any of the crystallographic point groups [95]. The breathing motions demonstrated by protein molecules are oscillations of the protein's symmetry emanating from the center of symmetry of the molecule. These vibrations could potentially be a source and receiver of multipole EMF. Symmetrical and oscillatory nature of proteins, which constitute enzymes, exhibits unique features that have the potential for interaction via external multiple EMF coupling.

6.4.2.6 PHYSICAL SIGNALS IN INTERMOLECULAR COMMUNICATION

Progress to understand the intercellular interactions of microorganisms has been linked to the investigation of prokaryotic signaling molecules; however, there is increasing evidence of physically mediated communication for some events, including cell division, adaptation and stress conditions

[96]. The hypothesis that electromagnetic forces have a fund mental role in organization and transport of entities is supported by indirect and direct measurements of the electromagnetic fields around living cells.

3.4.2.7 ELECTROMAGNETIC CELL FUNCTIONS

The electromagnetic fields serve as mediators for the interconnection of the organism with the environment as well as between organisms. Electric dipole and multipole moments are common to every biological structure and macromolecule. Oscillating multipole EMF may be generated as a result of interaction of these dipoles and multipoles with electromagnetic emitters and transceivers [97]. Thus the fields produced by the organisms play an important role in the coordination and communication of physiological systems and informational interactions in addition to energetic interactions which play a significant role [98]. The endogenous physiological EM rhythms control and determine the growth and differentiation of cells and are essential for spatiotemporal organization at the subcellular, cellular and organism level [70]. With the recent development of the "nanosized voltmeter" using a voltagedependent fluorescent nanosensor (E-PEBBLE), the first complete three-dimensional profiling throughout the entire volume of living cells was accomplished. The results indicated that the endogenous electric fields generated penetrate much deeper into the cytosol and non-membrane regions than previously estimated. These measurements support the picture of an electrically complex environment inside the cell [87].

Ions are the transducers of information in the regulation of cell structure. Modification in the interfacial structure of cell membrane alters its ionic composition and constitutes electrochemical information transfer. This alters biochemical and mechanical transport properties of the membrane that is interpreted by the cell as requiring a change in its function which could trigger specific enzyme activity [62,73]. Thousands of chemical reactions are carried out simultaneously and successively in different cellular compartments and are closely coordinated and linked together. The importance of vibrational coherence in the form of electrical and mechanical oscillations has been proven through the experiments [99]. It has been shown for instance that endogenous electric fields exhibiting coherent behavior can have a dominant effect on directed transport of molecules and electrons such that the probability to reach the target is enhanced in comparison with random thermal motion alone [97].

6.4.2.8 QUANTUM PHYSICS AND COHERENCE IN BIOLOGY

Coherence is a fundamental property of a quantum field in which coherent quanta give rise to an order extending over a long distance within which there is a finite probability of finding the system in this order-related state [100]. It is demonstrated in an organism by the movements that are fully coordinated at macroscopic to the molecular levels [90]. The metabolic functioning of living systems has revealed nanomechanical and electrical oscillations in the frequency range of 0.4 to 1.6 kHz, that were found in the yeast, *S. cerevisiae* using atomic force microscopy. If metabolic function was chemically inhibited, the oscillations ceased. It was concluded that the oscillations were consistent with cellular metabolism of molecular motors and may be part of a communication pathway or pumping mechanism by which the yeast cell supplements the passive diffusion of nutrients and/or drives transport of chemicals across the cell wall [101–103]. Physical signal transmission were also found in bacterial cells, where growth-promoting/regulating phonons or sonic vibrations, were effectively transmitted over a distance of at least 30 cm in air, through 2.5 mm plastic barrier, as well as a 2 mm iron plate to distant cultures [104]. Further, sound waves generated from a speaker at specific frequencies promoted colony formation under non-permissive stress conditions [105].

Remarkably, it has been found that even biological events traditionally considered chemically based, such as the lock-and-key model for olfaction, may actually rely more fundamentally on quantum scale atomic processes of inelastic electron tunneling from the donor to a receptor for critical discrimination [106,107]. For example in photosynthesis, light energy is ultimately transduced into chemical and electronic energy through the apparatus of the photosynthetic reaction center. Here the excitation of a chlorophyll molecule by the photon's energy initiates a series of charge-transfer processes from the antenna pigments to the reaction center via quantum coherence energy transfer [108]. The first steps are so fast that quantum dynamics of the nuclear motion needs to be accounted for as well as electron tunneling [109]. The wave-like characteristics of this energy transfer can explain the extreme efficiency that allows the light harvesting complex to sample vast areas of phase space to find the most efficient path [110].

Most notably, it was discovered that all living biological systems emit ultra-weak photons, or biophotons, which exhibit very unique physical characteristics during spontaneous emission and delayed luminescence. The hyperbolic decay and oscillations of these electromagnetic emissions or biophotons, in the optical regime have been observed experimentally and are indicative of coherent emission in accordance with multimodal laser theory. Coherent electromagnetic radiation strongly suggests the capacity for electromagnetic pathways in intercellular communication [111]. Groups of molecules cannot emit independently from each other because the distance between cells is smaller than the wavelength of the radiation they emit. Since they are coupled by a common radiation field, they will always be coherent [112]. Inside a coherent region or domain, energy travels in a wave-like fashion, whereas in non-coherent domains the energy propagates in a diffusive manner [72]. This coupling field consists of interference patterns reflecting the structure of the antenna system, i.e., groups of molecules, to which it is feedback coupled. Any field has a coherence space-time in which coherent states may exist by having a region where the phase is defined. Outside this region, the phase information is lost, but within it, the interference patterns are formed and a particle loses its classical pictures. Thus the particles and fields within the coherence region must be considered as an indivisible whole [112]. Gurwitsch first discovered coherent emission of ultraweak luminescence on the tips of onions roots in the 1920's. Modern interpretations of biophotonics conceptualize organisms as biological lasers of optically coupled emitters and absorbers operating at the laser threshold. A technical systems such as a laser, has a fixed coherence region or volume, while organisms may have a multitude of different coherence volumes, which can exist simultaneously and can overlap and demonstrate dynamic properties. The physical components of an organism is coupled with what can be described as a highly coherent, holographic, biophoton field, which has been proposed to be the basis of biological communication at all levels of organization. The components of the organism are seen to be connected in such a way by phase relations of the field that they are instantly informed about each in real-time. The coherent states appear to be fundamental for biological systems since they enable optimization of organization, information quality, pattern recognition and regulation of biochemical and morphogenetic processes [112]. It has been proposed that enzyme dynamics are an outcome of the coherent electromagnetic structure of living systems. Enzymes exhibit selective interactions with specific molecules which strongly suggest the existence of a coherent medium since the molecules no longer interact through random collisions. Classically enzymes are depicted as chemical polymers, however upon applying quantum electrodynamics (QED) principles an enzyme is projected as a coherent domain of its component monomers bound by electrodynamic as opposed to chemical attraction [72].

In various biophotonic experiments with cultures of the unicellular alga *Acetabularia acetabulum* exposed to variety of influences such as varying salt concentrations, chloroform, and temperature modulation, it was concluded that the delayed luminescence was not solely a function of the primary delayed photochemical fluorescence events of the photosynthetic apparatus. However, it demonstrated global correlations and information about the organization of streaming motility of the chloroplast and the cytoplasmic structure of the cell [113]. The cytoskeleton is an important milieu for providing coherent events being the basis for acoustic/photonic transmission. In established A. acetabulum cultures the individual cells form extensive electromechanical interactions where phase boundaries and mechanical tensions play an important role, which may be closely connected with biochemical changes and ultimately in a collective biophoton emission pattern [114].

6.4.2.9 BIOELECTROMAGNETICS FOR NON-CHEMICAL COMMUNICATION AND SIGNALING

A long history of extensive research on intercellular communication is found in the literature, which has primarily focused on receptor-based chemical signaling, molecular mechanisms, cell recognition, and cell surface receptors; however very few studies have focused on light-mediated interactions of cells, tissues and whole organisms [115]. Kaznacheyev and colleagues in Russia performed over 12,000 experiments in studying distant intercellular communication from two physically separated living tissues or cultures. They used two hermetically sealed vessels attached to each other via an interchangeable window composed of glass or quartz, where each vessel contained an identical culture. One of the vessel's cells was treated with a specific toxin, i.e., virus, chemical or radiation, while keeping the neighboring culture physically isolated from it. If a quartz window was used, so as to allow UV in addition to the visible and IR range of photons, approximately 75% of the physically isolated cultures began exhibiting toxin specific morphological stress and cell death 12 h after the directly exposed neighbor. However no effect was found if glass was used in the window to block the UV radiations indicating that biophoton signals passing through the quartz window were responsible for the specific morphological response [116–121]. By implementing a photomultiplier tube (PMT), they observed that normal functioning cells emit a uniform photon flux, while with the introduction of a toxin the radiation flux which intensifies at periodic intervals which depend on the different exposed toxin [120]. The harmonic relationship between the UV, visible and IR bands and their phase orientation has been suggested as a potential mechanism of intercellular communication [122] since the existence of coherent fields gives rise to destructive and constructive interference patterns in the space between living cells [123]. The biocommunication in these mutual interference regions leads to an optimized signal/noise ratio as the wave patterns achieve maximum destructive interference or compensation. Once the coherent superposition of modes of biophoton fields breaks down, one expects an increase in biophotonic emission, which was confirmed by Schamhart and Wijk [124], by examining the delayed luminescence of tumor cells as they lose their coherence and capacity for destructive interference by exhibiting exponential as opposed to hyperbolic decay [123]. The importance of biophotons in inter- and intracellular communication has been further confirmed through many other experiments that have been listed in the Table 1.

TABLE 1: Overview of biophotonic and distant intercellular interactions (D.I.) experiments, delayed luminescence (D.L.), and spontaneous emission (S.E.).

Culture	Experiment	Effect	Reference
Daphnia	D.I. & S.E.	Established destructive interference found at natural population density	[125]
D. tertiolecta	D.I. & D.L.	Changes in external environment demonstrated dose/intensity dependent decay curves	[126]
P. elegans	D.I. w/E-Field	E-field stimulated distant culture's photonic activity and synchronization	[127,128]
Gonyaulax sp.	D.I.	Established destructive interference and synchronization of photon pulses	[129]
XC tumor cells	D.I.	Dense cell culture stimulated growth rate of isolated culture via optical contact	[116]
Epithelial cells	D.I. w/H_2O_2	Reduction in protein, increased nuclear activation, and structural damage	[130]
E. coli	D.I.	Synchronized growth parameters when in optical contact of Vis-IR.	[96]
S. cerevisiae	D.I.	Stimulation of cellular subdivision via optical coupling with culture of same type	[131]
P. fluorescens	D.I.	Long range interactions of an isolated culture diminished adhesion between cells of another culture	[132]
V. costicola	D.I.	Isolated treated culture stimulated growth of second culture of same species	[133]
Fibroblasts	D.I. w/Viruses	Three viral effects transferred to 72–78% of distant isolated cells	[134]
	D.I. w/$HgCl_2$	Effects transferred to 78% of distant isolated cells	
	D.I. w/Rad	UV radiation effects transferred to 82% of distant isolated cells	
L. pekennisis	S.E.	Measured coherent emission from 200–800 nm which differed between male and female specimens	[135]

6.4.2.10 ENDOGENOUS EMF MODELING

Atoms to molecules to macromolecules, the process of modeling these interactions gets increasingly more complex. Biological systems behave like a macroscopic quantum system [112] therefore quantum mechanics is used to describe them. Modern quantum theory in biology has introduced the non-local property of interconnectedness, where the emphasis is no longer on isolated objects, but on relations, exchanges and interdependences on processes, fields and wholes [136].

The ability to detect, interpret and meaningfully interact with the endogenous bioelectromagnetic systems of living organisms could lead to dramatic advancements in modern biological sciences and engineering applications. However, in the case of biophotonic, distant interaction, and multipolar EMF experiments, where there is a destructive interference of EM signals, it becomes exceedingly difficult to directly measure phase conjugated or completely compensated EM fields in superposition. The decomposition of an electromagnetic field into scalar potential functions [137,138] is a traditional mathematical apparatus to describe EMFs at the complete destructive field interference. A conventional wisdom in engineering is that potentials have only mathematical, not physical significance. For instance, classical electrodynamic theory regards the complete cancellation of two fields as an absence of any field or effect. However, besides the case of quantum theory, where it is well known that the potentials are physical constructs, there are a number of physical phenomena - both classical and quantum mechanical, which possess physical significance as global-to-local operators or gauge fields, in precisely constrained topologies, such as the Aharonov-Bohm and Altshuler-Aronov-Spivak effects, the topological phase effects of Berry, Aharonov, Anandan, Pancharatnam, Chiao and Wu, the Josephson effect, the quantum Hall effect, the De Haas - Van Alphen effect, and the Sagnac effect [139]. In particular, the Aharonov-Bohm effect theoretically emphasized the importance of potentials rather than the force fields [140,141]. It was later experimentally demonstrated that interfering electromagnetic potentials could produce real effects on the phase via the magnetic vector potential (A-field) of charged particle systems even though the magnitude of the force field was zero around the charged particles [142]. Due to the relative phase factor of two interfering charges, the scalar field can transfer information, even though there is no transport of electromagnetic energy [143]. Furthermore, it appears that information is encoded as frequencies of alternating magnetic vector potential, and should be possible to control chemical reactions in vitro and in vivo through the interaction of magnetic vector potential with chemical potential [100].

The mathematics to describe the decomposition of an electromagnetic field or wave into two scalar potential functions was advanced by Whittaker at the turn of the century [137,138], which later became the basis for superpotential theory [144,145]. Maxwell's linear theory is of U(1) symmetry form, with Abelian commutation relations, but it can be extended to include physically meaningful A–field effects by its reformulation in SU(2) and higher symmetry forms. The commutation relations of the conventional classical Maxwell theory are Abelian. When extended to SU(2) or higher symmetry forms, Maxwell's theory possesses non-Abelian commutation relations, and addresses global, i.e., nonlocal in space, as well as local phenomena with the potentials used as local-to-global operators [139]. Success has been achieved in developing theoretical models for topological criteria for multiple coupled oscillators and higher group symmetry manifolds based on both classical and quantum electromagnetism to explain several phenomena in microbiology, nanoscience and metamaterials [146–150].

The application of these extended, higher topological mathematical models and quantum theories into biophysics and biophotonics may help elucidate the embedded or internal dynamics of the scalar potential functions that comprise the electromagnetic fields that destructively interfere between coupled biological systems or cultures. Despite the overwhelming complexity of modeling interdependent coherent electromagnetic interactions in complex biological systems, there exist both theoretical and empirical evidence that establishes spatial and temporal topology of fundamental geometric superposition, and interdependent relationships, such as multipolar influences, can uniquely affect biological systems.

6.4.2.11 ROLE OF WATER

Water is well known to be an anomalous substance and plays a great role in living organisms. Due to the critical role water plays in biochemical and biological reactions, many studies have focused on the effects of magnetic and electromagnetic fields on water molecules [51]. These experiments have shown that water previously exposed to electrical, magnetic, electromagnetic, acoustic or vibrating fields keeps the acquired biological activity for extended periods of time [151]. Liquid water is clearly a very complex system when considering the complexity of molecular clusters, gas-liquid and solidliquid surfaces, reactions between the materials and the consequences of physical and electromagnetic processing [152].

The investigation of indirect magnetic field effects have shown that magnetically treated water has changes in light absorption, specific electrical conductivity, magnetic susceptibility, Raman spectrum, index of light refraction, surface tension and viscosity. The exposure of water to a static magnetic field is connected with the energy influence of the field on the water and biostructures. Markov [153] has also shown that static magnetic fields influence the speed of protoplasm movement, the miotic activity, and the quantity of pigments such as chlorophyll a, b and organic acids in plants. Water stores and transmits information concerning solutes, by means of its hydrogen-bonded network. The conditioning of water via permanent magnetic and electromagnetic oscillating fields has been found to be stimulatory or inhibitory depending on the residence time of the water. *S. cerevisiae* exhibited the strongest influence by measuring a growth rate increase of ~60% after exposing the culture media to 15–30 seconds of a 100 kHz EMF at 2 μT. Longer exposure times that were inhibitory, could become stimulatory after dilution suggesting the existence of active agent(s) generated by the field exposure. Increases in toxicity after applying a biocide compared to a biocide+EMF indicates an enhanced cell wall permeability [154].

Ultra high dilutions are special preparations of a specific compound dissolved in a medium (usually water) that undergo dramatic dilutions (usually thirty 1:100 dilutions) that exceed Avogadro's number such that the final dilution is void of any original dissolved molecules. Each dilution step is accompanied by some activation force, usually mechanical succussion (shock wave) or vigorous mixing. However, other experiments have used sonication, high-voltage electromagnetic pulses, passive or active resonant circuits. The experimental results indicate that "pure" water samples can retain specific information regarding a "donor" substance which can be quantitatively measured via thermoluminescence, delayed luminescence, excess heat-of-mixing/microcalorimetry, changes in pH and conductivity, alterations to FTIR spectra, enzymatic activity, and modulation of chemical, biochemical, and biological processes usually in accord with the donor substance. These experiments have been carried out with biological bioassays with dinoflagellates comparing succussed media, and modulation to Ca^{2+} channel affinity by non-thermal microwave exposure, as well as investigating physico-chemical effects on purely chemical systems using ultra-high dilution of lithium chloride, sodium chloride, mercuric chloride, and mercuric iodide [155–168].

It has been proposed that the water molecules respond to incident EMF exposure and form metas water states [164]. The experiments with thermoluminescence, microcalorimetry, and conductivity measurements indicate molecular cluster formation, most likely originating from the hydrogen bond network. The evolution of these physico-chemical parameters with time suggests a trigger effect on the formation of molecular aggregates following the potentization procedure [159]. The various initial perturbations initiate development of a set of chain reactions of active oxygen species in water. Energy in the form of high-grade electronic excitations is released in reactions, which can support non-equilibrium state of an aqueous system [169]. Within these solutions, the molecular aggregates or clusters consisting of water molecules are connected by hydrogen bonds, in far from equilibrium conditions, which can remain in, or move away from their uns equilibrium state dissipating energy from the external environment in the manner Prigogine has described "dissipative structures" [170]. The lifetime of a particular cluster, containing specific water molecules will be not much longer than the life of individual hydrogen bonds, i.e., nanoseconds, but clusters can continue forever although with constant changing of their constituent water molecules [152]. However, the primacy of hydrogen bonds for the molecular aggregate structures is not essential, as the formation of H-bonded molecules are considered coherence domains in water by Coherent Quantum Electrodynamic Theory, where the H-bond dynamics are transferred to the origin of their pair potentials interacting with zero-point fluctuations of the A-field [171]. The existence of these physicochemical and biological effects from water should elevate water from its traditional role as a passive space-filling solvent in organisms, to a position of singular importance, the full significance of which is yet to be fully elucidated [143].

6.5 ELECTROMAGNETIC APPLICATIONS FOR PRODUCTION OF ALGAE BIOFUELS

The application of exogenous electromagnetic influences has been used for various commercial applications and an overview is given in Table 2.

TABLE 2: Overview of existing application of bioelectromagnetic fields.

	Influence	Application	Reference
Biomedical	PEMF	Chronic wound healing, and non-union fracture healing	[172]
		Chronic wound healing	[173]
		Treatment of osteonecrosis	[174]
		Treatment of pressure ulcers in spinal-cord injuries	[175]
		Treatment of osteoarthritis of the knee	[176]
		Treatment of grade I & II ankle sprains	[177]
		Treatment of venous leg ulceration	[178]
Agricultural	SMF	Treated water to stimulate germination in Pinus tropicalis seeds	[179]
		Treated chickpea seeds increased germination, seedling and root length & size	[180]
		Treated water increased plant height, branch number, and shoot dry weight	[181]
		Treated wheat seeds increased germination, yields, and protein	[182]
		Treated rice seeds and water increased rate and % of germination	[183]
		Treated barley seeds and water increased length and weight	[184]
	OMF	Treated tomato seeds for increased growth, yields, and disease resistance	[185]

The electric field pulses, or electroporation, have been traditionally implemented in metabolic engineering for gene transformation. Direct electroporation of a cyanobacterium *Synechococcus elongates*, introduced the enzyme *Clostridial hydrogenase*, which may lead to the development of a variety of hydrogenases for hydrogen production, coupled to photosynthesis in cyanobacteria for bioenergy production [186]. In addition to membrane-permeabilizing effects, it can also induce biochemical and physiological changes in plant protoplasts, such as stimulating protein and DNA synthesis, and cell division and differentiation [187]. Alternatively, electroporation can also be used as a process for cell membrane modification for enhanced oil/lipid extraction from microalgae for biodiesel. A preliminary study found 20% increase in oil yield, shorter extraction time, and 2/3 less solvent used without affecting the composition of extracted fatty acids compared to chemical solvent alone [33].

Although most electrochemical and electromagnetic effects mentioned thus far have been focused on biological responses, an integrated biorefining system also requires process engineering technologies for harvesting algae for instance. An electrochemical process using direct electric current, called electroflocculation, has a long history as a wastewater treatment technology for solid/liquid, and liquid/liquid separation [188]. This technology combines the use of a sacrificial electrode that dissolves to coagulate suspended particles (electrocoagulation) along with the use of electrolysis, which produces H_2 microbubbles that float the aggregates or flocs to the surface (electroflotation) for easy removal from the water. Electroflocculation is a promising technology for harvesting microalgae biomass since it has several advantages over other conventional processes. The efficiency of particle/biomass separation in electroflocculation is over 90% and this technology does not require moving parts, and consumes relatively little energy (0.3 kWh m^{-3}) with substantially lower capital costs [189]. In fact, a more recent study shows a 99.5% removal of total suspended solids (TSS) and chlorophyll a (algae) by applying 0.55 kWh m^{-3} for 15 minutes [190].

Bioelectrochemical denitrification is a novel technology being used for the treatment of ammonium and nitrate-containing wastewater by means of denitrifying bacteria and hydrogen gas produced on the cathode by the electrolysis of water. The denitrifying microorganisms are usually immobilized as a biofilm on graphite or a stainless steel cathode. A nitrate removal efficiency of 98% was observed at 20 mA when phosphate was used as a buffer. The studies suggested that the application of bioelectroreactors could be used for reduction and oxidation treatments of ammonium and nitrate-containing wastewaters [191–193].

In many cases, real-time monitoring of cultures is critical for productive and efficient cultivation/fermentation in which the optical density, pH, and dissolved gas levels may not elucidate the underlying bioprocesses occurring, especially when evaluating electrochemical or electromagnetic interactions. Pulsed Amplitude Modulation Fluorometry or PAM fluorometry is a special method for measuring fluorescence from photosynthetic organisms for real-time culture monitoring of the photosynthetic apparatus. It uses the characteristics of the fluorescence emitted by chlorophyll a as a probe for the biophysics and biochemical events occurring in the electron transport chain of Photosystem I & II. These measurements are a unique indicator of photosynthesis and provide information about the maximum photosynthetic efficiency (by a dark-adapted sample), the effective photosynthetic efficiency (under constant illumination), and the non-photochemical quenching (heat dissipation). These parameters indicate what fraction of the photon energy absorbed by the organism is used for photochemistry, dissipated as heat, and re-emitted as fluorescence [194]. Papazi and his colleagues found that PAM Fluorometry in conjunction with traditional biomass analysis was able to show how extremely high CO_2 concentrations impacted the photosynthetic apparatus, which stimulated intense biomass production in the microalgae, *Chlorella minutissima* [195].

The use of the biophotonic method of delayed luminescence (DL) has been used for quality control applications with fruits and vege s. A study with tomato fruits revealed marked changes due to different harvesting maturities. It was found that tomatoes exhibited DL measurements related to color and respiration as well as significant differences in soluble solids content and dry matter percentage. Therefore, DL values are directly related to tomato harvest maturity. Qualitative traits can depend on harvest maturity, thus suggesting that delayed luminescence could be used as a nondestructive indicator of fruit quality [196].

In addition to fluorescent measurements, the fast, non-invasive measurement of biological cells by dielectric spectroscopy, or impedance spectroscopy, is currently being utilized to determine cellular parameters, such as living cell volume, cell number distribution over cell cycle phase, cell length, internal structure, complex permittivity, and intracellular and extracellular media and morphological factors. The electrical and morphological properties of the cell membrane are assumed to represent sensitive parameters of the cellular state [197]. It has been demonstrated to be a powerful method for dielectric monitoring of biomass and cell growth in ethanol fermentation and the extension of the scanning dielectric microscope is a promising tool for dielectric imaging of biological cells [198]. The real-time monitoring of yeast cell division by measuring the dielectric dispersion can enable to tracking of cell cycle progression using an electromagnetic induction method [199]. Recently, online monitoring of lipid storage in microorganisms (yeasts) was conducted which found that using dielectric spectroscopy data, the change in capacitance divided by the characteristic frequency being used showed a clear shift from the growth phase to the lipid accumulation phase, which could be of use for technical control of intracellular biopolymer or oil accumulation, as well as enzyme overproduction [200]. Moreover, it has been established that there exists a connection between D.L. and impedance spectroscopic parameters, which explore related structures and mechanisms in living samples [201]. By applying the knowledge, gained from biophotonic and bioelectromagnetic experiments, it may be possible to detect, interpret and interact with the endogenous coherent electromagnetic signals that are correlated with regulation, communication, and organization of biological systems since oscillation dynamics are of essential importance in intercellular and intracellular signal transmission and cellular differentiation [70]. These signals may initially give us real-time insight into the internal dynamics of an organism or culture, which may precede the physically/chemically observable events.

Induction of specific cellular response to biophotonic signals could perhaps be achieved to stimulate a desired biological effect such as enhancement of lipid or enzyme synthesis or metabolite modulation using electromagnetic fields instead of an external stress or a biochemical initiator. Electromagnetic bioprocesses such as electroflocculation and electroporation can be used for algal harvesting and biomass processing. The use of static and oscillating electromagnetic fields has a potential for the enhancement of cell proliferation, metabolite production and cell cultivation for biomass production. After extraction, fermentation of the algae feedstock, using applied electric field parameters, can be designed for enhanced substrate utilization and higher ethanol/butanol yields. Any residual biomass may then be used for enhanced production of methane from anaerobic digestion using specific frequencies of microwaves reported by Banik et al. [5], who showed how the EM exposure parameters could be used for potential bioenergy/biofuel applications.

The application of electromagnetic coupling to electrochemical biological pathways, which have been studied and commercialized for biomedical applications can be introduced into bioengineering. Here investigations into the electrochemical impedance properties for triggering biochemical cascades of desired signaling pathways in microorganisms for bioenergy applications deserve significant attention. The application of exogenous EMF influences may synergistically couple with endogenous electric fields for enhancing directed mass transport in cells. It is conceivable that any cell could be stimulated, inhibited, or made to exhibit passive response, depending upon the appropriate choice of frequencies and amplitudes of the excitation signals employed [62]. The induction of mitosis for cell proliferation, as well as the stimulation of enzymatic pathways associated with energy metabolism and storage such as lipid accumulation, needs modeling and more experimentation. Such electrochemical processes may also be relevant for accelerating enzymes, such as Rubisco, in the carboxylation pathway of photosynthesis to enhance specific binding of CO_2 and limiting photorespiration to enhance overall system efficiency in microalgae or plants. The greatest challenge may be the evaluation of the proper dosimetry for modulation of the desired biochemical cascade [1].

The introduction of the complex topology of multipolar electromagnetic fields may provide an enhanced coupling effect to complex, interdependent biological systems. Such systems may be tailored to uniquely control endogenous electromagnetic processes and communication for cellular functioning and organization. Furthermore, the bioproducts, generated by engineered multipolar hybrid biosystems have additional properties. For example biofuel/bioenergy production processes potentially can have higher productivities through better substrate utilization and conversion and shorter processing times.

The use of electrochemical/electromagnetic triggering of specific metabolic pathways could be coupled with biophotonic analysis, where rapid screening and fine tuning of a desired effect could be devised. Such bioelectromagnetic and biophotonic monitoring could also be of significant

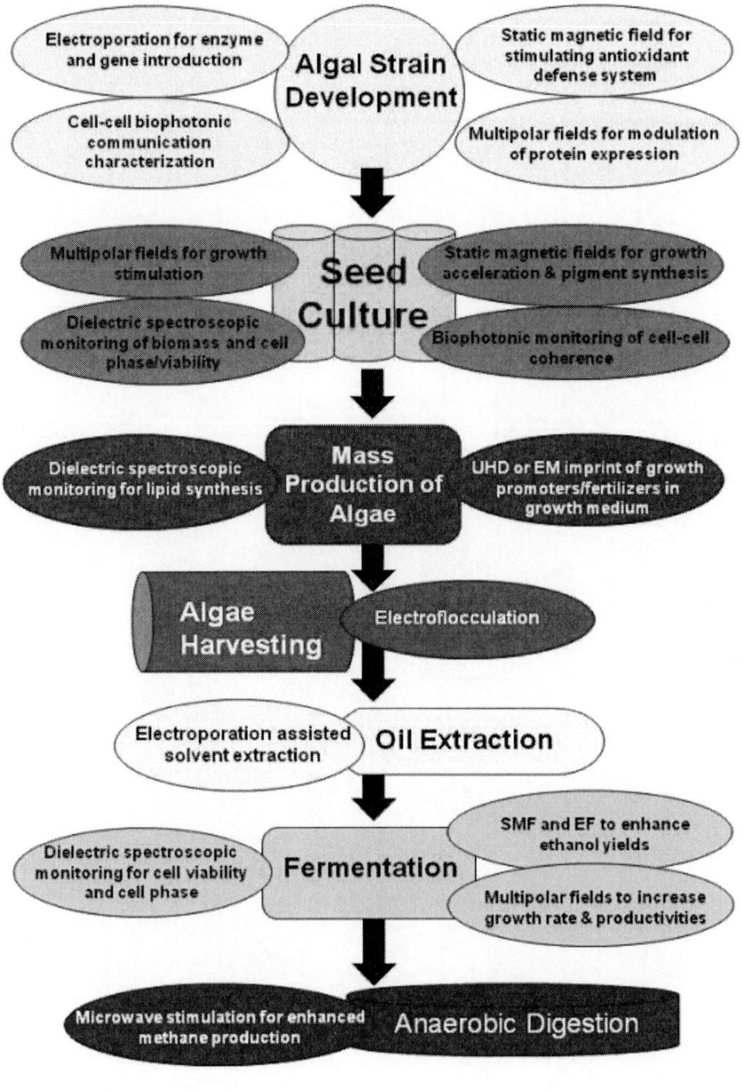

FIGURE 6: Integrated biostimulation/biofuel production system.

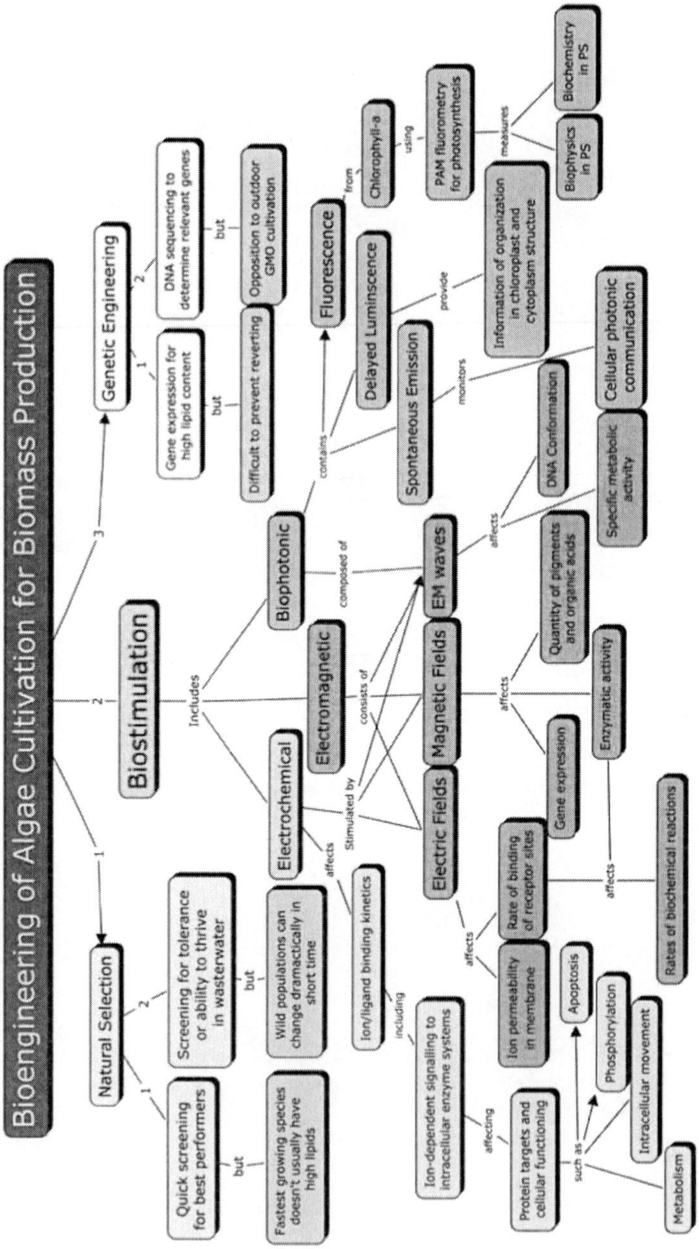

FIGURE 7: Bioengineering of algae cultivation.

interest to metabolic and genetic engineering, by incorporating and correlating electrochemical and endogenous electromagnetic signals with gene expression and enzymatic activity. The pervasive utilization of water in the cultivation of microorganisms particularly with algae, suggests possible application of the principles discovered in ultra-high dilution and activation studies to enhance and modulate biological responses. Water used in the growth medium for cultivation may be imprinted by various methods (electromagnetic information transfer probably being the most convenient) with specific information on relevant organic and inorganic nutrients as well as biochemical growth promoters to enhance growth characteristics, while decreasing demand for the potentially large amounts of the donor substance.

The combination of these separate disciplines, could blossom into a new integrative bioengineering approach that incorporates the diverse specializations of molecular biology, biochemistry, electrochemistry, biophysics, and quantum physics that could open up significant biotechnological progress of engineering of living systems for bioprocessing, bioconversion, biofuel and bioenergy applications (Figures 6 and 7).

6.6 CONCLUSIONS

Traditional cultivation and manipulation of biological systems have consisted of natural selection and genetic engineering modalities. Recently metabolic engineering and synthetic biology are gaining wide attention from the scientific community due to their immense potential in altering the metabolism in living systems especially microbes for medical, agricultural, industrial and environmental applications. However, genetic manipulation of microbes and living systems for agricultural and environmental applications may affect the ecosystem adversely as the changes in the species are permanent and inherited. In case of bioelectromagnetic stimulation, system reacts more in a transient fashion. The changes even if inherited are not sustained by the species for long thus they might be safer over genetic manipulation.

This review provided a broad spectrum of potentially useful bioeffects on microorganisms that are currently or potentially valuable in biotechnology and bioenergy. At this point it is difficult to ascertain exactly how economically feasible these emerging methods and potential technologies will be due to a variety of unknown factors from the nature and scalability of the bioeffects to the electronic design and efficiency for large-scale implementation. However, it is the aim to stimulate interest in the field and invite scientists with new ideas into the long standing discipline of bioelectromagnetics that modern biology is only recently starting to understand. In the new horizon of biologically derived fuels and materials, advancements in the area of biostimulation could impact the direction of biotechnology towards an energetic approach that may boost the potential for emerging biotechnologies such as microalgae based biofuel and biomass production.

Biofuels, bioenergy and carbon capture are considered to be the current priorities for the entire global community. The International Energy Agency (IEA) has reported that the world's primary energy need is projected to grow by 55% between 2005 and 2030, at an average annual rate of 1.8% per year. Fossil fuels are the main source of primary energy and if the governments around the world stick to current policies, the world will need almost 60% more energy in 2030 than today. Transportation is one of the fastest growing sectors using 27% of the primary energy. At the present staggering rates of consumption, the world's fossil oil reserve will be exhausted in less than 45 years. Considering the negative impacts of utilizing fossil fuel energy sources, many countries have already mandated the use of biofuels and set the targets to replace significant quantities of fossil derived fuels. Second and third generation biofuels such as lignocellulosic ethanol and algae biofuels are considered to be the viable alternatives as they do not compete with food needs. Bioelectromagnetic stimulation of microbes particularly with microalgae provides a new extended domain of disciplines and methodologies for cultivation, harvesting and processing of biomass for production of biofuels, bioenergy and added value bioproducts. Though this technology is promising, lots of research efforts are needed in future to exploit its commercial potential for biotechnology and biofuel applications.

REFERENCES

1. Pilla, A.A.; Markov, M.S. Bioeffects of weak electromagnetic fields. Rev. Environ. Health 1994, 10, 155–169.
2. Velizarov, S. Electric and magnetic fields in microbial biotechnology: Possiblities, limitations, and perspectives. Electro- Magnetobiol. 1999, 18, 185–212.
3. Chisti, Y. Biodiesel from microalgae. Biotechnol. Adv. 2007, 25, 294–306
4. Rai, S. Causes and mechanism(s) of ner bioeffects. Electromagn. Biol. Med. 1997, 16, 59–67.
5. Banik, S.; Bandyopadhyay, S.; Ganguly, S.; Dan, D. Effect of microwave irradiated Methanosarcina barkeri DSM-804 on biomethanation. Bioresour. Technol. 2005, 97, 819–823.
6. Del Re, B.; Bersani, F.; Agostini, C.; Mesirca, P.; Giorgi, G. Various effects on transposition activity and survival of Escherichia coli cells due to different ELF-MF signals. Radiat. Environ. Biophys. 2004, 43, 265–270.
7. Dutta, S.; Verma, M.; Blackman, C. Frequency-dependent alterations in enolase activity in Escherichia coli caused by exposure to electric and magnetic fields. Bioelectromagnetics 1994, 15, 377–383.
8. Justo, O.R.; Pérez, V.H.; Alvarez, D.C.; Alegre, R.M. Growth of Escherichia coli under extremely low-frequency electromagnetic fields. Appl. Biochem. Biotechnol. 2006, 134, 155–163.
9. Belyaev, I.Y.; Alipov, Y.D.; Matronchik, A.Y. Cell density dependent response of E. Coli cells to weak ELF magnetic fields. Bioelectromagnetics 1998, 19, 300–309.
10. Chang, Y.H.D.; Grodzinsky, A.J.; Wang, D.I.C. Augmentation of mass transfer through electrical means for hydrogel-entrapped Escherichia coli cultivation. Biotechnol. Bioeng. 1995, 48, 149–157.
11. Cellini, L.; Grande, R.; Campli, E.D.; Bartolomeo, S.D.; Giulio, M.D.; Robuffo, I.; Trubiani, O.; Mariggio, M.A. Bacterial response to the exposure of 50 Hz electromagnetic fields. Bioelectromagnetics 2008, 29, 302–311.
12. Zrimec, A.; Jerman, I.; Lahajnar, G. Alternating electric fields stimulate ATP synthesis in Escherichia coli. Cell. Mol. Biol. Lett. 2002, 7, 172–174.
13. Zavalin, A.; Collins, W.E.; Morgan, S. In A Compensation Zone of Multipolar System of EM Fields Stimulates Bacterial Growth, Proceeds of the 24th Meeting of Bioelectromagnetics Society, Quebec, Canada, 2002; pp. 8–9.
14. Zavalin, A.; Lensky, V.; McCarrol, P.; Westbrook, R.; Collins, W.E.; Morgan, S. Biostimulation of microorganisms exposed to multipolar systems of mutually compensated EMF. Bioelectromagnetics 2009, in review.
15. Erygin, G.D.; Pchedlkina, V.V.; Kulikova, A.K.; Rurinova, N.G.; Bezborodov, A.M.; Gogolev, M.N. Influence on microorganism growth and development of nutrient medium treatment with magnetic field. Prikl. Biokhim. Mikrobiol. 1988, 24, 257–263.
16. Ramon, C.; Martin, J.T.; Powell, M.R. Low-level, magnetic-field-induced growth modification on Bacillus subtilis. Biolelectromagnetics 1987, 8, 275–282.

17. Moore, R.L. Biological effects of magnetic fields: Studies with microorganisms. Can. J. Microbiol. 1979, 25, 1145–1151.
18. Hönes, I.; Pospischil, A.; Berg, H. Electrostimulation of proliferation of the denitrifying bacterium Pseudomonas stutzeri. Bioelectrochem. Bioenerg. 1998, 44, 275–277.
19. Kerns, G.; Bauer, E.; Berg, H. Electrostimulation of cellulase fermentation by pulsatile electromagnetically induced currents. Bioelectrochem. Bioenerg. 1993, 32, 89–94.
20. Grosse, H.-H. Electrostimulation during fermentation. Bioelectrochem. Bioenerg. 1988, 20, 279–285.
21. Thiemann, W.; Wagner, E. Die Wirkung eines homogenen Magnetfeldes auf das Wachstum von Micrococcus denitrificans. Z. Naturforsch. 1970, 25b, 1020–1023.
22. Lei, C.; Berg, H. Electromagnetic window effects on proliferation rate of Corynebacterium glutamicum. Bioelectrochem. Bioenerg. 1998, 45, 261–265.
23. Rao, T.B.M.L.R.; Sonolikar, R.L.; Saheb, S.P. Influence of magnetic field on the performance of bubble columns and airlift bioreactor with submersed microorganisms. Chem. Eng. Sci. 1997, 52, 4155–4160.
24. Bustard, M.; Rollan, A.; McHale, A.P. The effect of pulse voltage and capacitance on biosorption of uranium by biomass derived from whiskey distillery spent wash. Bioprocess Eng. 1998, 18, 59–62.
25. Utsunomiya, T.; Yamane, Y.-I.; Watanabe, M.; Sasaki, K. Stimulation of porphyrin production by application of an external magnetic field to a photosynthetic bacterium, Rhodobacter sphaeroides. J. Biosci. Bioeng. 2003, 95, 401–404.
26. Hirano, M.; Ohta, A.; Abe, K. Magnetic field effects on photosynthesis and growth of the cyanobacterium spirulina platensis. J. Ferment. Bioeng. 1998, 86, 313–316.
27. Li, Z.-Y.; Guo, S.-Y.; Lin, L.; Cai, M.-Y. Effects of electromagnetic field on the batch cultivation and nutritional compostion of Spirulina platensis in an air-lift photobioreactor. Bioresour. Technol. 2007, 98, 700–705.
28. Pakhomov, A.G.; Akyel, Y.; Pakhomova, O.N.; Stuck , B.E.; Murphy, M.R. Current state and implications of research on biological effects of millimeter waves: A Review of the literature. Bioelectromagnetics 1998, 19, 393–413.
29. Singh, S.S.; Tiwari, S.P.; Abraham, J.; Rai, S.; Rai, A.K. Magnetobiological effects on a cyanobacterium, Anabaena doliolum. Electromagn. Biol. Med. 1994, 13, 227–235.
30. Wang, H.-Y.; Zeng, X.-B.; Guo, S.-Y.; Li, Z.-T. Effects of magnetic field on the antioxidant defense system of recirculation-cultured Chlorella vulgaris. Bioelectromagnetics 2008, 29, 39–46.
31. Takahaski, F.; Kamezaki, T. Effect of magnetism of growth of Chlorella. Hakkokogaku 1985, 63, 71–74.
32. Yamaoka, Y.; Takimura, O.; Fuse, H.; Kamimura, K. Effect of magnetism on growth of Dunaliella salina. Res. Photosynth. 1992, 3, 87–90.
33. Sommerfeld, M.; Chen, W.; Hu, Q.; Giorgi, D.; Navapanich, T.; Ingram, M.; Erdman, R. Application of Electroporation for Lipid Extraction from Microlalgae; Algae Biomass Summit: Seattle, WA, USA, 2008.
34. Nimitan, E.; Topola, N. Influence of magnetic fields on the dehydrogenase activity of Saccharomyces cerevisiae. Analele S. Univ. Al. I. Cuza' din Iasi. Serie noua. Biologie 1972, 18, 259–264.

35. Perez, V.; Reyes, A.; Justo, O.; Alvarez, D. Bioreactor coupled with electromagnetic field generator: Effects of extremely low frequency electromagnetic fields on ethanol production by Saccharomyces cerevisiae. Biotechnol. Prog. 2007, 23, 1091–1094.
36. Fiedler, U.; Grobner, U.; Berg, H. Electrostimulation of yeast proliferation. Bioelectrochem. Bioenerg. 1995, 38, 423–425.
37. Mehedintua, M.; Berg, H. Proliferation response of yeast Saccharomyces cerevisiae on electromagnetic field parameters. Bioelectrochem. Bioenerg. 1997, 43, 67–70.
38. Nakanishi, K.; Tokuda, H.; Soga, T.; Yoshinaga, T.; Takeda, M. Effect of electric current on growth and alcohol production by yeast cells. J. Ferment. Bioeng. 1998, 85, 250–253.
39. Grundler, W.; Keilmann, F.; Fröhlich, H. Resonant growth rate response of yeast cells irradiated by weak microwaves. Phys. Lett. A 1977, 62, 463–466.
40. Russel, D.N.; Webb, S.J. Metabolic response of Danaüs archippus and Saccharomyces cerevisiae to weak oscillatory magnetic fields. Int. J. Biometereol. 1981, 25, 257–262.
41. Schaarschmidt, B.; Lamprecht, L.; Müller, K. Influence of a magnetic field on the UV-sensitivity in yeast. Z. Naturforsch. 1974, 29c, 447–448.
42. Pereira, M.R.; Nutini, L.G.; Fardon, J.C.; Cook, E.S. Cellular respiration in intermittent magnetic fields. Proc. Soc. Exp. Biol. Med. 1967, 124, 573–576.
43. McCabe, A.; Barron, N.; McHale, L.; McHale, A.P. Increased efficiency of substrate utilization by exposure of the thermotolerant yeast strain, Kluyveromyces marxianus IMB3 to electric-field stimulation. Biotechnol. Tech. 1995, 9, 133–136.
44. Goodman, E.M.; Greenebaum, B.; Marron, M.T. Effects of extremely low frequency electromagnetic fields on physarum polycephalum. Radiat. Res. 1976, 66, 531–540.
45. Marron, M.T.; Goodman, E.M.; Greenebaum, B. Effects of weak electromagnetic fields on Physarum polycephalum: Mitotic delay in heterokaryons and decreased respiration. Experientia 1978, 34, 589–591.
46. Marron, M.T.; Goodman, E.M.; Greenebaum, B.; Tipnis, P. Effects of sinusoidal 60-Hz electric and magnetic fields on ATP and oxygen levels in the slime mold, Physarum polycephalum. Bioelectromagnetics 1986, 7, 307–314.
47. Genkov, D.; Cvetkova, A.; Atmadzov, P. The effect of the constant magnetic field upon the growth and development of T. vaginalis. Folia Med. (Plovdiv). 1974, 16, 95–99.
48. Dihel, L.E.; Smith-Sonneborn, J.; Middaugh, C.R. Effects of extremely low frequency electromagnetic field on the cell division rate and plasma membrane of Paramecium tetraurelia. Biolelectromagnetics 1985, 6, 61–71.
49. Tabrah, F.L.; Guernsey, D.L.; Chou, S.-C.; Batkin, S. Effect of alternating magnetic fields (60–100 Gauss, 60 Hz) on Tetrahymena pyriformis. J. Life Sci. 1978, 8, 73–77.
50. Pirt, M.W.; Pirt , S.J. Production of biomass and starch by Chlorella in chemostat culture J. Appl. Chem. Biotechnol. 1977, 1917, 643–650.
51. Binhi, V.N. Theoretical concepts in magnetobiology. Electromagn. Biol. Med. 2001, 20, 43–58.
52. Nakasono, S.; Saiki, H. Effect of ELF magnetic fields on protein synthesis in Escherichia coli K12. Radiat. Res. 2000, 154, 208–216.

53. Nakasono, S.; Laramee, C.; Saiki, H.; McLeod, K.J. Effect of power-frequency magnetic fields on genome-scale gene expression in Saccharomyces cerevisiae. Radiat. Res. 2003, 160, 25–37.
54. Gao, W.; Liu , Y.; Zhou, J.; Pan, H. Effects of a strong static magnetic field on bacterium Shewanella oneidensis: An assessment by using whole genome microarray. Bioelectromagnetics 2005, 26, 558–563.
55. Fologea, D.; Vassu-Dimov, T.; Stoica, I.; Csutak, O.; Radu, M. Increase of Saccharomyces cerevisiae plating efficiency after treatment with bipolar electric pulses. Bioelectrochem. Bioenerg. 1998, 46, 285–287.
56. Belyaev, I.Y.; Alipov, Y.D.; Shcheglov, V.S. Chromosome DNA as a target of Resonant interaction between Esherichia coli cells and low-intensity millimeter waves. Electro-Magnetobiol. 1992, 11, 97–108.
57. Engstrom, S.; Markov, M.; McLean, M.; Holcomb, R.; Marko, J. Effects of non-uniform static magnetic fields on the rate of myosin phosphorylation. Bioelectromagnetics 2002, 23, 475–479.
58. Smith, S.; McLeod, B.R.; Liboff, A.R.; Cooksey, K. Calcium cyclotron resonance and diatom mobility. Bioelectromagnetics 1987, 8, 215–227.
59. Blackman, C.; Benane, S.G.; House, D.E.; Elliott, D.J. Importance of alignment between local DC magnetic field and an oscillating magnetic field in response to brain tissue in vitro and in vivo. Bioelectromagnetics 1990, 11, 159–167.
60. Reese, J.; Frazier, M.E.; Morris, J.E.; Buschbom, R.L.; D.L., M. Evaluation of changes in diatom mobility after exposure to 16-Hz electromagnetic fields. Bioelectromagnetics 1991, 12, 21–25.
61. Blackman, C.; Blanchard, J.P.; Benane, S.G.; House, D.E. Effect of AC and DC magnetic field orientation on nerve cells. Biochem. Biophys. Res. Commun. 1996, 220, 807–811.
62. Pilla, A.A.; Kaufman, J.J.; Ryaby, J.T. Electrochemical kinetics at the cell membrane: A physicochemical link for electromagnetic bioeffects. In Mechanistic Approaches to Interaction of Electric and Electromagnetic Fields with Living Systems; Blank, M, Findl, E., Ed.; 1987; pp. 39–61.
63. Lensky, V. Generation of multipolar electromagnetic energy. US Patent Application: 20,080,112,111, 2006.
64. Rein, G. Utilization of a cell culture bioassay for measuring quantum field generated from a modified caduceus coil. In Proceedings of the 26th International Society of Energy Conversion and Engineering Conference, Boston, MA, USA, 1991; pp. 400–403.
65. Ho, M.-W.; French, A.; Haffegee, J.; Saunders, P.T. Can weak magnetic fields (or potentials) affect pattern formation?. In Bioelectrodynamics and Biocommunication; Ho, M.-W., Popp, F.A., Warnke, U., Ed.; World Scientific: Singapore, 1994; pp. 195–212.
66. Rein, G. Bioinformation within the biofield: Beyond bioelectromagnetics. J. Altern. Complem. Med. 2004, 10, 59–68.
67. Repacholi, M.H.; Greenebaum, B. Interaction of static and extremely low frequency electric and magnetic fields with living systems: Health effects and research needs. Bioelectromagnetics 1999, 20, 133–201.

68. Liboff, A.R. Interactions between Electromagnetic Fields and Cells; Plenum Press: New York, NY, USA, 1985.
69. Pazur, A.; Schimek, C.; Galland, P. Magnetoreception in microorganisms ad fungi. Cent. Eur. J. Biol. 2007, 2, 597–659.
70. Kaiser, F. External signals and internal oscillation dynamics: Biophysical aspects and modelling approaches for interactions of weak electromagnetic fields at the cellular level. Bioelectrochem. Bioenerg. 1996, 41, 3–18.
71. McLeod, B.R.; Liboff, A.; Smith, S. Electromagnetic gating in ion channels. J. Theor. Biol. 1992, 158, 15–32.
72. Del Guidice, E.; De Ninno, A.; Fleischmann, M.; Mengoli, G.; Milani, M.; Talpo, G.; Vitiello, G. Coherent quantum electrodynamics in living matter. Electromagn. Biol. Med. 2005, 24, 199–210.
73. Pilla, A.A. Electrochemical information transfer at living cell membranes. Ann. N. Y. Acad. Sci. 1974, 238, 149–167.
74. Wojtczak, L.; Romanowski, S. Simple model of intermembrane communiation by means of collective excitations modified by an electric field. Bioelectrochem. Bioenerg. 1996, 41, 47–51.
75. Pilla, A.A.; Muehsam, D.J.; Markov, M.S.; Sisken, B.F. EMF signals and ion/ligand binding kinetics: Prediction of bioeffective waveform parameters. Bioelectrochem. Bioenerg. 1999, 48, 27–34.
76. Pilla, A.A.; Muehsam, D.J.; Markov, M.S. A dynamical systems/Larmor precession model for weak magnetic field bioeffects: Ion binding and orientation of bound water molecules. Bioelectrochem. Bioenerg. 1997, 43, 239–249.
77. Markov, M.S.; Pilla, A.A. Weak static magnetic field modulation of myosin phosphorylation in a cell-free preparation: Calcium dependence. Bioelectrochem. Bioenerg. 1997, 43, 233–238.
78. Binhi, V.N. Interference of ion quantum states within protein explains weak magnetic field's effect on biosystems. Electro- Magnetobiol. 1997, 16, 203–214.
79. Solov'yov, I.A.; Chandler, D.E.; Schulten, K. Magnetic field effects in Arabidopsis thaliana Cryptochrome-1. Biophys. J. 2007, 92, 2711–2726.
80. Ritz, T.; Adem, S.; Schulten, K. A model for photoreceptor-based magnetoreception in birds. Biophys. J. 2000, 78, 707–718.
81. Ahmad, M.; Galland, P.; Ritz, T.; Wiltschko, R.; Wiltschko, W. Magnetic intensity affects cryptochrome-dependent responses in Arabidopsis thaliana. Planta 2006, 225, 177–180.
82. Hunter, P. A quantum leap in biology. EMBO 2006, 7, 971–974.
83. Wiltschko, R.; Wiltschko, W. Magnetoreception. Bioessays 2006, 28, 157–168.
84. Wiesenfeld, K.; Moss, F. Stochastic resonance and the benefits of noise: From ice ages to crayfish and SQUIDs. Nature 1995, 373, 33–36.
85. Kruglikov, I.L.; Dertinger, H. Stochastic resonance as a possible mechanism of amplification of weak electric signals in living cells. Bioelectromagnetics 1994, 15, 539–547.
86. Pollack, G.H. The role of aqueous interfaces in the cell. Adv. Colloid Interface Sci. 2003, 103, 173–196.
87. Tyner, K.M.; Kopelman, R.; Philbert, M.A. "Nanosized voltmeter" enables cellular-wide electric field mapping. Biophys. J. 2007, 93, 1163–1174.

88. Collins, P.J. Liquid Crystal's Delicate Phase of Matter; Princeton University Press: Princeton, NJ, USA, 1990.
89. Ho, M.-W. The Rainbow and the Worm-The Physics of Organisms, 2nd ed.; World Scientific Publishing Co.: Singapore, 1998; p. 272.
90. Ho, M.-W. Bioenergetics and the coherence of organisms. Neuronetwork World 1995, 5, 733–750.
91. Del Guidice, E.; Doglia, S.; Milani, M.; Smith, C.W.; Vitiello, G. Magnetic flyx quantization and Josephson behaviour in living systems. Phys. Scr. 1989, 40, 786–791.
92. Costato, M.; Milani, M.; Spinoglio, L. Quantum mechancs: A breakthrough into biological system dynamics. Bioelectrochem. Bioenerg. 1996, 41, 27–30.
93. Fröhlich, H.; Kremer, F. Coherent excitations in biological systems. In Biological Coherence and Response to External Stimuli; Fröhlich, H., Ed.; Springer-Verlag: Berlin, Germany, 1983.
94. Merlino, A.; Sica, F.; Mazzarella, L. Approximate values for force constant and wave number associated with a low-frequency concerted motion in proteins can be evaluated by a comparison of X-ray structures J. Phys. Chem. B 2007, 111, 5483–5486.
95. Goodsell, D.; Olson, A. Structural symmetry and protein function. Annu. Rev. Biophys. Biomol. Struct. 2000, 29, 105–153.
96. Trushin, M.V. Studies on distant regulation of bacterial growth and light emission. Microbiology 2003, 149, 363–368.
97. Pokorný, J.; Hašek, J.; Jelínek, F. Electromagnetic field on microtubules: Effects on transfer of mass particles and electrons. JBP 2005b, 31, 501–514.
98. Presman, A.S. Electromagnetic Fields and Life; Plenum Press: New York, NY, USA, 1970; p. 336.
99. Pokorný, J.; Hašek, J.; Jelínek, F. Endogenous Electric field and organization of living matter. Electromagn. Biol. Med. 2005a, 24, 185–197.
100. Smith, C.W. Quanta and coherence effects in water and living systems. J. Altern. Complem. Med. 2004, 10, 69–78.
101. Cifra, M.; Vaniš, J.; Kučera, O.; Hašek, J.; Frýdlova, I.; Jelínek, F.; Šaroch, J.; Pokorný, J. Electrical Vibrations of Yeast Cell Membrane. In Progress In Electromagnetics Research Symposium, Prague, Czech Republic, 2007; pp. 215–220.
102. Cifra, M.; Pokorný, J.; Jelínek, F.; Hašek, J. Measurements of Yeast Cell Electrical Oscillations around 1 kHz, In Progress In Electromagnetics Research Symposium, Cambridge, MA, USA, 2008; pp. 780–784.
103. Pelling, A.E.; Sehati, S.; Gralla, E.B.; Valentine, J.S.; Gimzewski, J.K. Local nanomechanical motion of the cell qall of Saccharomyces cerevisiae. Science 2004, 305, 1147–1150.
104. Matsuhashi, M.; Pankrushina, A.N.; Endoh, K.; Watanabe, H.; Ohshima, H.; Tobi, M.; Endo, S.; Mano, Y.; Hyodo, M.; Kaneko, T.; Otani, S.; Yoshimura, S. Bacillus carboniphilus cells respond to growth-promoting physical signals from cells of homologous and heterologous bacteria. J. Gen. Appl. Microbiol. 1996, 42, 315–323.
105. Matsuhashi, M.; Pankrushina, A.N.; Takeuchi, S.; Ohshima, H.; Miyoi., H.; Endoh, K.; Murayama, K.; Watanabe, H.; Endo, S.; Tobi, M.; Mano, Y.; Hyodo, M., Kobayashi, T.; Kaneko, T.; Otanu, S.; Yoshimura, S.; Harata, A.; Sawada, T. Production

of sound waves by bacterial cells and the response of bacterial cells to sound. J. Gen. Appl. Microbiol. 1998, 44, 49–55.
106. Brookes, J.C.; Hartoutsiou, F.; Horsfield, A.P.; Stoneham, A.M. Could humans recognize odor by phonon assisted tunneling. Phys. Rev. Lett. 2007, 98, 1–4.
107. Turin, L. A method for the calculation of odor character from molecular structure. J. Theor. Biol. 2002, 216, 367–385.
108. Leegwater, J.A. Coherent versus incoherent energy transfer and trapping in photosynthetic antenna complexes. J. Phys. Chem. 1996, 100, 14403–14409.
109. Frauenfelder, H.; Wolynes, P.G.; Austin, R.H. Biological Physics. Rev. Mod. Phys. 1999, 71, 419–430.
110. Engel, G.S.; Calhoun, T.R.; Read, E.L.; Ahn, T.-K.; Mancal, T.; Cheng, Y.-C.; Blankenship, R.E.; Fleming, G.R. Evidence for wavelike energy transfer through quantum coherence in photosynthetic systems. Nature 2007, 446, 782–786.
111. Popp, F.A. On the coherence of ultraweak photon emission from living tissues. In Photon Emission from Biological Systems; Jeżowska-Trzebiatowska, B., Kochel, B., Sławiński, J., Stręk, W., Eds.; World Scientific Publishing Co., Inc.: Singapore, 1987.
112. Bischof, M. Introduction to integrative biophysics. In Integrative Biophysics- Biophotonics, Popp, F.A., Beloussov, L.V., Eds.; Kluwer Academic Publishers: Dordrecht, The Netherlands, 2003; pp. 1–116.
113. Musumeci, F. Physical basis and application of delayed luminescence. In Integrative Biophysics- Biophotonics, Popp, F.A., Beloussov, L.V., Eds.; Kluwer Academic Publishers: Dordrecht, The Netherlands, 2003; pp. 203–230.
114. van Wijk, R.; van Wijk, E.P.A. Oscillations in ultraweak photon emission of Acetabularia acetabulum (L.). Indian J. Exp. Biol. 2003, 41, 411–418.
115. Trushin, M.V. Distant non-chemical commuication in various biological systems. Riv. Biol./Biol. Forum 2004, 97, 409–442.
116. Vilenskaya, L. Kirkin, A.F. Non-chemical distant interactions between cell in a culture. Parapsychol. USSR 1982, 5, 25–29.
117. Kaznacheyev, V.P. Sverkhslabye Izlucheniia v Mezhkletochnykh Vzaimodeistviiakh (in Russian). Nauka: Novosibirsk, USSR, 1981.
118. Kaznacheyev, V.P. Electromagnetic bioinformation in intercellular interactions. PSI Res. 1982, 1, 47–76.
119. Kaznacheyev, V.P.; Shurin, S.P.; Mikhailova, L.P.; Ignatovish, N.V. Apparent information transfer between two groups of cells. Psychoenerg. Syst. 1974a, 1, 37–38.
120. Kaznacheyev, V.P.; Mikhailova, L.P.; Shurin, S.P. Informational interactions in biological systems caused by electromagnetic radiation of the optical range. Parapsychol. USSR 1974b, 5, 13–24.
121. Kaznacheyev, V.P.; Shurin, S.P.; Mikhailova, L.P.; Ignatovish, N.V. Distant intercellular interactions in a system of two tissue cultures. Psychoenerg. Syst. 1976, 1, 141–142.
122. Bearden, T.E. Gravitobiology; Tesla Book Co: Chula Vista, CA, USA, 1991.
123. Popp, F.A. Biophotons- background, experimental results, theoretical approach and applications. In Integrative Biophysics-Biophotonic; Popp, F.A., Beloussov, L., Eds.; Kluwer Academic Publishers: Dordrecht, The Netherlands, 2003; pp. 387–438.

124. Schamhart, D.H.J.; Wijk, R.V. Photon emission and the degree of differentiation. In Photon Emission from Biological Systems; Jeżowska-Trzebiatowska, B., Kochel, B., Sławiński, J., Stręk, W., Eds.; World Scientific: Singapore, 1987.
125. Galle, M.; Neurohr, R.; Altmann, G.; Popp, F.A.; Nagl, W. Biophoton emission from Daphnia magna: A possible factor in the self-regulation of swarming. Cell. Mol. Life Sci. 1991, 47, 457–460.
126. Zrimec, A.; Drinovec, L.; Berden-Zrimec, M., Influence of Chemical and Physical Factors on Long-Term Delayed Fluorescence in Dunaliella tertiolecta. Electromagn. Biol. Med. 2005, 24, 309–318.
127. Chang, J.J.; Popp, F.-A.; Yu, W.D. Research on cell communication of P. elegans by means of photon emission. Chin. Sci. Bull. 1995, 40, 76–79.
128. Chang, J.J. Biological effects of electromagnetic fields on living cells. In Integrative Biophysics- Biophotonics, Popp, F.A., Beloussov, L., Eds.; Kluwer Academic Publisher: Dordrecht, The Netherlands, 2003; pp. 231–259.
129. Popp, F.A.; Chang, J.J.; Gu, Q.; Ho, M.W. Bioelectrodynamics and Biocommunication. Ho, M.W., Popp. F.A., Warnke, U., Eds.; World Scientific: Singapore, 1994; pp. 293–317.
130. Farhadi, A.; Forsyth, C.; Banan, A.; Shaikh, M.; Engen, P.; Fields, J.Z.; Keshavarzian, A. Evidence for non-chemical, non-electrical intercellular signaling in intenstinal epithelial cells. Bioelectrochemistry 2007, 71, 142–148.
131. Musumeci, F.; Scordino, A.; Triglia, A.; Blandino, G.; Milazzo, I. Intercellular communication during yeast cell growth. Europhys. Lett. 1999, 47, 736–742.
132. Nikoleav, Y.A. Distant Interaction in the bacterium Pseudomonas fluorescens as a factor of adhesion regulation. Mikrobiologiia 2000, 69, 365–361.
133. Nikoleav, Y.A. Distant Interaction between bacterial cells (In Russian). Mikrobiologiia 1992, 61, 1066–1071.
134. Kaznacheyev, V.P.; Mikhailova, L.P.; Shurin, S.P. Informational interactions in biological systems caused by electromagnetic radiation of the optical range. In Parapsychology in the USSR; Vilenskaya, L., Ed.; 1982; Vol. 5, pp. 13–24.
135. Chang, J.J.; Popp, F.A.; Yu, W.D. Biocommunication and bioluminescence of Lampyride. In Biophotonics: Non-equilibrium and Coherent Systems in Biology, Biophysics and Biotechnology; Bioinform Services CO: Moscow, Russia, 1995; pp. 267–280.
136. Stapp, H.P. Nonlocal character of quantum theory. AmJPh 1997, 65, 300–304.
137. Whittaker, E.T. On the partial differential equations of mathematical physics. Math. Ann. 1903, 57, 333–355.
138. Whittaker, E.T. On an expression of the electromagnetic field due to electrons by means of two scalar potential functions. Proc. London Math. Soc. 1904, 1, 367–372.
139. Barret, T.W. Topological Foundations of Electromagnetism, 196th ed.; World Scientific Publishing Co: Singapore, 2008; Vol. 26.
140. Aharonov, Y.; Bohm, D. Significance of electromagnetic potentials in the quantum theory. Phys. Rev. 1959, 115, 485–491.
141. Aharonov, Y.; Bohm, D. Further considerations on electromagnetic potentials in the quantum theory. Phys. Rev. 1961, 123, 1511–1524.

142. Osakabe, N.; Matsuda, T.; Kawasaki, T.; Endo, J.; Tonomura, A.; Yano, S.; Yamada, H. Experimental confirmation of Aharonov-Bohm effect using a toroidal magnetic field confined by a superconductor. Phys. Rev. A. 1986, 34, 815–822.
143. Hyland, G.J. Bio-Electromagnetism. In Integrative Biophysics-Biophotonics; Popp, F.A., Beloussov, L.V., Eds.; Kluwer Academic Publsiher: Dordrecht, The Netherland, 2003; pp. 117–148.
144. Phillips, M. "Classical electrodynamics" in principles of electrodynamics and relativity. In Encyclopedia of Physics; Flugge, S., Ed.; Springer Verlag: Berlin, Germany, 1962; Vol. 4.
145. Bearden, T.E. Energy from the Vacuum: Concepts and Principles; Cheniere Press: Santa Barbara, CA, USA, 2002.
146. Šrobár, F.; Pokorný, J. Topology of mutual relationships implicit in the Fröhlich model Bioelectrochem. Bioenerg. 1996, 41, 31–33.
147. Guillot, J.-C.; Robert, J. The multipolar Hamiltonian in QED for moving atoms and ions. J. Phys. A: Math. Gen. 2002, 35, 5023–5039.
148. Shahverdiev, E.M. Chaos synchronization in the multi-feedback ikeda model. In Chaotic Dynamics: Adaptation and Self-Organizing Systems, arXiv:nlin/0405004v1; Ithaca: New York, NY, USA, 2004; pp. 1–5.
149. Raab, R.E.; De Lange, O.L. Multipole Theory In Electromagnetism: Classical, Quantum, and Symmetry Aspects, with Applications; Oxford University Press: Oxford, UK, 2005; p. 235.
150. Conti, D.; Tomassini, A. Special symplectic six-manifolds. QJMat 2007, 58, 297–231.
151. Lobyshev, V.I. Water is a sensor to weak forces including electromagnetic fields of low intensity. Electromagn. Biol. Med. 2005, 24, 449–461.
152. Chaplin, M.F. The memory of water: An overview. Homeopathy 2007, 96, 143–150.
153. Markov, M.S. In Direct and Indirect Action of Constant Magnetic Field on Biological Subjects, Sixth International Conference on Magnet Technology (MT-6); ALFA Bratislava: Bratislava, Czechoslovakia, 1977; pp. 384–389.
154. Goldsworthy, A. Biological effects of physically conditioned water. Water Res. 1999, 33, 1618–1626.
155. Tschulakow, A.V.; Yan, Y.; Klimek, W. A new approach to the memory of water. Homeopathy 2005, 94, 241–247.
156. Elia, V.; Niccoli, M. Thermodynamics of extremely diluted aqueous solutions. Ann. N. Y. Acad. Sci. 1999, 879, 241–248.
157. Rey, L. Thermoluminescence of ultra-high dilutions of lithium chloride and sodium chloride. Physica A 2003, 323, 67–74.
158. Lenger, K.; Bajpai, R.P.; Drexel, M. Delayed Luminescence of high homeopathic potencies on sugar globuli. Homeopathy 2008, 97, 134–140.
159. Elia, V.; Baiano, S.; Duro, I.; Napoli, E.; Niccoli, M.; Nonatelli, L. Permanent physico-chemical properties of extremely diluted aqueous solutions of homeopathic medicines. Homeopathy 2004, 93, 144–150.
160. Assumpção, R. Electrical impedance of HV plasma images of high dilutions of sodium chloride. Homeopathy 2008, 97, 129–133.

161. Sukul, N.C., Ghosh, S., Sukul, A., Sinhababu, S.P. Variation in fourier transform infrared spectra of some homeopathic potencies and their diluent media. J. Altern. Complem. Med. 2005, 11, 807–812.
162. Sinitsyn, N.; Petrosyan, V.L.; Yolkin, V.A.; Devyatkov, N.D.; Gulaev, Y.V.; Betskii, O.V. Particular role of an "MM wave-water medium" system in the nature. Biomed. Electron. 1998, 1, 5–23.
163. Geletyuk, V.I.; Kazachenko, V.N.; Chemeris, N.K.; Fesenko, E.E. Dual effects of microwaves on signle Ca2+ activated K+ channels in cultured kidney cells Vero. FEBS Lett. 1995, 359, 85–88.
164. Fesenko, E.E.; Gluvstein, A.Y. Changes in the state of water, induced by radio-frequency electromagnetic fields. FEBS Lett. 1995, 367, 53–55.
165. Sukul, N.C.; De, A.; Sukul, A.; Sinhababu, S.P. Potentized Mercuric chloride and Mercuric iodide enhance α-amylase activity in vitro. Homeopathy 2002, 91, 217–220.
166. Citro, M.; Smith, C.W.; Scott-Morley, A.; Pongratz, W.; Endler, P.C. Transfer of information from molecules by means of electronic amplification-preliminary results. In Ultra-High Dilution; Endler, P.C., Schulte, J., Eds.; Kluwer Academic Publishers: Dordrecht, The Netherland, 1994; pp. 209–214.
167. Cardella, C.; De Magistris, L.; Florio, E.; Smith, C.W. Permanent changes in the physicochemical properties of water following exposure to resonant circuits. J. Sci. Explor. 2001, 15, 501–518.
168. Jerman, I.; Ružič, R.; Krašovec, R.; Škarja, M.; Mogilnicki, L. Electrical transfer of molecule information into water, its storage, and bioeffects on plants and bacteria. Electromagn. Biol. Med. 2005, 24, 341–353.
169. Voeikov, V.L. The possible role of active oxygen in the memory of water. Homeopathy 2007, 96, 196–201.
170. Elia, V.; Napoli, E.; Germano, R. The "Memory of Water": An almost deciphered enigma. Dissipative structures in extremely dilute aqueous solutions. Homeopathy 2007, 96, 163–169.
171. Arani, R.; Bono, I.; Del Guidice, E.; Preparata, G. QED coherence and the thermodynamics of water. Int. J. Mod. Phys. B 1995, 9, 1813–1841.
172. Strauch, B.; Patel, M.K.; Navarro, J.A.; Berdichevsky, M.; Yu, H.-L.; Pilla, A.A. Pulsed magnetic fields accelerate cutaneous wound healing in rats. Plast. Reconstr. Surg. 2007, 120, 425–430.
173. Kloth, L.C.; Berman, J.E.; Sutton, C.H. Effect of pulsed radio frequency stimulation on wound healing: A double-blind pilot study. In Electricity and Magnetism in Biology and Medicine; Bersani, F., Ed.; Plenum: New York, NY, USA, 1999; pp. 875–878.
174. Aaron, R.K.; Lennox, D.; Bunce, G.E.; Ebert, T. The conservative treatment of osteonecrosis of the femoral head: A comparison of core decompression and pulsing electromagnetic fields. Clin. Orthop. 1989, 249, 199–208.
175. Salzberg, C.A.; Cooper, S.A.; Cooper, P.; Perez, P. The effects of non-thermal pulsed electromagnetic energy on wound healing; a double blind pilot study. OWM 1995, 41, 42–48.
176. Zizic, T.M.; Hoffman, K.C.; Holt, P.A.; Hungerford, D.S.; O'Dell, J.R.; Jacobs, M.A.; Lewis, C.G.; Deal, C.L.; Caldwell, J.R.; Cholewczynski, J.G. The treatment

of osteoarthritis of the knee with pulsed electrical stimulation. J. Rheumatol. 1995, 22, 1757–1761.
177. Pennington, G.M.; Danley, D.L.; Sumko, M.H. Pulsed non-thermal, high-frequency electromagnetic energy (Diapulse) in the treatment of grade I and grade II ankle sprains. Mil. Med. 1993, 158, 101–104.
178. Kenkre, J.E.; Hobbs, F.D.; Carter, Y.H. A randomized controlled trial of electromagnetic therapy in the primary care management of venous leg ulceration. Fam. Pract. 1996, 13, 236–241.
179. Morejón, L.P.; Castro palacio, J.C.; Velázquez Abad, L.; Govea, A.P. Stimulation of Pinus tropicalis M. seeds by magnetically treated water. Int. Agrophysics 2007, 21, 173–177.
180. Vashisth, A.; Nagarajan, S. Exposure of seeds to static magnetic field enhances germination and early growth characteristics in chickpea (Cicer arietinum L.). Bioelectromagnetics 2008, 29, 571–578.
181. Ozdemir, S.; Dede, O.H.; Koseoglu, G. Electromagnetic water treatment and water quality effect on germination, rooting, and plant growth on flower. Asian J. Wat. Environ. Pollut. 2005, 2, 9–13.
182. Pietruszewski, S. Effects of magnetic biostimulation of wheat seeds on germination, yields, and proteins. Int. Agrophysics 1996, 10, 51–55.
183. Carbonell, M.V.; Martinez, E.; Amaya, J.M. Stimulation of germination in rice (Oryza sative L.) by a static magnetic field. Electro- Magnetobiol. 2000, 19, 121–128.
184. Martinez, E.; Carbonell, M.V.; Amaya, J.M. A static magnetic field of 125 mT stimulates the initial growth stages of barley. Electro- Magnetobiol. 2000, 19, 271–277.
185. Souza, A.D.; Garcia, D.; Sueiro, L.; Gilart, F.; Poras, E.; Licea, L. Pre-sowing magnetic treatments of tomato seeds increase the growth and yield of plants. Bioelectromagnetics 2006, 27, 247–257.
186. Miyake, M.; Asada, Y. Direct electroporation of clostridial hydrogenase into cyanobacterial cells. Biotechnol. Tech. 1997, 11, 787–790.
187. Joersbo, M.; Brunstedt, J. Electroporation: Mechanism and transient expression, s transformation and biological effects in plant protoplasts. Physiol. Plant. 1991, 81, 256–264.
188. Holt, P.K.; Barton, G.W.; Mitchell, C.A. The future for electrocoagulation as a localised water treatment technology. Chemosphere 2005, 59, 355–367.
189. Poelman, E.; De Pauw, N.; Jeurissen, B. Potential of electrolytic flocculation for recovery of micro-algae. Resour. Conserv. Recycl. 1997, 19, 1–10.
190. Azarian, G.H.; Mesdaghinia, A.R.; Vaezi, F.; Nabizadeh, R.; Nematollahi, D. Algae removal by electro-coagulation process, application for treatment of the effluent from an industrial wastewater treatment plant. Iran. J. Publ. Health 2007, 36, 57–64.
191. Islam, S.; Suidan, M.T. Electrolytic denitrification: Long term performance and effect of current intensity. Water Res. 1998, 32, 528–536.
192. Kuroda, M.; Watanabe, T.; Umedu, Y. Simultaneous oxidation and reduction treatments of polluted water by a bio-electro reactor. Water Sci. Technol. 1996, 34, 101–108.
193. Kuroda, M.; Watanabe, T.; Umedu, Y. Simultaneous COD removal and denitrification of wastewater by bio-electro reactors. Water Sci. Technol. 1997, 35, 161–168.

194. Schreiber, U. Pulse-amplitude modulation (PAM) fluorometry and saturation pulse method: An overview. In Chlorophyll a Fluorescence: A Signature of Photosynthesis; Papageorgiou, G.C., Govindjee, Eds.; Springer: Dordrecht, The Netherlands, 2004.
195. Papazi, A.; Makridis, P.; Divanach, P.; Kotzabasis, K. Bioenergetic changes in the microalgal photosynthetic apparatus by extremely high CO2 concentrations induce an intense biomass production. Physiol. Plant. 2008, 132, 338–349.
196. Triglia, A.; La Malfa, G.; Musumeci, F.; Leonardi, C.; Scordino, A. Delayed luminescence as an indicator of tomato fruit quality. J. Food Sci. 1998, 63, 512–515.
197. Gheorghiu, E. Measuring living cells using dielectric spectroscopy. Bioelectrochem. Bioenerg. 1996, 40, 133–139.
198. Asami, K.; Yonezawa, T.; Wakamatsu, H.; Koyanagi, N. Dielectric spectroscopy of biological cells. Bioelectrochem. Bioenerg. 1996, 40, 141–145.
199. Asami, K.; Gheorghiu, E.; Yonezawa, T. Real-time monitoring of yeast cell division by dielectric spectroscopy. Biophys. J. 1999, 76, 3345–3348.
200. Maskow, T.; Rollich, A.; Fetzer, I.; Ackermann, J.-U.; Harms, H. On-line monitoring of lipid storage in yeasts using impedance spectroscopy. J. Biotechnol. 2008, 135, 64–70.
201. Lanzanò, L.; Grasso, R.; Gulino, M.; Bellia, P.; Falciglia, F.; Scordino, A.; Tudisco, S.; Triglia, A.; Musumeci, F. Corresponding measurements of delayed luminescence and impedance spectroscopy on acupuncture points. Indian J. Exp. Biol. 2008, 46, 364–370.

This chapter was originally published under the Creative Commons Attribution License. Hunt, R. W., Zavalin, A., Bhatnagar, A., Chinnasamy, S., and Das, K. C. Electromagnetic Biostimulation of Living Cultures for Biotechnology, Biofuel and Bioenergy Applications. International Journal of Molecular Science 2009: 10; 4515–4558.

CHAPTER 7

CATALYTIC TRANSFORMATIONS OF BIOMASS-DERIVED ACIDS INTO ADVANCED BIOFUELS

JUAN CARLOS SERRANO-RUIZ, ANTONIO PINEDA, ALINA MARIANA BALU, RAFAEL LUQUE, JUAN MANUEL CAMPELO, ANTONIO ANGEL ROMERO, and JOSE MANUEL RAMOS-FERNÁNDEZ

7.1 INTRODUCTION

Fossil fuels are the primary source of energy, chemicals and materials for our modern society. Petroleum, natural gas and coal supply most of the energy consumed worldwide and their massive utilization has allowed our society to reach high levels of development in the past century. However, these natural resources are highly contaminant, unevenly distributed around the world and they are in diminishing supply. These important concerns have stimulated the search for new well-distributed and non-contaminant renewable sources of energy such as solar, wind, hydroelectric power, geothermal activity, and biomass. This shift toward a renewable-based economy is currently spurred by governments which have established ambitious targets to replace an important fraction of fossil fuels with renewable sources within next 20 years [1] and [2]. In this sense, biomass is considered the only sustainable source of organic carbon currently available on earth and, consequently, it is the ideal substitute for petroleum in the production of fuels, chemicals and carbon-based materials [3].

Transportation sector of our society heavily relies on petroleum which accounts for essentially all (96%) of the transportation energy. This high reliance on petroleum is especially relevant since transportation is the largest and fastest growing energy sector, and it is responsible for almost one third of the total energy consumed in the world [4]. A large fraction of the extracted petroleum (70–80%) is consumed in form of transportation fuels (e.g. diesel, gasoline and jet fuels) in an attempt to cover this enormous demand for transportation energy. Consumption of petroleum is currently estimated to be around 80 millions of barrels per day, with projections to increase this amount by 30% within the next 20 years [4]. With these aspects in mind, an eventual displacement of petroleum by biomass will necessarily involve development of new technologies for large-scale production of fuels from this resource, the so-called biofuels.

The liquid biofuels most widely used today are bioethanol and biodiesel which are obtained from edible biomass sources such as sugar cane or corn and vegetable oils, respectively. An exponential increase in the consumption of such biofuels has taken place in the past few years [5]. Two are the main driven forces for this rapid expansion of bioethanol and biodiesel: (i) the simple and well-known technologies for their production (e.g. fermentation of sugars and transesterification of triglycerides with methanol) that has accelerated scale up of technologies and subsequent commercialization; and (ii) the partial compatibility of these biofuels with existing transportation infrastructure of diesel and gasoline which has allowed an easy penetration of these biofuels in the current fuel market.

Even though biodiesel and bioethanol (denoted as conventional biofuels) are produced by simple and mature technologies and are already commercially available, they possess a number of important drawbacks that seriously limit their further implementation in current transportation infrastructure [6]. For example, bioethanol is slightly corrosive and it has to be used in form of dilute mixtures with gasoline (e.g. E blends) in existing spark-ignition vehicles; it contains less energy per volume than gasoline (leading to lower fuel economy of vehicles running on E blends); and it induces water absorption in the fuel when added to gasoline thereby increasing risk of phase-separation episodes and engine damages. The corrosive nature of biodiesel also obligates to use it dilute with petrol-based diesel (e.g. B blends) and its higher cloud point compared to regular diesel increases the risk of plugging filters or small orifices at cold temperatures. Furthermore, this issue is compounded for the new generation of diesel engines which operate at higher injection pressures and with nozzles with a lower diameter.

These important limitations of conventional biofuels have stimulated the search for new technologies that allow production of high energy-density, infrastructure-compatible fuels (i.e. advanced biofuels) which could be easily implemented in the existing hydrocarbons-based transportation infrastructure (e.g. engines, fueling stations, distribution networks and petrochemical processes). In the past few years, these strong incentives have favored a dramatic change in funding directions from projects involving biodiesel and bioethanol to those aimed to the synthesis of advanced biofuels [7]. Relevant examples of advanced biofuels include higher alcohols (C4–C7) which possess energy density and polarity properties similar to gasoline [8]; and liquid hydrocarbon fuels (e.g. green hydrocarbons) which are chemically identical to those currently used in the transportation fleet [6], [9] and [10].

When operating with biomass resources, the structural and chemical complexity of feedstocks is an important issue. One of the most common strategies to overcome biomass complexity involves previous conversion into simpler fractions that are more easily transformed in subsequent processes. Thus, complex biomass resources can be converted into simpler compounds or platform molecules, which can subsequently serve as starting materials for a number of valuable products [11]. These platform molecules are carefully selected in base of a number of indicators such as the availability of commercial technologies for their production from biomass sources, and the platform potential of these compounds for the simultaneous production of fuels and chemicals in biorefineries [11]. This selected group of biomass platform molecules include sugars (glucose, xylose), polyols (sorbitol, xylitol, glycerol), furans (hydoxymethyl furfural or HMF, furfural), acids (lactic acid, levulinic acid, succinic acid) and alcohols (ethanol).

In the present paper we explore the potential of two of these platform molecules (e.g. lactic acid and levulinic acid) for the production of advanced biofuels. As represented in Fig. 1, the very different chemical composition of these molecules compared to final products (e.g. hydrocarbons or higher alcohols) suggests that deep chemical transformations will be required during catalytic processing of these resources. As is common to all biomass derivatives, lactic acid (2-hydroxypropanoic acid) and levulinic acid (4-oxo-pentanoic acid) are highly oxygenated compounds and, consequently, their conversion into advanced biofuels will necessarily involve deoxygenation steps. This oxygen removal step increases energy density in the molecule and, simultaneously, achieves reduction of the chemical reactivity generating less-reactive intermediates that are more easily processed to final products with high yields. As a result, oxygen will be removed from biomass acids in form of H_2O and/or CO_x species (CO and CO_2) by hydrogenation, dehydration, Csingle bondO hydrogenolysis and decarboxylation/decarbonylation catalytic processes. This requisite deoxygenation step normally (although not always) involves consumption of large amounts of hydrogen which is expensive and typically derived from fossil sources (which negatively affects CO_2 footprint of the bioprocess). As will be shown in subsequent sections, efforts are currently being made to drastically reduce external hydrogen consumption during deoxygenation of biomass platform molecules by, for example, utilization of renewable sources of this gas such as formic acid (a by-product of levulinic acid production industrial process) [12] and [13].

The high oxygen content of biomass platform molecules (as compared to petroleum feedstocks) is not the only limitation to overcome when advanced fuel production technologies are envisaged. Platform molecules are typically derived (by chemical and biological routes) from biomass sugars which are compounds with a maximum number of carbon atoms limited to 6 (derived from glucose). Consequently, if targeted products are liquid hydrocarbon transportation fuels (e.g. C5–C12 for gasoline, C9–C16 for jet fuel, and C10–C20 for diesel applications) deoxygenation will necessarily have to be combined with additional reactions aimed to increase the molecular weight in the molecule (e.g. Csingle bondC coupling reactions) [14]. Among the numerous Csingle bondC bond forming routes that organic chemistry can offer us, there are some reactions with particular interest in biomass conversion processes [15] and [16]. Thus, well known reactions such as aldol condensation of carbonyl compounds, catalytic ketonic decarboxylation or ketonization of carboxylic acids,

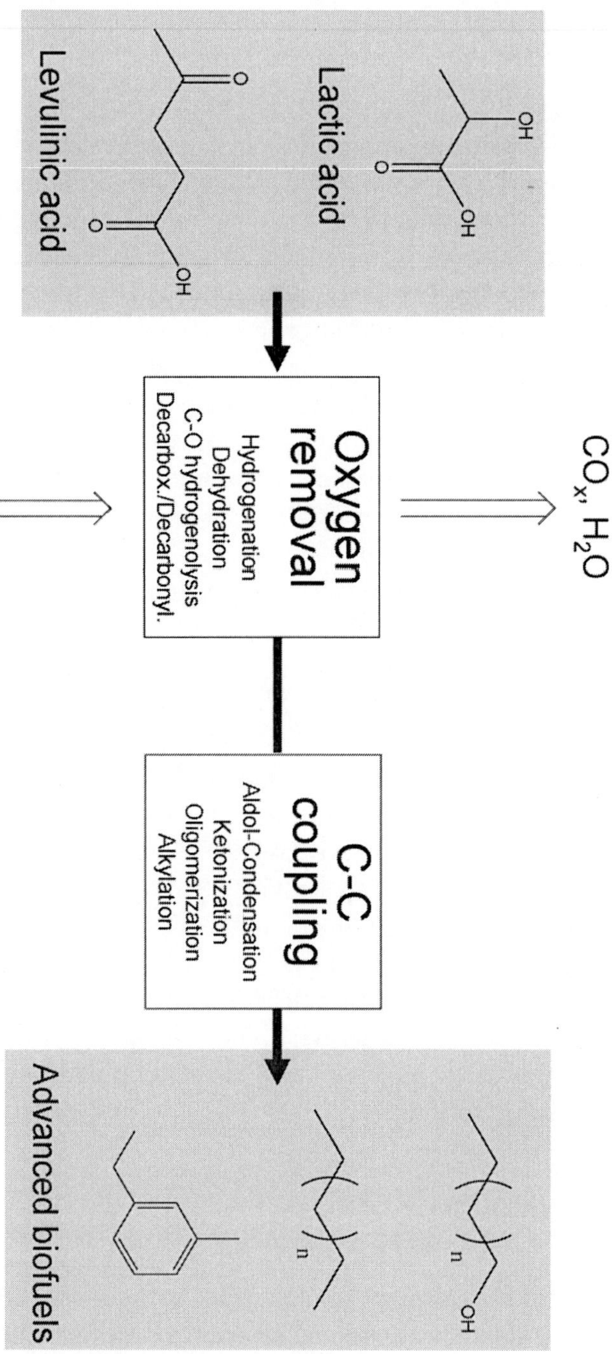

FIGURE 1: Scheme of the catalytic routes required to convert biomass-derived acids such as lactic acid and levulinic acid into advanced biofuels.

oligomerization of alkenes, and alkylations of hydrocarbons are especially indicated to increase molecular weight and to adjust structure of final advanced biofuel products.

Aldol condensation and ketonization represents two of the most relevant reactions for Csingle bondC coupling of biomass derivatives [15]. The former achieves effective and high-yield coupling between carbonyl-containing biomass intermediates at moderate reaction conditions. With regard to advanced biofuels, aldol condensation has been shown to be effective for Csingle bondC coupling of biomass-derived furanics (containing an aldehyde group) such as HMF and furfural with ketones such as acetone to produce diesel and jet fuel green hydrocarbon components [17], [18] and [19]. Ketonization, on the other hand, involves condensation of two molecules of carboxylic acid to produce a larger (2n − 1) symmetric ketone [20]. This reaction possesses a great potential for the catalytic upgrading of biomass since Csingle bondC coupling takes place with simultaneous oxygen removal (i.e., the reaction involves the removal of CO2 and water) from carboxylic acids, the latter of which are common intermediates in biomass conversion processes [21] and [22]. This reaction is typically catalyzed by inorganic oxides such as CeO_2, TiO_2, Al_2O_3 and ZrO_2 at moderate temperatures (300–425° C).

In the case of oligomerization and alkylation, they are especially indicated for upgrading of deoxygenated petroleum feedstocks (e.g. alkenes and hydrocarbons). However, relevant works by Dumesic and Corma's groups have recently demonstrated that they can be successfully employed for advanced biofuels production in the last stages of biomass conversion processes, that is, when oxygen content in bio-derived feedstocks is very low [23] and [24]. Next sections will provide examples on the application of the abovementioned Csingle bondC coupling processes for the transformation of lactic acid and levulinic acid into advanced biofuels for the transportation sector.

7.2 LACTIC ACID

Lactic acid (2-hydroxypropanoic acid) is the most widely occurring carboxylic acid in nature. Annual production reaches 120,000 tons per year, 90% of which is produced by bacterial fermentation of biomass sugars

[25], including pentoses [26]. The bacterial route affords lactic acid in high yields (e.g. 90%) although it possesses important drawbacks namely low reaction rates and troublesome separation/purification from the reaction broth of the lactic acid product. Lactic acid is obtained in form of calcium salt, and subsequent neutralization generates large amounts of residual $CaSO_4$ (1 kg per kg of lactic acid) which raises production costs and produces a waste disposal problem. Recently, a non-biological route for the conversion of aqueous sugars into lactic acid based on easily separable and recyclable solid zeolites has been developed [27]. This promising technology opens the possibility of producing lactic acid from biomass sugars at more competitive prices in near future thereby considerably increasing the platform potential of lactic acid. The classical market for lactic acid involves food and food-related applications; however, the development and commercialization of new applications in the field of polymers and chemicals has caused steady expansion of the lactic acid market since the early 1990s [28].

Lactic acid possesses a rich chemistry based on its two functional groups (e.g. single bondOH and single bondCOOH). Thus, a variety of transformations to useful compounds such as acetaldehyde [29] (via decarbonylation/decarboxylation), acrylic acid [30] (via dehydration), propanoic acid [31] (via reduction), 2,3-pentanedione [32] (via condensation) and polylactic acid (PLA) [33] (via self-esterification to dilactide and subsequent polymerization) has been described. All these transformations convert lactic acid in an attractive feedstock for the renewable chemicals industry [34].

Lactic acid, with its two adjacent functional groups concentrated in a small molecule of three carbon atoms, can be considered as a prototype of an over-functionalized biomass-derived molecule. This chemical structure determines its high reactivity as well as its natural tendency to decompose with temperature [35]. As indicated in the Introduction, an effective approach for the conversion of biomass derivatives into advanced biofuels involves a requisite oxygen removal step that helps to reduce reactivity leaving molecule more amenable for subsequent Csingle bondC coupling upgrading processes. Following this approach, lactic acid can be converted into hydrophobic C4–C7 alcohols suitable as high energy density gasoline-compatible liquid fuels for the transportation sector (Fig. 2) [31]. In this scheme, lactic acid is first deoxygenated to generate two reactive intermediates, namely propanoic acid and acetaldehyde, by means of dehydration–hydrogenation and decarbonylation/decarboxylation processes, respectively. These intermediates were detected at low lactic acid conversions indicating that they are primary products in the synthesis [36]. Importantly, these intermediates are less reactive than lactic acid but still preserve oxygen functionality for subsequent Csingle bondC coupling upgrading. Thus, acetaldehyde can undergo self-coupling by aldol-condensation to generate butanal (after hydrogenation of the corresponding C4 unsaturated aldol-adduct), whereas propanoic acid is self-coupled into 3-pentanone via ketonization. As shown in Fig. 2, successive aldol condensations between acetaldehyde (which is present in the reactor in high amounts) and butanal and 3-pentanone products generate C6 and C7 ketones. There are several important aspects of this process: (i) lactic acid is processed solved in water which the classical medium in which this molecule is obtained after microbial fermentation of sugars; (ii) the number of reactions leading from lactic acid to C4–C7 carbonyl compounds can be carried out in a single reactor by employing a multifunctional and water-stable Pt/Nb_2O_5 catalyst in which niobic support plays a crucial role catalyzing dehydration, decarboxylation/decarbonylation and Csingle bondC coupling reactions; and (iii) C4–C7 carbonyl compounds (precursors of the corresponding alcohols by simple hydrogenation) are stored in a spontaneously separating from water organic layer accounting for 50% of the carbon in the lactic acid feed.

7.3 LEVULINIC ACID

Levulinic acid (4-oxopentanoic acid) is a high-boiling point, water-soluble biomass-derived acid that crystallizes at room temperature. Levulinic acid contains two reactive functional groups (single bondCdouble bond; length as m-dashO and single bondCOOH) that provides, as in the case of lactic acid, a rich chemistry to this compound [37]. Levulinic acid occupies a prominent place in the selected list of biomass platform molecules [11] since it is simply and inexpensively produced from lignocellulose wastes (paper mill sludge, urban waste paper, agricultural residues) by acid dehydration of C6 sugars [38]. Interestingly, equimolar amounts of formic acid

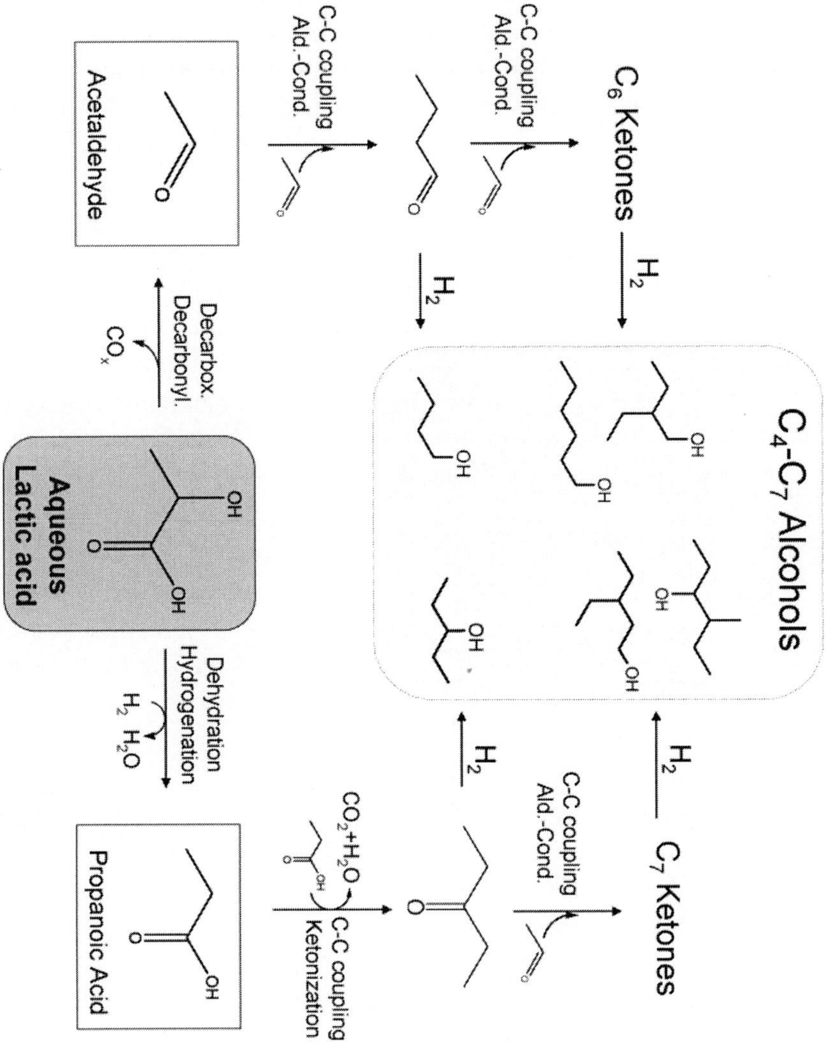

FIGURE 2: Scheme of the catalytic processing of lactic acid into C4–C7 alcohols suitable as advanced liquid biofuels.

are co-produced along with levulinic during the dehydration process. As will be described below, this formic acid can be used as a renewable internal source of hydrogen in the conversion of levulinic acid into advanced biofuels. Dehydration of sugars to levulinic acid is typically accompanied by unwanted polymerization reactions that produce intractable black insoluble materials denoted as humins. These humins are typically burnt in the industrial process to generate heat and electricity. Interestingly, C5 sugars such as xylose (the main component of hemicellulose which typically accounts for 20–30% of lignocellulose) can also serve as a source of levulinic acid. The process is not as straightforward as in the case of hexoses since it involves previous dehydration of xylose to furfural, subsequent hydrogenation to furfuryl alcohol and final hydrolysis of the latter to levulinic acid [39]. However, this route would benefit from an easier deconstruction of amorphous hemicellulose compared to highly recalcitrant crystalline cellulose (the main component of lignocellulose and the source of C6 sugars) leading to potential operation at milder acidic conditions at which formation of humins is more controlled.

Among the different processes that have been developed for the large-scale continuous production of levulinic acid, the most promising technology has been patented by the Biofine Corporation [40] and [41]. This approach utilizes a double-reactor system that minimizes the formation of by-products and the resulting separation problems. Lignocellulose is fed to the reactor along with a solution of sulfuric acid. The by-products of the process (furfural and formic acid) are condensed and separately collected, while solid humins are removed from the levulinic acid solution and collected as combustible wastes. This technology allows production of levulinic acid at low cost (0.06–0.18 €/kg) [42], making this compound suitable for use as a platform molecule. There are currently several plants operating with tons of waste cellulosic materials per day, both in the U.S. [42] and in Europe [43].

As shown in Fig. 3, levulinic acid possesses great potential for the production of advanced biofuels of diverse classes. Most of the routes involve intermediate formation of γ-valerolactone (GVL), a stable and water-soluble biomass derivative with potential to be blended with gasoline as well as to serve as a precursor of polymers and fine chemicals [44] and [45]. The reduction of levulinic acid to GVL is thus a process with interest, and several routes involving different catalysts and hydrogen sources for this reduction have been explored in recent years. The most simple route involves utilization of carbon-supported noble metal catalysts under hydrogen pressure [46], [47] and [48] which achieves near quantitative yields of GVL at mild temperatures (e.g. 150 °C) with slight deactivations with time on stream. The utilization of mild temperature conditions and non-acidic supports such as carbon is crucial to direct synthesis through 4-hydroxypentanoic acid which subsequently undergoes highly favorable internal esterification to the five member ring GVL compound [49]. The utilization of higher temperatures and/or acidic catalysts promotes dehydration of levulinic acid to α-angelica lactone. This compound polymerizes readily over acidic surfaces leading to loss of carbon as coke and severe catalyst deactivation.

As remarked above, the formic acid co-produced in the sugar dehydration process can serve as a internal source of hydrogen for the reduction of levulinic acid to GVL [13], [50], [51], [52], [53], [54] and [55]. These technologies take advantage of the easy decomposition of formic acid into CO_2 and H_2 to generate an in situ source of this gas within the reaction system. Interestingly, the same materials used for reduction of levulinic acid to GVL are also able to catalyze decomposition of formic acid. In this sense, excellent GVL yields (96%) have been reported by Deng and co-workers in two different works involving homogenous [13] and heterogeneous [51] Ru-based catalysts. This ability of Ru to quantitatively convert levulinic acid into GVL via formic acid decomposition has been utilized to expand this route to hexoses thereby allowing production of GVL from biomass sugars in a one-pot process and without hydrogen requirements [52]. The process, however, requires additional utilization of an acidic medium which serves as a catalyst for sugar dehydration, and GVL yields obtained, limited by the sugar dehydration process, are typically modest (50%).

Since industrial manufacturing of levulinic acid involves treatment of biomass with aqueous sulfuric acid, it would be interesting to find a catalyst that can efficiently transform aqueous sulfuric acid streams of levulinic acid and formic acid into GVL without the need for previous and waste-producing neutralization steps. In this sense, aqueous sulfuric acid solutions of levulinic and formic acids, obtained after acid hydrolysis of

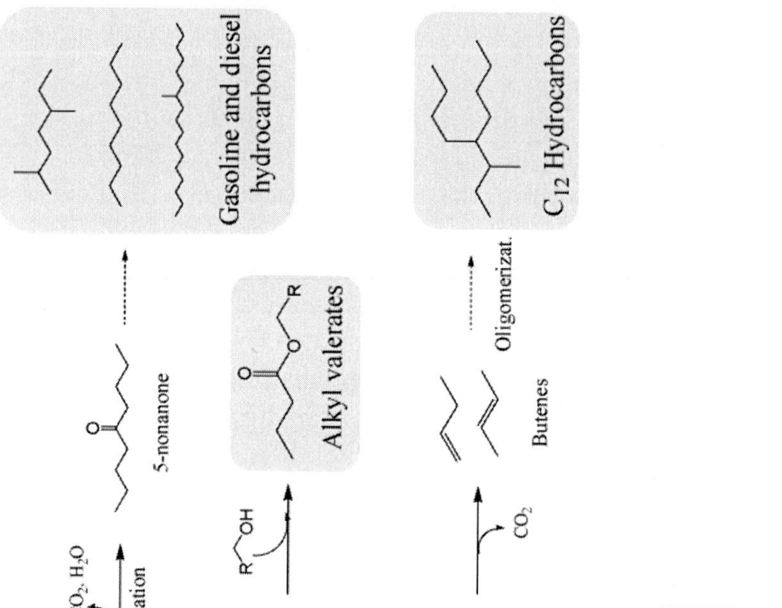

FIGURE 3: Main catalytic routes for the conversion of levulinic acid into advanced biofuels.

solid cellulose, can be transformed into GVL with acceptable yields over Ru/C and Ru-Re/C catalysts [53] and [56]. Importantly, formation of GVL allows the design of strategies for the recycling of most of the sulfuric acid utilized for biomass depolymerization and sugar dehydration processes. Thus, alkylphenol solvents, with superior abilities to selectively extract GVL from aqueous sulfuric acid solutions, have been recently proposed for this task [57].

As summarized in Fig. 3, GVL presents high versatility to synthesize advanced transportation biofuels of diverse classes. The most direct route involves conversion into methyltetrahydrofuran (MTHF) via hydrogenation to 1,4-pentanediol over metal catalysts at moderate temperatures (250 °C) and subsequent dehydration of the diol to yield the cyclic ether [42]. The process takes advantage of the natural tendency of 1,4 diols to undergo dehydration/cyclation with temperature ($\Delta G = -73$ kJ/mol for 1,4-pentanediol at 250 °C) to afford MTHF from levulinic acid with high yields (83%). MTHF is a hydrophobic molecule which, unlike ethanol, can be blended with gasoline up to 60% (v/v) without adverse effects on engine performances or gas mileage and can be distributed by existing pipeline for hydrocarbons without water contamination. MTHF is one of the components of the so-called P-series fuels which are approved by the US DOE for use in gasoline vehicles.

One route that is gaining interest in recent years involves transformation of GVL into pentanoic acid. The process involves acid-catalyzed ring opening of GVL to pentenoic acid and subsequent hydrogenation of the latter over bifunctional (metal and acid) catalysts at moderate temperatures and hydrogen pressures [48] and [58]. The formation of pentanoic acid achieves reduction of the oxygen content of levulinic acid thereby producing a less-reactive intermediate which is more appropriate for new upgrading strategies to larger compounds. For example, Lange and coworkers [59] have used this route to produce the so-called valeric biofuels (i.e. alkyl valerates). Valeric biofuels can be used in conventional engines without any modification since they present similar energy-density, polarity and volatility-ignition properties than hydrocarbon fuels. The process is flexible in that by varying the alkyl chain length the fuels can be adapted to fit in both gasoline and diesel engines. The main drawback of this technology lies in the need for external alcohol source for esterification.

Alternatively, liquid hydrocarbon fuels appropriate for gasoline and diesel applications can be produced via 5-nonanone, the ketonization product of pentanoic acid (Fig. 3). Interestingly, 5-nonanone can be produced in high yields (70%) from aqueous GVL over a single bed of Pd/Nb_2O_5 catalyst in which niobic support catalyzes GVL ring opening and pentanoic ketonization reactions [48]. Nonanone yield can be increased to almost 90% by using a double-bed reactor configuration with Pd/Nb2O5 + $Ce_{0.5}Zr_{0.5}O_2$ operating at two different temperature zones (325 and 425 °C) which allows for optimum control of reactivity [53]. As shown in Fig. 4, the C9 ketone, which is obtained in high yields stored in an organic layer that spontaneous separates from water, can be upgraded to liquid hydrocarbon fuels by means of well-known petroleum-based chemistry. For example, by consecutive cycles of hydrogenation/dehydration over a bifunctional metal-acid catalyst such as Pt/$NbPO_4$, 5-nonanone can be converted into n-nonane which possesses excellent cetane number, lubricity and cloud point properties to be used as a blender agent for winter diesel applications. Alternatively, 5-nonanol, obtained by the hydrogenation of the C9-ketone, can be dehydrated and isomerized in a single step over an USY zeolite catalyst to produce a mixture of branched C9 alkenes with the appropriate molecular weight and structure for use in gasoline after hydrogenation to the corresponding alkanes [53]. Larger hydrocarbons such as those required for diesel vehicles can be produced from the nonanone stream by means of oligomerization reactions of the previously formed C9-alkenes over an acid catalyst such as Amberlyst 70 [60]. This process allows conversion of approximately half of the mass of GVL into C18 alkenes which retain more than 90% of its energy content.

Aviation requires fuels with high energy density (to allow storage of large amounts of fuels in tanks with a size determined by aircraft design) and with extremely low cloud points (to ensure operational use at high altitude temperatures). Branched hydrocarbons in the C9–C16 range meet those requirements and, consequently, routes for the production of these compounds from biomass sources are highly valuable. Recently, a promising route to upgrade aqueous solutions of GVL into jet fuels through the formation of C4 alkenes has been developed by Bond et al. [23] (Fig. 3). In this process, GVL undergoes decarboxylation at elevated pressures (e.g. 36 bar) over a inexpensive silica/alumina catalysts, producing a clean gas

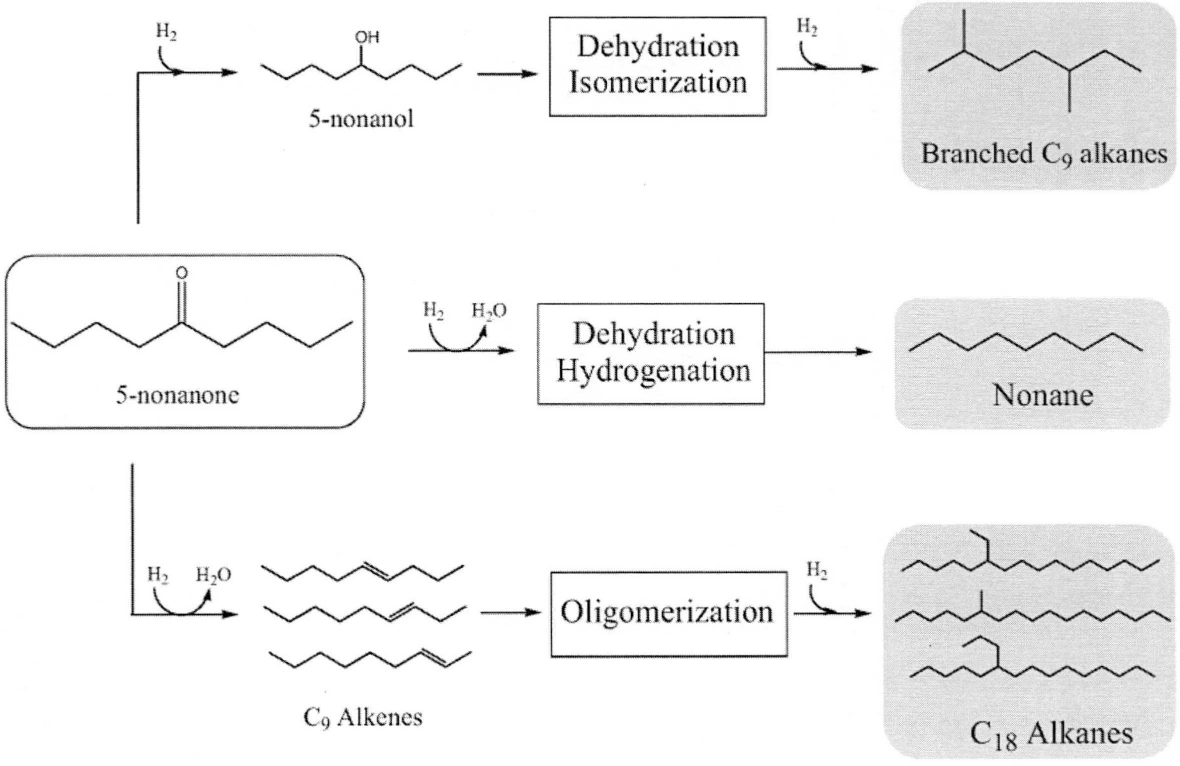

FIGURE 4: 5-Nonanone as a platform for the production of liquid hydrocarbon fuels for gasoline and diesel applications.

stream composed of butenes isomers and CO_2. This gaseous stream is then passed over an acidic catalyst (H-ZSM5, Amberlyst) that achieves oligomerization of butenes yielding a distribution of branched alkenes centered at C12 suitable to be used as jet fuels after hydrogenation. This technology presents important economic and environmental advantages: (a) minimum amounts of external hydrogen are required in the process, (b) precious metal catalysts are not required, (c) a gas stream of pure CO_2 is produced at the elevated pressures, thereby permitting effective utilization of sequestration or capture technologies to mitigate greenhouse gas emissions.

H_2 is the typical reagent employed for levulinic acid deoxygenation to advanced biofuels, and efforts are currently aimed to obtain this gas from biomass sources. Electricity, based on renewable wind or solar power, could be alternatively used for electrochemical deoxygenation of biomass to fuels. Nilges and co-workers have recently developed an approach for the electrochemical conversion of levulinic acid into octane [61]. As shown in Fig. 3, the process involves the combination of electroreduction of aqueous levulinic acid into pentanoic acid and subsequent oxidative Csingle bondC coupling of the latter (e.g. Kolbe reaction) to octane. Apart from the spontaneous separation of octane from the water medium, the use of electricity gives this process a number of green characteristics including the use of mild conditions (room temperature and water solutions), the replacement of any reducing chemical agent by immaterial electrons, and the minimization of waste generation. Although this preliminary work still has to solve some issues such as the effect of potential impurities in levulinic acid (from biomass depolymerization steps) on the electrochemical reactions and the design of more efficient electrodes for large-scale applications, it represents an interesting approach for converting the unused electricity generated during overproduction cycles (typical of fluctuating electricity production profiles of wind and photovoltaics) into a storable biofuel form.

Deoxygenation of levulinic acid to advanced biofuels can be carried out without utilization of hydrogen by means of a pure thermal pyrolysis treatment. The process is denoted as thermal deoxygenation (TDO) and involves previous formation of calcium levulinate and heating of the latter to temperatures ranging 350–450 °C under inert atmosphere [62] (Fig. 3). At these conditions, calcium levulinate simultaneously condenses (by ketonization) and deoxygenates (by internal cyclation and dehydration) leading to the production of a broad product distribution of cyclics and aromatics with very low oxygen content. Production of aromatics, which are valuable components of gasoline and jet fuels, can be maximized by operating at higher temperatures. The process has been recently improved by addition of equimolar amounts of calcium formate which serves as an in situ hydrogen source allowing a deeper TDO process leading to the formation of a petroleum-like oil which could be further processed in existing refinery facilities [63].

7.4 CONCLUSIONS

Our society is highly dependent on fossil fuels, which are non-renewable and contribute to global warming. The conversion of biomass into fuels for the transportation sector can help to partially alleviate this reliance. Biodiesel and bioethanol, the main biofuels used today, present serious compatibility issues which can be overcome by the production of advanced biofuels such as higher alcohols and green hydrocarbons which are fully compatible with our existing hydrocarbons-based transportation infrastructure.

However, working with complex biomass feedstocks is difficult and approaches based on the formation of simpler and more stable intermediate derivatives, denoted as platform molecules, have been shown to be effective for efficient biomass conversion to fuels and chemicals. Lactic acid and levulinic acid are two of these relevant biomass derivatives that can be transformed into advanced biofuels by a number of catalytic routes involving deoxygenation reactions combined with Csingle bondC coupling processes. The present paper offers a state of the art overview of the most relevant catalytic strategies available today for this paradigmatic conversion.

REFERENCES

1. Directive 2003/30/EC of the European Union parliament Off. J. Eur. Union (2003) http://ec.europa.eu/energy/res/legislation/doc/biofuels/en_final.pdf
2. President Bush State on the Union Address. White House (2007) http://usgovinfo.about.com/b/2007/01/23/bush-delivers-his-seventh-state-of-the-union-address.htm
3. A.J. Ragauskas et al. Science, 311 (2006), p. 484
4. US Energy Information Administration (EIA) International Energy Outlook (2010) http://www.eia.doe.gov/oiaf/ieo (accessed December 2011)
5. R. Luque, L. Herrero-Davila, J.M. Campelo, J.H. Clark, J.M. Hidalgo, D. Luna, J.M. Marinas, A.A. Romero Energy Environ. Sci., 1 (2008), p. 542
6. J.R. Regalbuto Science, 325 (2009), p. 822
7. J.R. Regalbuto Biofuels Bioprod. Bioref., 5 (2011), p. 495
8. P. Durre Biotechnol. J., 2 (2007), p. 1525
9. N. Savage Nature, 474 (2011), p. S9
10. J.C. Serrano-Ruiz, J.A. Dumesic Energy Environ. Sci., 4 (2011), p. 83
11. J.J. Bozell, G.R. Petersen Green Chem., 12 (2010), p. 539
12. H. Heeres, R. Handana, D. Chunai, C.B. Rasrendra, B. Girisuta, H.J. Heeres Green Chem., 11 (2009), p. 1247
13. L. Deng, J. Li, D.M. Lai, Y. Fu, Q.X. Guo Angew. Chem. Int. Ed., 48 (2009), p. 6529
14. D. Simonetti, J.A. Dumesic ChemSusChem, 1 (2008), p. 725
15. E.F. Iliopoulou Curr. Org. Synth., 7 (2010), p. 587
16. D.M. Alonso, J.Q. Bond, J.A. Dumesic Green Chem., 12 (2010), p. 1493
17. G.W. Huber, J. Chheda, C. Barret, J.A. Dumesic Science, 308 (2005), p. 1447
18. R.M. West, Z.Y. Liu, M. Peter, J.A. Dumesic ChemSusChem, 1 (2008), p. 417
19. R. Weingarten, J. Cho, W.C. Conner Jr., G.W. Huber Green Chem., 12 (2010), p. 1423
20. M. Renz Eur. J. Org. Chem. (2005), p. 979
21. C.A. Gartner, J.C. Serrano-Ruiz, D.J. Braden, J.A. Dumesic J. Catal., 266 (2009), p. 71
22. E.L. Kunkes, D.A. Simonetti, R.M. West, J.C. Serrano-Ruiz, C.A. Gartner, J.A. Dumesic Science, 322 (2008), p. 417
23. J.Q. .Bond, D. Martin-Alonso, D. Wang, R.M. West, J.A. Dumesic Science, 327 (2010), p. 1110
24. A. Corma, O. de la Torre, M. Renz, N. Villandier Angew. Chem., 123 (2011), p. 2423
25. R. Datta, M. Henry J. Chem. Technol. Biotechnol., 81 (2006), p. 1119
26. M. Ilmen, K. Koivuranta, L. Ruohonen, P. Suominen, M. Penttila Appl. Environ. Microbiol., 73 (2007), p. 117
27. M.S. Holm, S. Saravanamurugan, E. Taarning Science, 328 (2010), p. 602
28. R. Datta Hydrocarboxylic acids Kirk–Othmer Encyclopedia of Chemical Technology, vol. 14John Wiley & Sons (2004) pp. 114–134
29. W.S.L. Mok, M.J. Antal, M. Jones J. Org. Chem., 54 (1989), p. 4596
30. R.A. Sawicki, 1998. U.S. Patent No. 4,729,978.
31. J.C. Serrano-Ruiz, J.A. Dumesic Green Chem., 11 (2009), p. 1101
32. G.C. Gunter, D.J. Miller, J.E. Jackson J. Catal., 148 (1994), p. 252

33. L. Lunt Polym. Degrad. Stab., 59 (1998), p. 145
34. C.H. Christensen, J. Rass-Hansen, C.C. Marsden, E. Taarning, K. Egeblad ChemSusChem, 1 (2008), p. 283
35. C.H. Fisher, E.M. Filachione Properties and Reactions of Lactic Acid – A Review U.S. Department of Agriculture (1951)
36. J.C. Serrano-Ruiz, J.A. Dumesic ChemSusChem, 2 (2009), p. 581
37. B.V. Timokhin, V.A. Baransky, G.D. Eliseeva Russ. Chem. Rev., 68 (1999), p. 73
38. D.W. Rackemann, W.O.S. Doherty Biofuels Bioprod. Bioref., 5 (2011), p. 198
39. J.P. Lange, W. van de Graaf, R. Haan ChemSusChem, 2 (2009), p. 437
40. S.W. Fitzpatrick, 1997. U.S. Patent No. 5,608,105.
41. S.W. Fitzpatrick, P.J. Bilski, J.L. Jarnefeld, 1990. World Patent 8,910,362.
42. J.J. Bozell, L. Moens, D.C. Elliott, Y. Wang, G.G. Neuenscwander Resour. Conserv. Recycl., 28 (2000), p. 227
43. M.H.B. Hayes Nature, 443 (2006), p. 144
44. I.T. Horvath, H. Mehdi, V. Fabos, L. Boda, L.T. Mika Green Chem., 10 (2008), p. 238
45. J.P. Lange, J.Z. Vestering, R.J. Haan Chem. Commun. (2007), p. 3488
46. Z.P. Yan, L. Lin, S.J. Liu Energy Fuels, 23 (2009), p. 3853
47. P.P. Upare, J.-M. Lee, D.W. Hwang, S.B. Halligudi, Y.K. Hwang, J.-S. Chang J. Ind. Eng. Chem., 17 (2011), p. 287
48. J.C. Serrano-Ruiz, D. Wang, J.A. Dumesic Green Chem., 12 (2010), p. 574
49. J.C. Serrano-Ruiz, R.M. West, J.A. Dumesic Annu. Rev. Chem. Biomol. Eng., 1 (2010), p. 79
50. H. Mehdi, V. Fábos, R. Tuba, A. Bodor, L.T. Mika, I.T. Horvath Top. Catal., 48 (2008), p. 49
51. L. Deng, Y. Zhao, J. Li, Y. Fu, B. Liao, Q.X. Guo ChemSusChem, 3 (2010), p. 1172
52. H. Heeres, R. Handana, D. Chunai, C. Borromeus, C. Rasrendra, B. Girisuta, H.J. Heeres Green Chem., 11 (2009), p. 1247
53. J.C. Serrano-Ruiz, D.J. Braden, R.M. West, J.A. Dumesic Appl. Catal., B, 100 (2010), p. 184
54. E.I. Gürbüz, D.M. Alonso, J.Q. Bond, J.A. Dumesic ChemSusChem, 4 (2011), p. 357
55. D. Kopetzki, M. Antonietti Green Chem., 12 (2010), p. 656
56. D.J. Braden, C.A. Henao, J. Heltzel, C. Maravelias, J.A. Dumesic Green Chem., 13 (2011), p. 1755
57. D.M. Alonso, S.G. Wettstein, J.Q. Bond, T.W. Root, J.A. Dumesic ChemSusChem, 4 (2011), p. 1078
58. P.M. Ayoub, J.P. Lange, Processes for converting levulinic acid into pentanoic acid, WO 2008/142 127 patent to Shell Internationale, 2008.
59. J.P. Lange, R. Price, P.M. Ayoub, J. Louis, L. Petrus, L. Clarke, H. Gosselink Angew. Chem. Int. Ed., 49 (2010), p. 4479
60. D.M. Alonso, J.Q. Bond, J.C. Serrano-Ruiz, J.A. Dumesic Green Chem., 12 (2010), p. 992
61. P. Nilges, T.R. dos Santos, F. Harnisch, U. Schröder Energy Environ. Sci., 5 (2012), p. 5231

62. T.J. Schwartz, A.R.P. van Heiningen, M.C. Wheeler Green Chem., 12 (2010), p. 1353
63. P.A. Case, A.R.P. van Heiningen, M. Clayton Wheeler Green Chem., 14 (2012), p. 85

Serrano-Ruiz, J. C., Pineda, A., Balu, A. M., Luque, R., Campelo, J. M., Romero, A. A., and Ramos-Fernández, J. M. Catalytic Transformations of Biomass-Derived Acides into Advanced Biofuels. Catalysis Today 2012:1; 162–168. Copyright © 2012. Elsevier B.V. All rights reserved. Reprinted with permission.

CHAPTER 8

USE OF ANION EXCHANGE RESINS FOR ONE-STEP PROCESSING OF ALGAE FROM HARVEST TO BIOFUEL

JESSICA JONES, CHENG-HAN LEE, JAMES WANG, and MARTIN POENIE

8.1 INTRODUCTION

Some strains of algae show promise as a sustainable source of biofuel due to their rapid growth, ability to grow on non-arable land, and high triacylglycerol (TAG) content. TAGs can be easily converted to biodiesel, which is compatible with current fuel infrastructure [1]. Biodiesel has a higher energy density than ethanol, yet it is relatively non-toxic, biodegradable, and produces lower exhaust emissions than petroleum-based fuels, making it is one of the most attractive forms of alternative energy [2,3]. Additionally, when derived from plant sources such as algae, biodiesel has the potential of a near-neutral carbon footprint [4], though considerable work must be done in order to realize this goal in an economically and environmentally-feasible manner [5].

While algae are promising, there are technical challenges that currently make it prohibitively expensive as a source of fuel. Algae grow at dilute concentrations, generally less than 1 g/L [6], so it must be concentrated before it can be processed. Most concentration processes require pumping the dilute algal suspension and may involve other energy intensive steps such as centrifugation [7], compressors for dissolved air flotation, or else treatment of large volumes of water with chemicals (flocculation). Centrifugation, for example, could account for 30% of processing costs [8].

Once algae have been harvested, there are additional processing steps that often include lysis and drying of the algae followed by extraction with organic solvents to obtain the neutral lipids, primarily TAGs, that can then be converted to biodiesel [3]. Drying can be expensive but it facilitates the interaction between solvent and algae to improve extraction efficiency. However, the extraction solvent must be removed and recovered prior to conversion of lipids to biodiesel and there are attending questions about pollution of the air and contamination of the biomass with solvents. This is also considered a large part of the cost in processing algae [5].

Once the lipids have been isolated, they can be converted to biodiesel by either acid- or base-catalyzed transesterification, typically with methanol, to yield fatty acid methyl esters (FAMEs) and glycerol [9]. The transesterification reaction is sensitive to water [10–14] but this is not normally a problem since oils, such as would be obtained by hexane extraction, normally contain little water. When the entire processes is analyzed for cost and energy expenditure, some current projections suggest that the energy spent in the cultivation, harvest, and extraction of oil from algae could be greater than that gained from the product [5] along with the possibility of significant environmental impact [15]. It is generally agreed that for algae to be economically feasible, improvements in technology will be needed.

Efforts to reduce cost and simplify processing of algal biomass to biofuel have led to studies of direct conversion of dry or even wet algal biomass to biodiesel. Previous studies have shown that algal lipids can be transesterified in situ by adding reagents to a dried sample of algae [11,16–19]. One study examined a two-step procedure where the acyl groups of component lipids were hydrolyzed with base and then re-esterified in excess sulfuric acid/methanol [20]. This procedure gave greater amounts FAME than obtained by lipid extraction followed by transesterification. Another study showed that direct acid-catalyzed transesterification of wet biomass can produce FAME yields similar to that of dried biomass, although FAME compositions differed [11].

In this study we show that anion exchange resins such as Amberlite can concentrate and dewater algae (i.e., harvest algae) and then be eluted with 5% sulfuric acid/methanol reagent. The eluted algae appear to dissolve in the sulfuric acid reagent and esterified fatty acids are converted to FAMEs (biodiesel). This one step harvesting and transesterification process can potentially eliminate many of the costly steps of processing algae to biofuel.

8.2 RESULTS AND DISCUSSION

8.2.1 ALGAE ARE HARVESTED AND CONCENTRATED ONTO AMBERLITE ANION EXCHANGE RESIN

Amberlite CG-400, a divinylbenzene-based resin containing quaternary ammonium groups (3.8 mmol/g [21]), has been previously shown to bind and concentrate two different species of green algae for removal from the water supply [22]. Here, we used Amberlite to concentrate and dewater biomass from two potentially high oil-producing green algal species, *Neochloris oleoabundans* [23] and KAS 603 [24]. For routine comparisons, algal suspension was passed through the column until saturation was achieved. Algae appeared to bind on contact, initially accumulating as a band at the top of the resin bed. As more algae were added, the upper green layer progressively expanded down to the bottom of the resin bed. Prior to the resin bed becoming solid green, the flow-through was clear and colorless. Once the resin bed became solid green, color quickly began to appear in the effluent and this was taken as the saturation point.

To determine the binding capacity of the resin for *Neochloris* and KAS 603, the OD680 of the algal suspension was measured before and after passing the algae through an Amberlite column. The concentration of algae in the initial suspension and the in flow-through were determined from algal OD680 absorbance versus dry cell weight (DCW) calibration curves generated for both types of algae. The flow-through included the solution from the algae algal suspension that went through the resin and the subsequent wash to remove unbound algae. The total amount of algae bound to the resin was then determined by subtracting the amount of algae in the flow-through from the total that was added to the column.

For all the experiments algal binding to the resin was tested at pH 7, although there was no significant difference in binding over the pH range of 6–9. Since for Amberlite, algal binding decreases with increasing ionic strength, binding studies for both algae were tested under freshwater conditions (salinity 5 psu). The algal suspensions were used at 0.4 g/L, the value that we typically obtain in our simple airlift photobioreactors. When more dilute algal suspensions were tested, saturation of the resin was still achieved but, as would be expected, the volume of algal suspension and time required to achieve saturation increased. Since algae can be entrained in the resin beads, our protocol included a washing step between binding and elution to ensure only algae that were actually bound to the resin were counted.

Figure 1(a) gives a schematic for resin-binding and elution experiments shown in Figures 1–3. Algal binding capacity was determined for each cycle of algal binding. Reusability of the resin was assessed in terms of loss of binding capacity over multiple cycles of algal binding and elution. These experiments were carried out either using only fresh sulfuric acid/methanol reagent to elute the algae (Series A, "fresh") or using sulfuric acid/methanol that was recycled after the first use (Series B, "used").

Although the flow through approach in Figure 1(a) gave repeatable values for saturation and was used for all experiments except that shown in Figure 4, we later found that longer contact times, which involved stirring excess algae with resin for up to 15 min, gave somewhat larger binding capacities than those obtained using the flow through method. However, while the binding capacity numbers would be improved with longer contact times, the patterns remain the same.

The results showed that the binding capacity for *Neochloris* was 37.0 mg/g resin whereas that for KAS 603 was 12.8 mg/g resin [Figure 1(b)]. In terms of efficiency of removal of algae from 1 L of 0.4 g/L suspension, 10 g of Amberlite will sequester 93% of the *Neochloris*, whereas 30 g of Amberlite will sequester 96% of the KAS603 during a single pass through the column. Although the difference in binding capacity could possibly be attributed to differences in surface charge density, studies to be detailed elsewhere show that the nature of the resin backbone, especially its hydrophobicity has a large impact on binding capacity [25]. In resins based on a methacrylate backbone, both *Neochloris* and KAS 603 showed similar

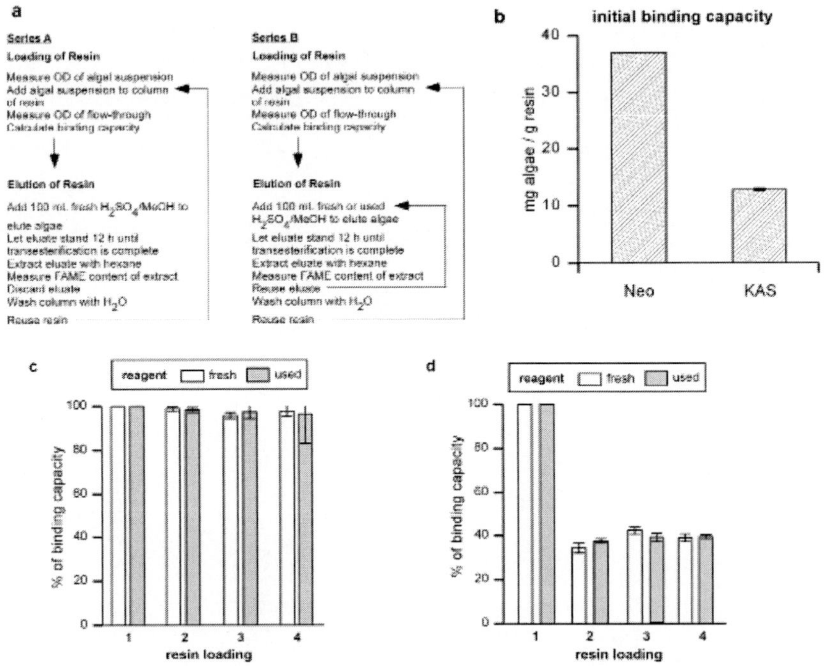

FIGURE 1: Algal binding capacity of Amberlite and resin reuse. (a) Experiment series for determining binding capacity and reusability of resin and transesterification reagent. (b) Binding capacities of Amberlite (mg algal dry weight / gram resin) were determined for Neochloris and KAS 603. (c) Neochloris or (d) KAS 603 were loaded onto resin columns and eluted with 100 mL of 5% sulfuric acid/methanol reagent. After washing the column with distilled water, the operation was repeated for three more cycles, eluting with either fresh methanol sulfuric acid reagent (fresh) or reusing the previously used reagent (used). The binding capacity of the resin was determined for each cycle of algal loading onto the resin.

binding capacities. Furthermore, binding capacity was greatly increased with the inclusion of glycol groups, which do not alter the charge characteristics of the resin but make it more hydrophilic.

After binding, algae were eluted with 100 mL of 5% sulfuric acid in methanol. This treatment visibly removed algae from the resin and regenerated the resin. It appears that most of the algae was dissolved by the sulfuric acid/methanol reagent as shown by filtering an eluate of

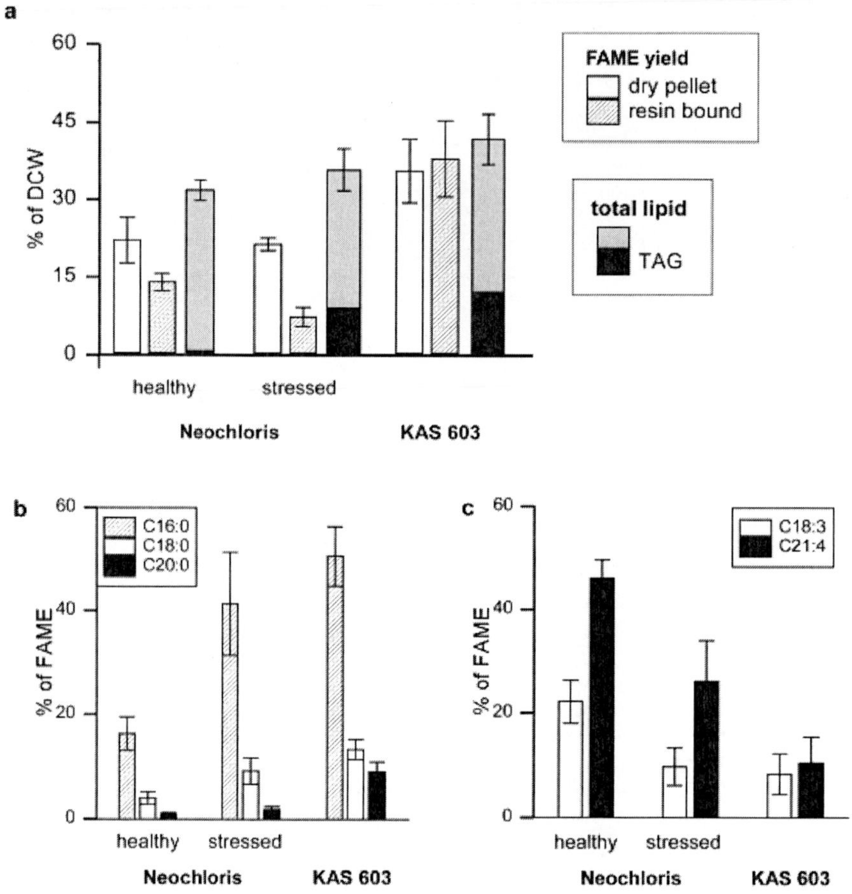

FIGURE 2: FAME yield and characterization by resin-bound transesterification. (a) FAMEs from normal or stressed (nitrogen-starved) Neochloris and KAS 603 were generated either by elution of resin-bound algae with 5% sulfuric acid/methanol reagent or by subjecting dry algal pellets subjected to sequential base hydrolysis followed by treatment with sulfuric acid in methanol to esterify free fatty acids. The FAME yields were expressed as percent of dry cell weight. The weight of a crude lipid extract from parallel batches of algae and the TAG content those extracts (determined by HPLC) are given for comparison. The FAME prepared from resin-bound algae were analyzed by HPLC/MS to determine the abundance of (b) saturated C16:0, C18:0, and C20:0 and (c) unsaturated C18:3 and C21:4 acyl constituents relative to total FAME derived.

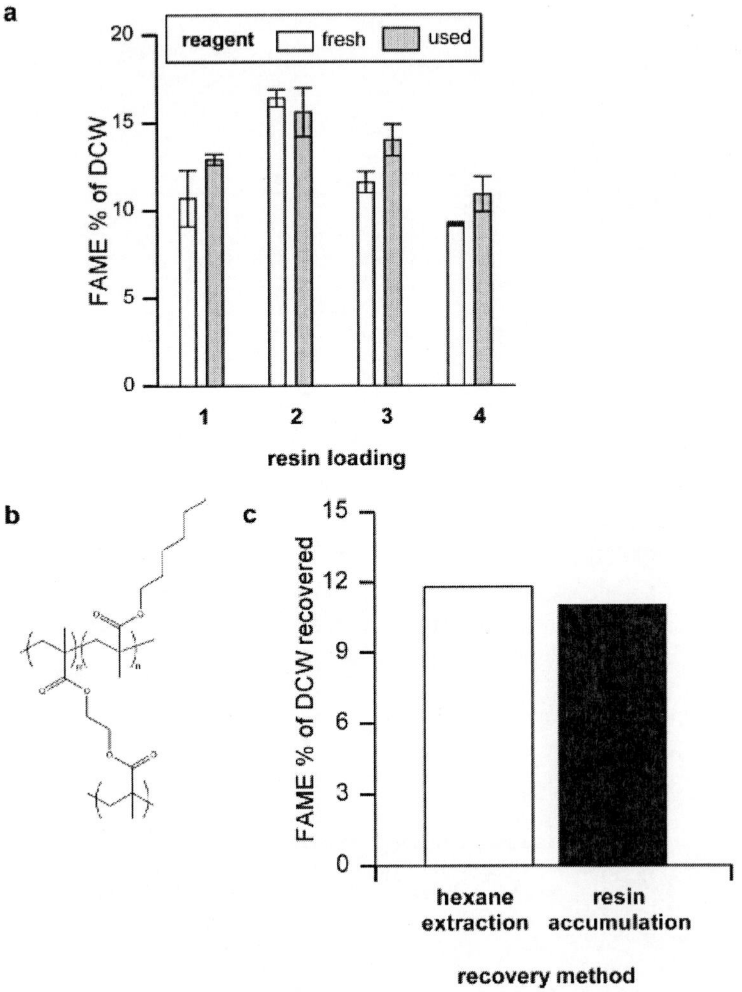

FIGURE 3: Efficiency of FAME production using recycled reagent, and FAME removal. (a) Algal were loaded onto Amberlite and eluted with 5% sulfuric acid/methanol four times in succession using either fresh reagent or reusing the old reagent after the initial elution. Twelve h after each elution the FAME was extracted with hexane, solvent was removed and the amount of FAME determined as a percentage of algal dry cell weight; (b) The sulfuric acid/methanol reagent containing algal FAME was either extracted with hexane or, for comparison, passed over a hydrophobic resin column composed of 80% EGDMA and 20% HMA; (c) FAME relative to total dry algal weight obtained by hexane extraction or after eluting FAME bound to the hydrophobic column.

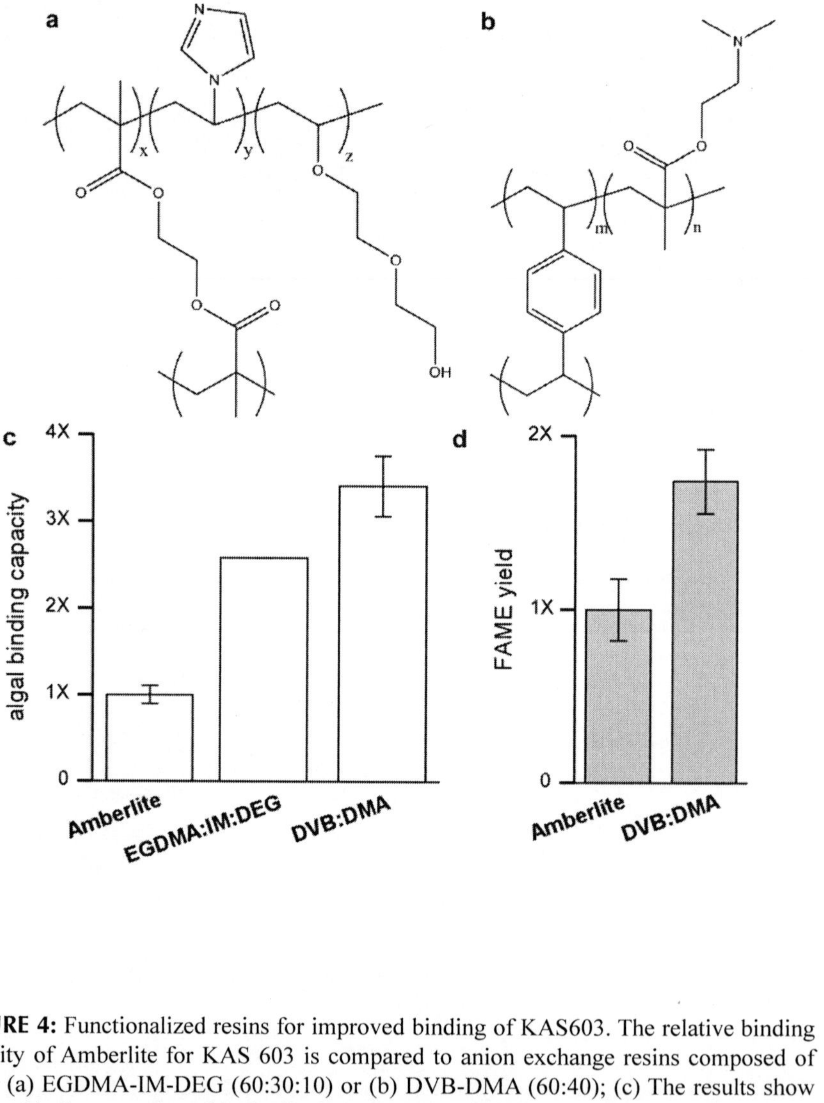

FIGURE 4: Functionalized resins for improved binding of KAS603. The relative binding capacity of Amberlite for KAS 603 is compared to anion exchange resins composed of either (a) EGDMA-IM-DEG (60:30:10) or (b) DVB-DMA (60:40); (c) The results show that the binding capacity of Amberlite is low compared to the other resins; (d) In addition to binding more algae, subsequent elution of either Amberlite or DVB-DMA shows that greater amounts of FAME are produced from the DVB:DMA resin.

Neochloris through a pre-weighed PTFE filter (Millipore Omnipore, pore size 0.1 μm). Only 6.1% of the original DCW was recovered on the filter.

In order to determine how completely the resin was regenerated, resin columns were loaded with *Neochloris* [Figure 1(c) "fresh"] or KAS 603 [Figure 1(d) "fresh"] and eluted with fresh portions of sulfuric acid/methanol four times in succession. For comparison, parallel experiments were carried out where, after the initial binding and elution using fresh sulfuric acid/methanol solution, the reagent was then recycled for the three remaining cycles of algal binding and elution for both *Neochloris* [Figure 1(c) "used"] and KAS 603 [Figure 1(d) "used"]. With each cycle, the amount of algae bound to the resin was determined based on the OD680 as described above. Using fresh reagent, *Neochloris* after the initial binding and elution, binding capacity dropped to 97% of its initial value and remained constant thereafter. Using fresh reagent with KAS 603, binding dropped to 39% of its initial value after the first cycle but remained constant thereafter. When the experiments were carried out using the same sulfuric acid/methanol reagent for each cycle, the results were comparable to those obtained using the fresh reagent [Figures 1(c) and 1(d) "used").

The results show that the sulfuric acid/methanol reagent can be used for multiple cycles before it is consumed or rendered ineffective. Eventually methanol will be consumed while products such as glycerol, FAME, and other molecules will build up in the recycled sulfuric acid/methanol reagent. Unused methanol, being quite volatile would be easy to remove from the mixture. FAME, as we detail below, can also be easily removed. It will be interesting to see if the recycled reagent can be further fractionated and if potentially high value compounds might also be isolated from the mixture. The drop in binding capacity of Amberlite for KAS 603 is not understood. It is not due to the action of the sulfuric acid/methanol reagent alone since pretreatment of the resin with sulfuric acid/methanol prior to initial binding of algae does not lower the binding capacity. A similar drop in binding capacity is not seen with the high capacity anion exchange resins based on a more polar methacrylate backbone [25]. Perhaps then the best explanation relates to the hydrophobicity of the divinylbenzene that interacts with hydrophobic components in the algae thereby either masking some of the charges on the resin or sterically interfering with binding.

8.2.2 CONVERSION OF ALGAL LIPIDS TO FAME

Since algae were eluted off the resin by 5% sulfuric acid in methanol, a reagent that catalyzes the transesterification of esterified fatty acid to FAMEs, tests were carried out to measure conversion of lipids in the eluate to FAME. To quantify FAME and other lipids, normal-phase HPLC was used in conjunction with an evaporative light scattering detector and mass spectrometry (HPLC-ELSD/MS;) [24]. The advantage here is that one can rapidly measure the amount of FAME generated, its fatty acid composition, and the amount of residual triacylglycerol starting material present in the reaction product as well as in crude total lipid extracts. In the reaction product, the presence of residual TAG is an indicator that the transesterification reaction did not reach completion. Preliminary tests carried out by extracting the sulfuric acid methanol eluate with hexane to obtain the products at various times after elution showed that 12 h at room temperature was sufficient to consume all the TAG. What remains in the extract are mainly saturated hydrocarbons, which have been characterized in more detail elsewhere [24], and FAME.

As a reference, algal lipids were converted to FAME by a two step procedure that entailed treatment of dry algal pellets with base to hydrolyze fatty acid esters followed by re-esterification of free fatty acids in sulfuric acid and methanol [26]. The results showed 21.7%, 20.9%, and 35.2% of dry weights were recovered as FAME for healthy *Neochloris*, stressed *Neochloris*, and KAS 603 respectively [Figure 2(a)]. Acid-catalyzed transesterification of resin-bound algae resulted in 13.6%, 6.9%, and 37.6% of dry weight recovered as FAME for healthy *Neochloris*, stressed *Neochloris*, and KAS 603 respectively. For comparison, FAME synthesis yields are shown alongside total lipid extract amounts [Figure 2(a)]. Crude lipid extract constituted 31.4% of total dry weight for healthy *Neochloris*, 35.4% for stressed *Neochloris*, and 41.3% for KAS 603. HPLC analysis of the crude lipid extract showed that TAG constituted 0.4%, 8.6%, and 11.7% of dry weight for healthy *Neochloris*, stressed *Neochloris*, and KAS 603, respectively.

For KAS 603 comparable yields of FAME were obtained using either method and the total FAME was close to the weight of total lipid. For

Neochloris, both methods generated substantially less FAME than total lipid and the resin-bound algae yielded only 60% of the FAME generated from the dried pellet. Clearly, substantially more FAME can be generated by direct transesterification than can be accounted for by TAG alone. For healthy *Neochloris*, there was hardly any TAG present in the extracts yet 15%–20% of the DCW could be recovered as FAME. For KAS 603, 10% of the DCW was present as TAG but nearly 40% of the DCW was recovered as FAME. These data suggest that much of the FAME is derived from polar lipids such as glycolipids and phospholipids.

One of the surprising results of this study is the finding that *Neochloris* accumulates high amounts of TAG when subjected to nitrogen deprivation [23], yet with only a modest increase in total lipid. While TAG increased nearly ten-fold, total lipid still constituted approximately 20% of dry weight, and FAME yield from dried biomass was comparable between healthy and stressed *Neochloris*. The most notable difference between FAME generated from healthy and stressed *Neochloris* was found in the fatty acid composition. This can be seen by analyzing positive mode APCI mass spectra of the various FAME reactions. Mass signatures were determined based on the fragmentation behavior of FAME standards. Fatty acyl groups in FAME were identified and quantified by positive mode APCI-MS. Figures 2(b) and 2(c) indicate a trend towards a higher degree of fatty acid saturation with increased TAG content. For example, healthy *Neochloris*, having little TAG content, yielded more C18:3 and C21:4 species and less C16:0, C18:0, and C20:0 than the other two algal groups. In contrast, stressed *Neochloris* and KAS 603, having a higher TAG content, yielded more C16:0, C18:0, and C20:0, and less C18:3 and C21:4 than healthy *Neochloris*. This shift in TAG fatty acid composition from unsaturated to saturated species with nitrogen starvation has been reported previously [21].

8.2.3 ACID VERSUS BASE ELUTION AND TRANSESTERIFICATION

Synthesis of FAMEs can be catalyzed by either base or acid. The base-catalyzed reaction is faster but can also generate soaps [4]. The sulfuric acid-catalyzed reaction, while slower, is also better at esterifying free fatty acids by dehydration [27]. Heat can speed up the acid-catalyzed reaction but there is a tradeoff between speed and the cost of heating. The previous results showed that 5% sulfuric acid in methanol could both elute algae from the resin and convert lipids to FAME. To test how well a base-catalyzed transesterification reagent would work, algae bound to Amberlite was eluted with either potassium hydroxide/methanol or sodium methoxide/methanol. We found that FAME yields from the sodium hydroxide/methanol reaction were undetectable and yields from the sodium methoxide/methanol reagent were 4% lower than using the sulfuric acid/methanol reagent. The reduced levels of FAME may be due to residual water, as shown by Griffiths [20], or to binding and depletion of the hydroxide or methoxide ions from solution by the anion exchange resin. For acid-catalyzed transesterification, we found that ethanol could be substituted for methanol with essentially identical results. Because of its low cost and wide availability, methanol is often used for biodiesel synthesis [4]. However, ethanol cost may decrease in the future as advances in bioethanol productions lead to increases in supply.

8.2.4 RECOVERY OF FAME FROM TRANSESTERIFICATION REAGENT USING HYDROPHOBIC RESIN

The results from Figure 1 showed that the sulfuric acid/methanol transesterification reagent could be reused multiple times for eluting algae from resin. We next tested whether recycling the sulfuric acid/methanol reagent led to a progressive decrease in FAME synthesis in comparison to fresh sulfuric acid/methanol [Figure 3(a)]. For the recycled sulfuric acid/methanol solution, each eluate was extracted with hexane and analyzed for FAME content before it was reused. The results showed that both fresh and recycled sulfuric acid/methanol solutions gave comparable amounts of FAME during four successive cycles of algal loading and elution.

Reuse of the sulfuric acid/methanol solution would minimize chemical costs, and removal of the FAME with each cycle should promote product formation. Hexane extraction of FAME is a relatively cumbersome and time consuming process leading us to explore a simple hydrophobic resin-based method to collect the FAME generated during each cycle. For comparison, the sulfuric acid/methanol eluate was either directly extracted with hexane or passed over a hydrophobic ethylene glycol dimethacylate:hexyl methacrylate (EGDMA:HMA) resin [Figure 3(b)] which was then eluted with hexane in a separate recovery step. The results show that FAME recovery from the resin (11.0% of DCW) was comparable to that obtained by direct hexane extraction [11.8% of DCW; Figure 3(c)]. While the hydrophobic column was effective at collecting FAME on the lab bench scale, we suspect that other methods, such as a hollow fiber membrane extractor [28] would be more useful commercially.

8.2.5 RESINS WITH HIGHER ALGAL BINDING CAPACITY FOR DIRECT TRANSESTERIFICATION

While these studies were initiated and largely carried through using Amberlite, it became apparent through efforts to develop better resins that the binding capacity of Amberlite for algae is generally quite low [25]. It is particularly low for KAS603, which gave the highest yields of FAME relative to its dried weight. A comparison of Amberlite to other functionalized resins we have generated shows that EGDMA-IM-DEG [Figure 4(a)] and DVB-DMA [Figure 4(b)] resins showed equilibrium binding of 2.6 and 3.4 times more KAS603 than Amberlite, respectively [Figure 4(c)]. Currently we have obtained binding capacities up to 150 mg algae per g resin. While these resins can be eluted cleanly with the sulfuric acid/methanol reagent so as to generate FAME [Figure 4(d)], they were not designed for this purpose and are potentially susceptible to attack by the strong acid. Future resin designs will need to address the need for chemical stability under harsh conditions.

8.2.6 EVALUATION

While we are not equipped to do a formal life cycle analysis and feel it would be premature given the improvements needed in resin technology, nevertheless a comparison of the proposed resin-based processing to current approaches seems in order. Figure 5 presents a general schematic for current algal processing schemes derived on published evaluations [5,29]. For the most part, the cost of growing algae would be the same. The use of resins could impact growth in terms of the need for nitrogen starvation or growth of monocultures as discussed below. Primarily, our one-step harvest to biodiesel approach would impact the most expensive steps in processing algae to biofuel, namely the harvesting and extraction steps.

Currently harvesting typically requires both pumping the algal suspension and some treatment such as centrifugation, flocculation or dissolved air flotation. Centrifugation and dissolved air flotation both have relatively large capital costs and substantial electricity. Where resins could make a difference is by eliminating the need for pumping water. While we have used resin beads for comparisons, we do not feel this is the best way to deploy them for harvesting. Resin beads are porous with a large internal volume that can trap water. The pores are typically 10–30 nm in diameter whereas the algae are on the order of 2–3 μm in diameter so they cannot enter the resin. Our measurements show that after the bulk water is removed, about 100 mg of water/gram resin is retained either inside the resin or entrained between the beads. A better alternative is to use thin films of resin where the internal volume is low compared to the surface area. We have proposed elsewhere a belt harvester type approach where a belt moves from the pond, where it picks up the algae to a vat which, in this case would contain sulfuric acid/methanol [25]. In that study we estimated that a bristled belt 1 m wide × 7.5 m long could collect 3 kg algae per one complete turn of the belt. Other modes are possible such as mesh bags containing nonporous resin-coated particles that float or are semisubmerged.

Once the algae is harvested and eluted, the conversion to biodiesel is direct. There is no need for lysing, further drying and solvent extraction, steps that are considered quite expensive [5]. Hexane extraction requires recovery of the hexane and a certain amount will be lost to the environment. Furthermore, hexane only recovers the neutral lipids (mainly TAG). Our experience at measuring TAG in pond-grown algae is that TAG levels rarely surpass 5% DCW and we have never seen them greater than 12% DCW. One of the advantages of this approach is that FAME is generated from both polar and neutral lipids thereby increasing the yield from the

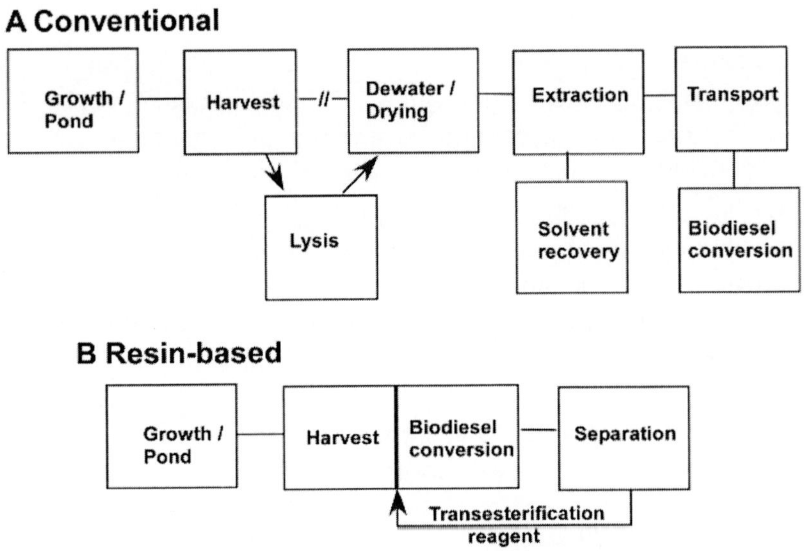

FIGURE 5: Sequence of steps to obtain fuel from algae. (a) A conventional sequence of steps is given for the obtaining biodiesel from algae. (b) A sequence of processing steps using a resin-based approach. Both sequences are essentially the same in terms of growth. Conventional harvesting involves pumping of algal suspension and some other treatment such as centrifugation, dissolved air flotation or flocculation with chemicals. For the resin-based approach the resin will collect algae leaving most of the water behind. One possible mechanism is to use a belt harvester employing a resin- coated belt. For the conventional approach a possible lysis step between harvest and extraction has been included since it can improve extraction efficiency. For solvent extraction it is usually necessary to dry the algae. In the resin-based approach drying and conventional solvent extraction is eliminated. In this scenario, biodiesel would be produced on site so the cost of transportation of the extracted oil to a biodiesel plant would also be eliminated. Since biodiesel is generated with or without employing resins, the cost of that step would be similar in either scenario.

algae. Our data suggest improvements of up to 4-fold were possible with an alga (KAS 603) that made 10% TAG.

The improved yield could affect several aspects of the algal growth process. Firstly, our data suggested that there was not so much to be gained by nitrogen starvation. Both starved and unstarved algae contained similar amounts of total lipid. Admittedly, the spectrum of fatty acids was different and this might be a factor. Secondly, resin-based conversion of total lipid to biodiesel might make it less important for growers to maintain a monoculture of high TAG-producing algae. Of course, high TAG production is a benefit but it is difficult to maintain monocultures in open ponds or in potentially other attractive scenarios where algae are grown in conjunction with wastewater treatment plants. Finally production of biodiesel onsite eliminates the transportation cost of moving algal oil to a biodiesel plant. While the biodiesel must be recovered from the sulfuric acid methanol, we believe this can be done using a low solvent or even solventless hollow fiber extraction system. In commercial biodiesel plants, the glycerol separates from the biodiesel layer and the separation is facilitated by water. This may be possible as well but more study is needed to determine how best to fractionate the spent sulfuric acid/ methanol mixture. There will be many cellular components in the final sulfuric acid/methanol solution, some of which may be valuable as side products.

8.2.7 FUTURE DIRECTIONS

Despite the potential advantages, more work is needed to optimize the resin approach. The use of resin beads is convenient for comparing properties such as binding capacity and fouling for both commercial and our own laboratory-synthesized resins. Here we used a commercially available strong cation exchange resin where sulfuric acid/methanol is used to elute the resin. It is entirely feasible to use a weakly basic anion exchange resin where raising the pH releases the algae from the resin [25]. At lower pH (7–8) these resins are positively charged and effectively bind algae. Raising the pH to ~pH 10 or above deprotonates the resin rendering it uncharged such that algae are released. Sodium methoxide in methanol, a strongly basic commercial transesterification reagent, would raise the pH, thereby releasing the algae and catalyzing the transesterification reaction. Here also, it would be critical to use a nonporous resin that did not entrain water.

Future directions are aimed at determining which approach, using strong or weak anion exchange resins is more effective in terms of binding capacity, fouling, and FAME yields. It will also be important to develop new modes for use of the resins such as thin film coatings or resin coated particles that do not entrain water. Finally, more studies of the transesterification reaction are in order since, as shown with *Neochloris*, the one step conversion to FAME was not quite complete and was further affected by the nitrogen starvation.

8.3 EXPERIMENTAL SECTION

Chloroform, ethanol, ethyl acetate, hexane, isopropanol, isooctane, methanol, and 2-ethoxyethanol (2-EE; Acros, Bridgewater, NJ, USA) were obtained from Fisher (Waltham, MA, USA) and were either HPLC or ACS reagent grade. Sulfuric acid and potassium hydroxide were reagent grade (Fisher). Sodium methoxide in methanol (0.5M), methyl oleate, methyl palmitate, di(ethylene glycol) vinyl ether (DEG), 2-dimethylamino methacrylate (DMA), divinylbenzene (DVB), ethylene glycol dimethacrylate (EGDMA), hexyl methacrylate (HMA), vinyl imidazole (IM), and azobisisobutyronitrile (AIBN) were obtained from Sigma-Aldrich (St. Louis, MO, USA). Triolein was obtained from Nuchek Prep (Elysian, MN, USA) and mineral oil (Squibb) was purchased locally.

All HPLC solvents were degassed and filtered through 0.5 micron PTFE filters (Ominpore, Waters Corp, Milford, MA, USA) prior to use. Isooctane was dried by storing over calcium hydride and filtered before use. Amberlite® ion exchange resin CG-400 (100–200 mesh, chromatographic grade; Mallinckrodt, St. Louis, MO, USA) was prepared by washing and settling in distilled water in order to remove fines, then dried at 55 °C.

8.3.1 ALGAL CULTIVATION AND HARVEST

Neochloris oleoabundans (UTEX LB 1185) was obtained from the University of Texas at Austin Algal Culture Collection and KAS 603, a saltwater species of *Chlorella*, was provided by Kuehnle Agro Systems. *Neochloris* was cultivated in freshwater Bold 3N (B3N) medium [30] using an airlift bioreactor illuminated with cool white fluorescent lights on a 12:12 light:dark cycle and aerated with ambient air using an oil-free diaphragm pump. KAS 603 was cultivated in the same way using f/2 saltwater medium [31]. For nitrogen starvation experiments, *Neochloris* was grown for 14 days in B3N media, harvested by centrifugation, resuspended in sodium nitrate-free B3N, and cultured for an additional 21 days. For resin binding and FAME synthesis, algae were harvested by centrifugation and resuspended in distilled water to concentration of 0.4 g/L.

8.3.2 ALGAL MASS QUANTIFICATION

Routine determination of algal dry cell weight (DCW) was obtained by measuring the optical density of chlorophyll at 680 nm (OD680) using a spectrophotometer (Shimadzu UV-265). The conversion of OD680 to DCW was accomplished by generating a dilution series for each species, recording the OD680, and then collecting samples onto pre-weighed cellulose acetate membranes (Pall Co., Port Washington, NY, USA), which were then dried in a vacuum oven (15 in. Hg., 60 °C) for 12 h before obtaining the final weights. To avoid optical filtering effects, samples were diluted, if need be, such that the OD680 was always less than 1.8.

8.3.3 ALGAL BINDING AND FAME GENERATION

For algae binding and FAME conversion studies, 10 mL polypropylene columns were loaded with 2 g of Amberlite resin and then washed with 1M hydrochloric acid followed by distilled water. To load the resin, 200 mL algal suspension containing 80 mg algae (a slight excess; based on the OD680 of the algae suspension) was passed through the resin. The OD680 of the flow through was then measured to determine the DCW of algae bound to the resin.

For transesterification and elution of algae from the resin, excess water was removed from column by vacuum, followed by the addition of 100 mL of transesterification reagent. For elution of algae and transesterification, reagents tested included 5% w/v sulfuric acid in either methanol or ethanol, 5% w/v potassium hydroxide in methanol, or 0.1 M sodium methoxide in methanol. The eluate was collected and the transesterification reaction continued at room temperature for 12 h. FAME was extracted sequentially with two 20 mL portions of hexane. The combined hexane extracts were then dried by rotary evaporation and resuspended in 1 mL hexane:isopropanol (3:1, v/v) for HPLC analysis. For comparison, FAME was generated directly from dried algal biomass, based on the method described by O'Fallon et al. [26]. Algal pellets containing approximately 30 mg algae in preweighed tubes were dried using a SpeedVac concentrator (Savant) for 12 h at 10^{-1} Torr with heat. Final tube weight was then measured to obtain an accurate algal DCW. The dried algal pellet was suspended in methanol and transferred to a glass centrifuge tube where methanol was added to a final volume of 5.3 mL. Lipids were then saponified by adding 0.7 mL of 10 N potassium hydroxide and heating at 55 °C for 1.5 h. Once the samples cooled, lipids were re-esterified by adding 0.6 mL of concentrated sulfuric acid to the same methanol suspension and heating again at 55 °C for 1.5 h. After cooling, FAME was extracted with 2 mL of hexane using a centrifuge to force the separation of the mixture into two layers. The upper hexane layer was then transferred to a vial for HPLC analysis.

8.3.4 TOTAL LIPID EXTRACTION FROM ALGAE

Total lipid was determined as described by Jones et al. [24] in order to assess the oil available for biodiesel synthesis for both algal species. Briefly, dried algal pellets (approximately 30 mg) were extracted with 20 mL of 2-EE for 30 min at 60 °C with continuous stirring. The solution was then filtered through 0.47 μm PTFE membrane (Millipore) and the residual biomass was extracted with a second 20 mL portion of 2-EE. After filtration, the two filtrates were combined, dried under vacuum, and weighed.

8.3.5 HYDROPHOBIC RESIN SYNTHESIS AND FAME RECOVERY

Hydrophobic resins used in this study were synthesized in bulk by combining 8 g of EGDMA and 2 g of HMA with 10 mL of toluene as the porogen. Resin synthesis was carried out in a round bottom flask fitted with an argon bubbler and heated to 60 °C with constant stirring. Polymerization was initiated by addition of 1 mol percent AIBN and continued until the mixture formed a brittle solid. The polymer was then dried at 55 °C for 12 h, scraped from the flask and ground by mortar and pestle. The crushed resin was then sized between #35 and #170 meshes to obtain beads of approximately 100–500 μm diameter.

As an alternative to solvent extraction, FAME generated by sulfuric acid/methanol transesterification was collected by passing the reaction mixture over a hydrophobic resin bed. Here, the 100 mL of sulfuric acid/methanol used to elute algae off the Amberlite resin was passed through a 10 mL polypropylene column containing 2 g of EDGMA-HMA resin. The resin was then eluted with 50 mL of hexane-ethyl acetate (3:1, v/v) solvent that was subsequently removed by rotary evaporation and the residue resuspended in 1 mL hexane-isopropanol (3:1, v/v) for HPLC analysis. For comparison, parallel samples of algae bound to Amberlite were eluted with 100 mL sulfuric acid/methanol and extracted with two 20 mL portions of hexane. After mixing and phase separation, the upper organic phase was recovered, dried by rotary evaporation, and the residue resuspended in 1 mL hexane-isopropanol (3:1, v/v) for HPLC analysis.

8.3.6 FUNCTIONALIZED RESIN SYNTHESIS AND COMPARISON WITH AMBERLITE

Functionalized weak anion exchange resins used in this study were synthesized in the manner stated above, using 6 g of EGDMA, 3 g of IM, and 1 g of DEG (EGDMA:IM:DEG), or 6 g of DVB and 4 g of DMA (DVB:DMA), with a solution of 5 mL of toluene (porogen) and 5 mL 3% acetic acid (dispersing agent). Steady-state binding capacity was determined by agitating 2 g of resin in a 500 mL flask containing 100 mL of KAS603 at 0.4 g/L concentration for 15 min. The suspension containing unbound algae was filtered off and resin transferred into columns. FAME was directly transesterified from the resin as stated previously using 100 mL of 5% sulfuric acid-methanol. For comparison, algae was bound and eluted from a parallel set of Amberlite resin. Because of the higher binding capacity of resin, columns were eluted for 3 consecutive cycles using the same transesterification reagent to ensure complete removal. After reaction for 12 h as room temperature, FAME was recovered by extracting twice with 20 mL of hexane. The recovered organic phase was dried by rotary evaporation, and the residue resuspended in 1 mL hexane:isopropanol (3:1, v/v) for HPLC analysis.

8.3.7 QUANTIFICATION AND CHARACTERIZATION BY HPLC-ELSD/MS

Lipid composition and FAME content were analyzed as previously described [24], using an HPLC (Surveyor LC Pump and Autosampler Plus, Thermo Finnegan, USA) coupled to both ELSD (Sedere Sedex 75) and quadrupole mass spectrometer (Thermo Finnigan MSQ) using a 10:1 line splitter (Analytical Instruments, USA). Xcalibur software controlled operation of the autosampler, pump, and mass spectrometer. ELSD analog data was acquired through an A/D data acquisition box (Agilent Technologies, SS420X) and RS232 PCI data acquisition card (Sea Level Systems, 7406S). Lipid standards and extracts were resolved using a YMC Pack PVA-Sil-NP column (250 mm × 4.6 mm I.D., 5 μm bead size) protected by a Waters Guard Pak™ guard column containing Nova-Pak™ silica inserts. The solvent program is given in Table 1. ELSD was run at 30 °C at gain setting 8. Mass spectrometer was run in APCI positive mode with probe temperature of 400 °C.

TABLE 1: Resolution of algal lipid classes using normal-phase HPLC. Normal-phase HPLC mobile phase gradient method using a three-solvent system of iso-octane (A), ethyl acetate (B), and isopropanol:methanol:water (3:3:1, v/v/v) + 0.1% acetic acid (C).

time (min)	flow rate (mL/min)	A (%)	B (%)	C (%)
0	1.5	100	0	0
5	1.5	98	2	0
15	1.5	75	25	0
19	1.5	20	80	0
24	1.5	0	100	0
32	1.3	0	50	50
38	1.0	0	15	85
43	1.0	0	0	100

52	1.0	0	100	0
54	1.0	0	100	0
60	1.5	90	10	0
64	1.5	100	0	0
74	1.5	100	0	0

8.4 CONCLUSIONS

Anion exchange resins can be used as a simple and inexpensive support for one-step algal harvest and biodiesel generation. The yields of FAME are greatly improved over methods that first isolate the TAG fraction since polar lipids also contribute to the FAME pool. Both the resin and the transesterification reagent can be reused for numerous cycles with the resultant FAME collected during each cycle. Although the basic principles have been demonstrated there is much room for improvement, especially in the design of resins. Although resins bind to algae based on surface charge, there are clearly additional issues that impact binding capacity, fouling and FAME conversion that are specific to different species of algae.

REFERENCES

1. Scott, S.A.; Davey, M.P.; Dennis, J.S.; Horst, I.; Howe, C.J.; Lea-Smith, D.J.; Smith, A.G. Biodiesel from algae: Challenges and prospects. Curr. Opin. Biotechnol. 2010, 21, 277–286.
2. Durrett, T.P.; Benning, C.; Ohlrogge, J. Plant triacylglycerols as feedstocks for the production of biofuels. Plant J. 2008, 54, 593–607.
3. Hossain, A.B.M.S.; Salleh, A.; Boyce, A.N.; Chowdhury, P.; Naqiuddin, M. Biodiesel fuel production from algae as renewable energy. Am. J. Biochem. Biotechnol. 2008, 4, 250–254.
4. Gerpen, J.V. Biodiesel processing and production. Fuel Process Technol. 2005, 86, 1097–1107.
5. Lardon, L.; Hélias, A.; Sialve, B.; Steyer, J.-P.; Bernard, O. Life-Cycle Assessment of Biodiesel Production from Microalgae. Environ. Sci. Technol. 2009, 43, 6475–6481.
6. Li, Y.; Horsman, M.; Wu, N.; Lan, C.Q.; Dubois-Calero, N. Biofuels from microalgae. Biotechnol. Prog. 2008, 24, 815–820.
7. Olaizola, M. Commercial development of microalgal biotechnology: From the test tube to the marketplace. Biomol. Eng. 2003, 20, 459–466.

8. Salim, S.; Bosma, R.; Vermue, M.H.; Wijffels, R.H. Harvesting of microalgae by bio-flocculation. J. Appl. Phycol. 2010, 1–7.
9. Schurchardt, U.; Sercheli, R.; Vargas, R.M. Transesterification of vegetable oils: A review. J. Braz. Chem. Soc. 1998, 9, 199–210.
10. Bikou, E.; Louloudi, A.; Papayannakos, N. The effect of water on the transesterification kinetics of cotton seed oil with ethanol. Chem. Eng. Technol. 1999, 22, 70–75.
11. Johnson, M.B.; Wen, Z. Production of biodiesel fuel from the microalga Schizochytrium limacinum by direction transesterification of algal biomass. Energy Fuels 2009, 23, 5179–5183.
12. Komers, K.; Machek, J.; Stioukal, R. Biodiesel from rapeseed oil and KOH, 2. Composition of solution of KOH in methanol as reaction partner of oil. Eur. J. Lipid Sci. Technol. 2001, 103, 359–362.
13. Kusdiana, D.; Saka, S. Effects of water on biodiesel fuel production by supercritical methanol treatment. Bioresour. Technol. 2004, 91, 289–295.
14. Ma, F.; Clements, L.D.; Hanna, M.A. The effects of catalyst, free fatty acids, and water on transesterification of beef tallow. Trans. ASAE 1998, 41, 1261–1264.
15. Clarens, A.F.; Resureccion, E.P.; White, M.A.; Colosi, L.M. Environmental life cycle comparison of algae to other bioenergy feedstocks. Environ. Sci. Technol. 2009, 44, 1813–1819.
16. Armenta, R.E.; Scott, S.D.; Burja, A.M.; Radianingtyas, H.; Barrow, C.J. Optimization of fatty acid determination in selected fish and microalgal oils. Chromatographia 2009, 70, 629–636.
17. Haas, M.J.; Wagner, K. SImplifying biodiesel production: The direct or in situ transesterification of algal biomass. Eur. J. Lipid Sci. Technol. 2011, 113, 1219–1229.
18. Johnson, E.A.; Liu, Z.; Salmon, E.; Hatcher, P.G. One-step Conversion of Algal Biomass to Biodeisel with Formation of an Algal Char as Potential Fertilizer. In Advanced Biofuels and Bioproducts; Lee, J.W., Ed.; Springer: New York, NY, USA, 2012.
19. Koberg, M.; Cohen, M.; Ben-Amotz, A.; Gedanken, A. Bio-diesel production directly from the microalgae biomass of Nannochloropsis by microwave and ultrasound radiation. Bioresour. Technol. 2011, 102, 4265–4269.
20. Griffiths, M.J.; Hille, R.P.; Harrison, S.T.L. Selection of direct transesterification as the preferred method for assay of fatty acid content of microalgae. Lipids 2010, 45, 2010.
21. Braithwaite, A.; Smith, F.J. Chromatographic Methods; Springer: London, UK, 1996.
22. Onyancha, D.; Mavura, W.; Ngila, J.C.; Ongoma, P.; Chacha, J. Studies of chromium removal from tannery wastewaters by algae biosorbents, Spirogyra condensata and Rhizoclonium hieroglyphicum. J. Haz. Mat. 2008, 158, 605–614.
23. Tornabene, T.G.; Holzer, G.; Lien, S.; Burris, N. Lipid composition of the nitrogen starved green alga *Neochloris* oleoabundans. Enz. Microb. Technol. 1983, 5, 435–440.
24. Jones, J.; Manning, S.; Montoya, M.; Keller, K.; Poenie, M. Extraction of algal lipids and their analysis by HPLC and mass spectrometry. J. Am. Oil Chem. Soc. 2012, 89, doi:10.1007/s11746-012-2044-8.

25. Jones, J.; Poenie, M. Resins that reversibly bind algae for harvesting and concentration. Environ. Prog. 2012, submitted for publication.
26. O'Fallon, J.V.; Busboom, J.R.; Nelson, M.L.; Gaskins, C.T. A direct method for fatty acid methyl ester synthesis: Application to wet meat tissues, oils, and feestuffs. J. Anim. Sci. 2007, 85, 1511–1521.
27. Chisti, Y. Biodiesel from microalgae. Biotechnol. Adv. 2007, 25, 294–306.
28. Seibert, F.; Poenie, M. Non-Dispersive Process for Insoluble Oil Recovery From Aqueous Slurries. U.S. Patent 2011/0174734 A1, 21 July 2011.
29. Sander, K.; Murthy, G.S. Life cycle analysis of algae biodiesel. Int. J. Life Cycle Assess. 2010, 15, 704–714.
30. Brown, M.R.; Bold, H.C. Comparative Studies of the Algal Genera Tetracystis and Chlorococcum; University of Texas Publication: Austin, UT, USA, 1964; p. 213.
31. Jeffrey, S.W.; LeRoi, J.M. Simple Procedures for Growing SCOR Reference Microalgal Cultures; UNESCO: Marseille, France, 1997.

This chapter was originally published under the Creative Commons Attribution License. Jones, J., Lee, C.-H., Wang, J., and Poenie, M. Use of Anion Exchange Resins for One-Step Processing of Algae from Harvest to Biofuel. Energies 2012: 5; 2608–2625.

CHAPTER 9

MICROALGAE ISOLATION AND SELECTION FOR PROSPECTIVE BIODIESEL PRODUCTION

VAN THANG DUONG, YAN LI, EKATERINA NOWAK, and PEER M. SCHENK

9.1 INTRODUCTION

Microalgae have been considered for biodiesel production, based on their ability to grow rapidly and to accumulate large amounts of storage lipids, primarily in the form of triacylglycerides (TAG). Microalgae are a group of mostly photoautotrophic microorganisms that includes both prokaryotic and eukaryotic species. These organisms can photosynthetically convert CO_2 and minerals to biomass, but some species also grow heterotrophically. Prokaryotic microalgae are cyanobacteria (blue-green algae) and eukaryotic microalgae include the nine phyla *Glaucophyta*, *Chlorophyta*, *Chlorarachniophyta*, *Euglenophyta*, *Rhodophyta*, *Cryptophyta*, *Heterokontophyta*, *Haptophyta* and *Dinophyta*. To date, about 2/3 of 50,000 species have been identified and are kept in collection by various algal research institutes [1]. For example, the largest collection at present is the Collection of Freshwater Algae at the University of Coimbra, Portugal, maintaining about 4000 strains and 1000 species of algae; the Culture Collection of Algae of the Göttingen University, Germany harbors 2213 strains and 1273 species of both freshwater and marine algae; the

Culture Collection of Algae in the University of Texas, USA, maintains 2300 strains of freshwater species; the National Institute for Environmental Studies in Japan is keeping 2150 strains with about 700 species of freshwater and marine algae [2]; the Australian National Algae Culture Collection (ANACC) maintains about 1000 strains of microalgae which were mostly isolated from Australian waters [3]. Although algae collections are maintained for many purposes (e.g., for pharmaceutical, food, energy and industrial products), only a few hundred strains have been investigated for chemical content and very few are cultivated in industrial quantities. To date, although there is mounting interest to develop microalgal biodiesel production, the cost for microalgal biomass production is currently much higher than from other energy crops [4]. Thus, selection of an energy and cost-efficient production model could play a very important role in achieving competitive biodiesel production. This includes the selection of high lipid-producing algae, suitable farming locations, efficient cultivation and harvesting methods and oil extraction procedures. Here, we focus on the first step, the selection of suitable high lipid-accumulating microalgae strains, a process that can be compared with the early domestication of current crop plants. In alignment with this purpose, this review aims to present a practical guide to several simple and robust methods for microalgae isolation and selection for traits that maybe most relevant for commercial biodiesel production.

9.2 ADVANCED MICROALGAE BIODIESEL PRODUCTION

The developments in the biodiesel industry have progressed dramatically in recent years. Developed countries have set priorities on biodiesel fuels for the transport and mechanical industry and established a Biodiesel Board for policy making and development. The production of biodiesel in the EU has been increasing from 1.9 million tons in 2004 to 3.2 million tons in 2005 and to 4.9 million tons in 2006 [5]. In 2011, biodiesel production rose to 22.117 million tones as reported by the European Biodiesel Board (EBB); the leading biodiesel producing countries in the area include Germany, France and Spain [6]. The United States is also developing biodiesel applications in many different industries. According to the US National Biodiesel Board, biodiesel production increased from 0.016 million tons in 1999 to 0.787 million tons in 2004 [7]. Production and sales were estimated to have tripled from 2004 to 2005 and to have reached 21.73 million tons of fuel in 2008 and more than 31.5 million tons in 2011, as reported by The U.S. National Biodiesel Board (TUSNBB) [8]. To date, research on biodiesel production from microalgae is enthusiastically attempted globally.

In comparison with other sources (e.g., animal fat, oleaginous grain crops and oil palm), there are remarkable advantages of biodiesel from microalgae as an alternative energy source for the future. Advantages include the following: (i) areal growth rate and oil productivity of microalgae per unit of land use are much higher than those of other biofuel crops; (ii) algae grow in a wide range of environments. Fresh, brackish and saline waters are ideal environments for growth of different algae species. Even in municipal and other types of wastewater, algae grow well by using inorganic (NH_4^+, NO_3^-, PO_4^{3-}) as well as organic sources of nutrients [9]; (iii) microalgae absorb CO_2 photosynthetically and convert it into chemical energy and biomass. The removal of CO_2 from the atmosphere (and possibly industrial flue gases) may play an important role in global warming mitigation by replacing fossil fuel emissions [9,10]. Producing 100 tonnes of algal biomass fixes roughly 183 tonnes of carbon dioxide from the atmosphere [4]; (iv) microalgae can provide raw materials for different types of fuels such as biodiesel, ethanol, hydrogen and/or methane which are rapidly biodegradable and may perform more efficiently than fossil fuels [11]; (v) products extracted from algal biomass can be used as sources for organic fertilizers or high value products, such as omega-3 fatty acids, sterols, carotenoids and other pigments and antioxidants, and could be amenable to a zero waste biorefinery concept [9]. Therefore, microalgae have been regarded as possibly the only route to sustainable displacement of high proportions of fossil oil consumption.

9.3 BIODIESEL CONVERSION FROM MICROALGAE

The process of isolation and selection of algae strains needs to consider the requirements of algal oil suitable for biodiesel production. Algal lipids occur in cells predominantly as either polar lipids (mostly in membranes) or lipid bodies, typically in the form of triacylglycerides. The latter are accumulated in large amounts during photosynthesis as a mechanism to endure adverse environmental conditions. Polar lipids usually contain polyunsaturated fatty acids which are long-chained, but have good fluidity properties. TAG in lipid storage bodies typically contain mostly saturated fatty acids which have a high energy contents, but, depending on the fatty acid profile of the algae strain, may lack fluidity under cold conditions. Provided the algal oil is low enough in moisture and free fatty acids, biodiesel is typically produced from TAG with methanol using base-catalyzed transesterification [12]. Most current feedstock for biodiesel production is based on plant oils produced from oil palm, soybean, cottonseed and canola, recycled cooking greases or animal fats from beef tallow or pork lard [13]. According to Fukuda, transesterification using base catalysts is 4,000 times faster than using acid catalysts [14]. Some common base catalysts used by industry are sodium hydroxide and potassium hydroxide. Use of lipase enzymes as a catalyst is efficient, however their use is limited because of the high costs [14]. The best temperature for the reaction is typically 60 °C under normal atmospheric pressure. If the temperature is higher, methanol will boil, lowering the efficiency [14]. During the transesterification process, saponification reactions can occur, forming soap. Thus, oil and alcohol must be dried. Finally, biodiesel is recovered by washing rapidly with water to remove glycerol and methanol [4]. The high potential of oil production from microalgae has attracted several companies to commercialize biodiesel from microalgae (e.g., MBD Energy Pty and Muradel Pty Ltd in Australia). Basically, algal biodiesel is produced after algae cultivation and harvesting, followed by oil extraction and its conversion by transesterification. Principally, microalgal oil can be directly used as fuel feedstock, based on the conventional process of biodiesel production, provided the fatty acid profile is favorable. But even algal oils with a high degree of saturation (e.g., similar as tallow) can be considered as a drop-in fuel (e.g., for B20 blends). In addition, scientists are also focusing on the conversion to higher value products. For example, thermal cracking is used for decomposition of triglycerides into hydrocarbons such as alkans, alkenes, and aromatic compounds [15,16].

9.4 ISOLATION AND SELECTION CRITERIA FOR MICROALGAE WITH POTENTIAL FOR BIODIESEL PRODUCTION

9.4.1 SAMPLING AND ISOLATION OF PURE CULTURES

Microalgae grow in most of the natural environments including water, rocks and soil, but interestingly also grow on and in other organisms. Their main habitats are freshwater, brackish and marine ecosystems. Microalgae can be found and collected not only in general aquatic ecosystems such as lakes, rivers and the oceans, but also in extreme environments such as volcanic waters and salt waters. Local microalgae species should be collected because it can be expected that they have a competitive advantage under the local geographical, climatic and ecological conditions. Our experience has shown that water and sediment samples from aquatic environments that undergo fluctuating and/or occasional adverse conditions provide a higher chance of isolating high lipid accumulating microalgae. Most likely these conditions would favor robust and opportunistic (fast-growing) algae with superior survival skills (e.g., by accumulation of storage lipids). Examples of these environments are tidal rock pools, estuaries and rivers.

Isolation is a necessary process to obtain pure cultures and presents the first step towards the selection of microalgae strains with potential for biodiesel production. Traditional isolation techniques include the use of a micropipette for isolation under a microscope or cell dilution followed by cultivation in liquid media or agar plates. Single cell isolation, based on traditional methods from the original sample is time-consuming and requires sterilized cultivation media and equipment, but the result of this elaborate process is always a pure culture that is usually easily identifiable. Another approach in the laboratory includes the enrichment of some microalgae strains by adding nutrients for algal growth. The most important nutrient sources for algal growth are nitrogen and phosphate. Some particular algae species may require trace minerals for their growth (e.g., silicon for diatoms). Soil water extract is an excellent source of nutrients for algae growth at this stage because this medium is easy to produce and satisfies nutrient intake of many algae strains. Although automatic isolation techniques have offered some advantages towards traditional methods (see below), single cell isolation by a micropipette (e.g., a glass capillary) is still a very effective method that can be used for a wide range of samples and is very cost-effective. An automated single cell isolation method that has been developed and widely used for cell sorting is flow cytometry [17]. This technique has been successfully used for microalgae cell sorting from water with many different algae strains [18], primarily based on properties of chlorophyll autofluorescence (CAF) and green autofluorescence (GAF) to distinguish algae species such as diatoms, dinoflagellates or prokaryotic phytoplankton.

Unlike for many agricultural crops, a targeted selection and domestication of microalgae strains is still in its infancy, while technology to economically grow microalgae with high lipid content is still being developed [4]. Each microalgae strain requires careful selection and optimization in order to increase lipid productivity with the aim to provide a cropping system with improved biofuel production and performance properties [19]. Each micro-environment may provide algae strains with very different properties. As opposed to only manipulating a few individual cultured algae strains in the laboratory for high lipid productivity, a more efficient and useful strategy to identify oleaginous microalgae would be an in-depth and systematic investigation of a whole taxonomic group of microalgae over a wide geographical and ecological distribution [20]. By correlating this with algal oil contents and optimal environmental growth conditions, a predictive tool for selecting optimal microalgae strains for biofuel production maybe developed. A bioinformatics approach could assist with the discovery of new algae isolates capable of biodiesel production and their phylogenetic grouping may suggest that potentially many more species have this ability. Typically, the steps involved in obtaining data for phylogenetic analysis include primers design (Table 1), DNA and/or RNA extraction, PCR amplification, denaturizing gradient gel electrophoresis and/or sequencing.

TABLE 1: Examples of 18S rDNA primers for the identification of microalgae by sequencing.

Primer name	Forward (5'–3')	Primer name	Reverse (5'–3')	Species	References
TH18S5'	GGTAAC-GAATTGTTAG	TH18S3'	GTCGGCATAGTTTATG	Thalassiosira pseudonana	[21]
P45	ACCTGGTT-GATCCTGC-CAGT	P47	TCTCAG-GCTCCCTCTCCGGA	Chlorella vulgaris	[22]
	GTCAGAGGT-GAAATTCTTG-GATTTA		AGGGCAGGGACGTA-ATCAACG	Dunaliella salina	[23]
SS5	GGTGATCCT-GCCAGTAGT-CATATGCTTG	SS3	GATCCTTCCGCAG-GTT CACCTACGGAAACC	Navicula sp. Chlorella sp.	[24]
	GAAGTCGTAA-CAAGGTTTCC		TCCTGGT-TAGTTTCTTTTCC	Chlamydomonas coccoides Tetraselmis suecica Nannochloris atomus	[25]
	CCAACCTG-GTTGATCCT-GCCAGTA		CCTTGTTAC GACTTCACCTTCCTCT	Nannochloropsis sp.	[26]

In 2010, seven microalgae genomes had been completed [27] and current efforts to obtain many other microalgal genome sequences will enhance gene-based biofuel feedstock optimization studies (e.g., by metabolic engineering). The accumulation of storage lipid precursors and the discovery of genes associated with their biosynthesis and metabolism is a promising topic for investigation. For example, genes encoding key enzymes involved in biosynthesis and catabolism of fatty acids/TAG and their regulation are currently not well understood (for a review of the lipid biosynthesis pathway in microalgae see Schuhmann et al. [28]). By providing insight into the mechanisms underpinning the relevant metabolic processes, efforts can be made to identify molecular markers for selection or to genetically manipulate microalgae strains to enhance the production of feedstock for commercial microalgal biofuels. To date, genetic engineering approaches have been successfully used to improve biofuel phenotypes only in *Chlamydomonas reinhardtii, Nannochloropsis gaditana* and *Phaeodactylum tricornutum* [29].

9.4.2 LIPID DETERMINATION

Lipid determination in qualitative and quantitative analysis is crucial for identification of suitable strains for biodiesel production. Conventional methods such as solvent extraction or gravimetric methods have been used by Bligh and Dyer [30]. Separation and profiling of lipid components require elaborate techniques in order to satisfy criteria of biodiesel quality and includes thin layer chromatography (TLC), gas chromatography-mass spectroscopy (GC/MS) and/or high pressure liquid chromatography (HPLC) [31]. These methods are time-consuming for lipid extraction and analysis, especially for a large number of samples. Thus, a rapid screening for lipid content in organisms or cells is necessary and important for high-throughput screening. Nile red (9-diethylamino-5-benzo[α] phenoxazinone), a lipophilic stain, maybe used for this purpose. It was first synthesized by Thorpe in 1907 by boiling Nile blue with sulfuric acid, and in the same year, Smith reported the use of Nile red for detecting lipids in human cells [32]. The application of Nile red for lipid staining in microorganisms such as bacteria, yeasts and microalgae is now a common practice that allows a rapid qualitative determination of lipids in cells (Figure 1) [33].

Although Nile red can be applied for rapid lipid screening, this method has not been successful in some particular microalgae species due to variables such as staining time, temperature, rigid cell walls, etc. [34]. Thus, Nile red dye concentrations applied for lipid staining are different for particular microalgae species. To improve staining efficiency some factors can be considered. For instance, microwaves applied for staining were first introduced by Leong and Milios and then improved by Chiu et al. in 1987 [34–36]. Microwave exposure time was optimized for processes of pretreatment and staining. Results of this research showed that microwave-assisted staining increased remarkably fluorescence intensity

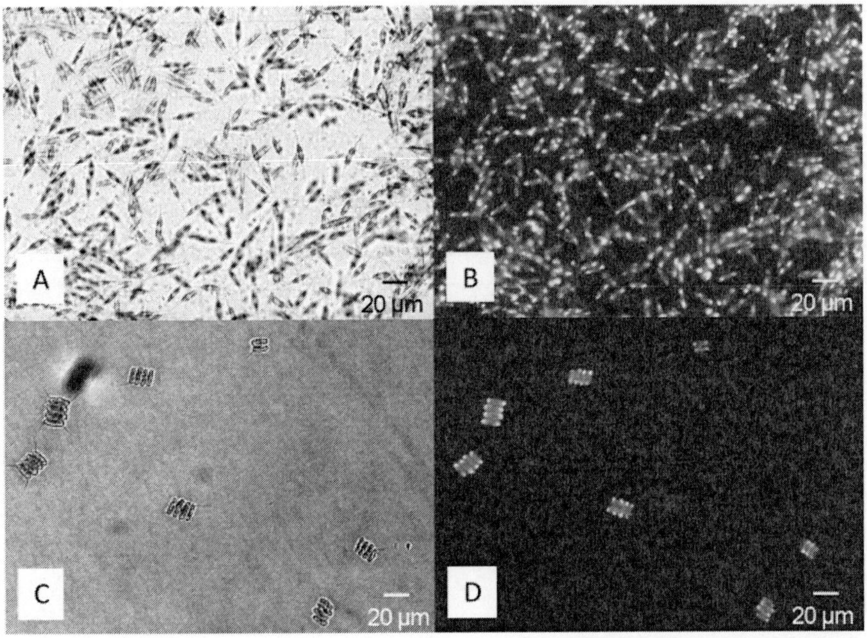

FIGURE 1: Cylindrotheca closterium and Scenedesmus sp. after Nile red staining under bright light (A,C) or fluorescence microscopy under blue light (B,D), respectively. Yellow dots show lipid bodies containing triacylglycerides; orange color indicates polar lipids and red shows autofluorescence from chlorophyll. Samples were obtained from a coastal rock pool (A,B); and a freshwater creek (C,D) in South East Queensland, Australia.

using a spectrofluorometer from 476 to 820 arbitrary units (a.u.) for Pseudochlorococcum sp. and from 662 to 869 a.u. for Chlorella zofingiensis after 50 s of microwave exposure in a pretreatment process and after 60 s of staining. Dimethyl sulfoxide (DMSO) has also been used for enhancing lipid staining effectiveness [33,34,37] and maybe used in low quantities instead of acetone as a solvent to allow viability of cells after staining. This means that Nile red staining can not only be used as a preliminary quantitative fluorometric assay for relative comparisons among closely-related strains, but potentially also for mutant screening and selection of high lipid-yielding strains.

9.4.3 CULTIVATION AND BIOMASS PRODUCTION

Microalgae strains with potentially high lipid content (e.g., as determined by Nile red staining) need to be cultured to increase biomass and directly compared to each other in larger cultivation systems to assess their potential as biodiesel feedstock. Initial tests of the most promising algae strains usually are carried out at laboratory-scale using culturing flasks and other vessels, such as hanging bags, under well-defined growth conditions. The test should follow a standard protocol over a certain growth period to allow direct comparisons between strains in terms of growth rate and lipid accumulation ("lipid productivity"). It should be noted though that a standard assay does not take into consideration the potential of certain microalgae under carefully optimized conditions. An example of this assay is the following that is routinely used by our laboratory to compare lipid productivity: Pure (but not axenic) algae strains are cultured in F/2 medium (fresh or seawater) until near stationary growth occurs (less than 20% growth in 24 h as determined by cell counts). An inoculum of 5 mL of this culture is then used to start growth in 20 mL fresh F/2 medium exactly at 2 h after the start of the light cycle. The culture is then grown and monitored by cell counting for 7 days after which medium is replaced with nutrient-free water. Nutrient starvation is conducted after that for another 2 days of cultivation to test the potential for rapid TAG accumulation. In addition, biomass is collected at the end of the experiment for lipid content analysis. This assay is useful for screening of growth and lipid producing capacity of microalgae, leading to selection of potentially useful strains. The best candidate strains with potential for biodiesel production should then be used to optimize parameters for rapid growth, lipid induction, harvesting/dewatering and oil extraction. While most of these parameters are typically optimized under small-scale laboratory conditions, it seems advisable to move towards larger size outdoor cultivation conditions as soon as possible, as these are typically quite different. Parameters, such as salinity, nutrient composition, pH and cell density can be controlled to some extent, but other factors such as temperature, irradiation and the co-cultivation of other organisms are much harder to control under outdoor conditions.

In summary, it is advantageous to isolate and screen a large number of local microalgae strains and test these under mid-scale outdoor conditions as soon as possible to be close to conditions that maybe expected for large-scale cultivation. Figure 2 provides a step-by-step overview of how microalgae may be rapidly isolated and selected for larger scale biodiesel production.

9.4.4 TESTING AT LARGER SCALE

At larger scale, two popular cultivation systems have been used for microalgal biomass and lipid production: open raceway ponds and closed photobioreactors. At present, open pond systems, especially large raceway ponds, are much more widely used, but bear the risk of attracting competing algae, grazers or viruses [38]. Although minimizing the cost of algae farming is one of highest priorities to achieve commercial algal biodiesel production, both systems require optimization of complex factors that satisfy high level production cultivation.

For example, these include irradiation, nutrients, temperature, dissolved O_2 and CO_2 contents, pH, salinity, water quality, mixing efficiency and harvesting ability. Culturing and environmental conditions affect algae productivity, lipid yield and fatty acid compositions. For example, a pilot study showed that high growth rates and lipid accumulation of *Chlorella sp.* could be achieved primarily by increasing nitrogen concentrations and nitrogen starvation, respectively [39]. Similarly, growth of hydrocarbon-producing *Botryococcus braunii* was strongly dependent on light, temperature, salinity [40], nutrient quantity and composition [41]. Total lipid, carotene and chlorophyll contents of *Navicula sp.* increased with increasing salinity of the medium from 0.5 to 1.7 M NaCl [42]. In some cases, wastewater (e.g., municipal wastewater) can be used as a nutrient source of microalgae growth. Pulz suggested that productivity of 60–100 mg dry weight L^{-1} d^{-1} or a biomass concentration of 1 g L^{-1} is achievable in open pond systems [43]. Table 2 lists some of the desirable traits for the selection process of microalgae with potential for biodiesel production.

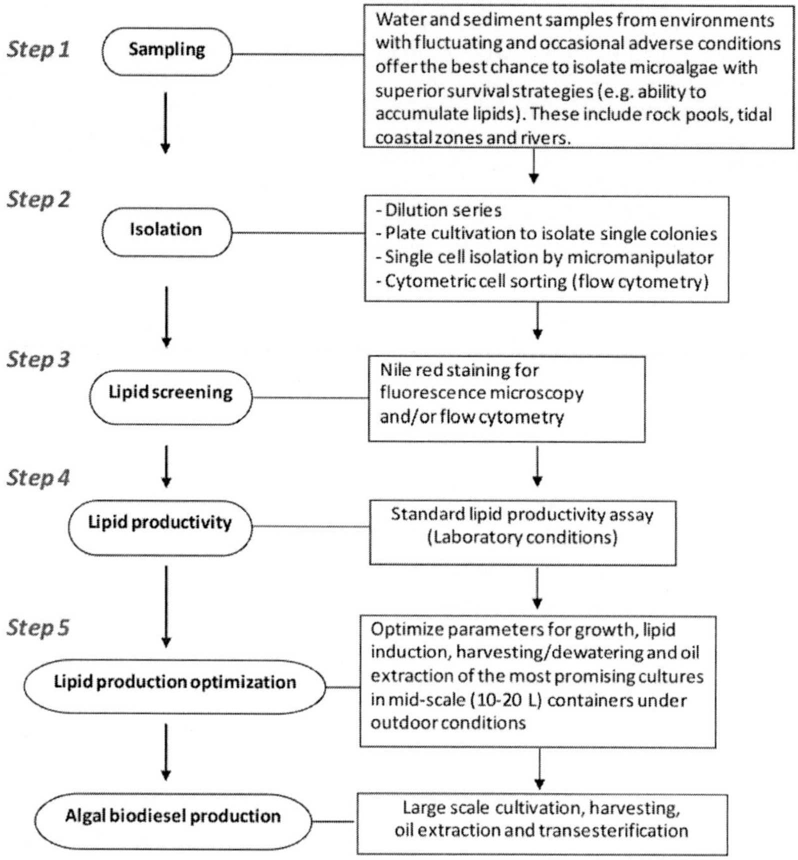

FIGURE 2: Suggested 5-step protocol for rapid selection of microalgae for biodiesel production. Step 1: Local sampling sites should be chosen where microalgae frequently undergo adverse conditions; Step 2: Dilution series in growth medium provide the simplest, most cost-effective and fastest method; Step 3: Nile red staining of near stationary cultures followed by visual inspection provides a simple and rapid screening for algae with high lipid accumulation ability; Step 4: A standardized growth assay in the laboratory can provide comparative data on lipid productivity; Step 5: Parameters can be directly optimized under outdoor conditions using mid-scale cultures as these are often very different to small-scale laboratory conditions.

TABLE 2: Checklist for desirable traits for microalgae selection with potential for biodiesel production and high-value byproducts.

Steps	Desirable traits
Screening	High oil
	High saturated fatty acids
	Low unsaturated fatty acids
	High omega 3 fatty acids
	Rapid and synchronized lipid production
	Radiation tolerance/pigment synthesis
	Antioxidants, sterols, carotenoids, astaxanthins and other pigments
	Low starch contents
	High protein contents
Cultivation	Rapid growth
	Salinity/freshwater tolerance
	High/low temperature tolerance
	Reduced antennal pigments (for improved photosynthesis in bioreactor)
	Flagella properties/possession
	Sheering resistance
Harvesting	Cell size and cell wall properties amenable for autoflocculation
	Sinking speed
	Foam fractionation properties
	Structure and cell wall properties
Extraction	Cell wall properties amenable for oil extraction
	Lipid extraction efficiency

9.5 LIPID CONTENT IN MICROALGAE

Many microalgae are capable of accumulating a large amount of lipids in the cells [10]. On average, the lipid contents typically range from 10 to 30% of dry weight (Table 3). Depending on the specific algae species and their cultivation conditions, however, microalgal lipid production may range widely from 2 to 75% [2]. In some extreme cases, it can reach 70%–90% of dry weight [4,5]. For instance, the freshwater green alga Botryococcus braunii can produce oil (including hydrocarbons) up to 86% of its dry cell weight [44]. This species is being considered as a possible source for biodiesel production in the near future [4], but has the major disadvantage of slow growth rates and a low tolerance for contamination. As a result, lipid productivities (lipid production per area or volume) of other microalgae, such as *Nannochloropsis, Chlorella, Tetraselmis* and *Pavlova sp.* are typically much higher [39,45]. Lipid productivity can be dramatically increased by external application of stress factors and is considered a survival strategy for microalgae under adverse conditions. Most notably these include nutrient deprivation, exposure to chemicals, changes in salinity, temperature, pH and/or irradiation [4,39,46]. The composition of fatty acids-containing lipids differs widely among species, but, as mentioned above, generally includes structural unsaturated polar lipids, as well as neutral storage lipids, mostly in the form of TAG. Significant fatty acids used for biodiesel include saturated fatty acids and polyunsaturated fatty acids (PUFAs) containing 14–18 carbon molecules, such as C14:0, C16:0, C16:1, C18:0, C18:1, C18:2, C18:3 fatty acids [41]. According to European requirements for biodiesel standards, some fatty acids should be excluded because of undesirable properties. For instance, methyl linolenate and fatty acid methyl esters with more than four double bonds are limited to 12% due to oxidation properties [47].

Table 3. Examples of lipid contents in some microalgae species [4,48].

Species	Total lipids (% dry weight)	PUFA (% total lipids)	PUFA (% dry weight)
Isochrysis galbana	25.6	17	4.3
Nanaochloropsis sp.	5.6	2.8	0.2
Chaetoceros calcitrans	11.8	8.7	0.9
Tetreselmis suecica	2.5	20.9	0.2
Skeletonema costatum	9.7	5.1	0.5
Phaeodactylum tricornutum		30	
Porphyridium cruentum	1.5	17.1	0.3
Crypthecodinium cohnii	20		
Botryococcus braunii	25.0–75.0		
Chlorella sp.	10.0–48.0		

It is expected that microalgae that offer a multiple product portfolio as part of a biorefinery, will be most applicable to large-scale commercial cultivation. In a microalgae screening process, besides fatty acids with properties relevant for biodiesel production, some high value products such as protein-rich biomass, omega-3 fatty acids, sterols, antioxidants, vitamins and pigments should also be taken into account. In particular, omega-3 fatty acids from microalgae have received significant attention as a high-value add product, as the current sources of fish oil are unsustainable due to depleting global fish stocks. A comparison of omega-3 fatty acid contents of different microalgae shows that these differ considerably between species (Table 4).

9.6 CULTIVATION AND LIPID EXTRACTION PROPERTIES OF MICROALGAE

High lipid productivity is not the only factor that should be considered early during strain selection. Outdoor cultivation should determine whether the selected microalgae are robust enough to withstand variable local climatic conditions and whether they can dominate a culture. This is particularly important for open pond systems where other algae strains, grazers or viruses may easily contaminate the culture. For this purpose, many phycologists recommend the use of local dominant species, even if their lipid productivity may not be as high as other species [43].

TABLE 4: Examples of potential microalgae species for omega-3 production [48].

Species	Eicosapentaenoic acid (EPA) (% of total fatty acids)	Docosahexaenoic acid (DHA) (% of total fatty acids)
Isochrysis galbana	0.9	
Nannochloropsis sp.	30.1	
Chaetoceros calcitrans	34	
Tetraselmis suecica	6.2	
Chaetoceros muelleri	12.8	0.8
Pavlova salina	19.1	1.5
Skeletonema costatum	40.7	2.3
Porphyridium cruentum	30.7	
Crypthecodinium cohnii		30
Chroomonas salina	12.9	7.1
Chaetoceros constriccus	18.8	0.6
Tetraselmis viridis	6.7	

Harvesting capability is another important feature of microalgae with biodiesel potential. Harvesting or dewatering can be best achieved through settling, flocculation or froth flotation [49,50]. For example, many microalgae settle under adverse conditions, and this can be tested under small scale conditions [51]. Lipid extraction efficiency from microalgae is dependent on residual water content after drying and in particular the structure of their cell wall. For example, *Nannochloropsis sp.* is regarded a highly productive microalga with strong potential for large-scale biodiesel production [43], but ideally requires pretreatment to open up the highly rigid cell walls for higher lipid extraction efficiency.

9.7 CONCLUSIONS

Development of biodiesel production from microalgae presents an important move to address the limitations posed by current first generation biodiesel crops. Microalgae, once developed for commercial biodiesel production, may offer many economical and environmental advantages. Current biodiesel production from microalgae is in the research phase, but is being developed to commercial scale in many countries. Finding promising microalgae for commercial cultivation is multi-facetted and challenging because particular microalgae strains have different requirements in terms of nutrients intake, environmental and culturing conditions and lipid extraction technology. However, diversity of lipid-producing microalgae species is one of the major advantages of this group of organisms that is likely to lead to selection of suitable algae crops to achieve algal biodiesel production in different regions. A combination of conventional and modern techniques is likely the most efficient route from isolation to large-scale cultivation (Figure 2). Careful initial analyses and far-sighted selection of microalgae with a view towards downstream processing and large-scale production with potential value-add products, is an important prerequisite to domesticate and develop algae crops for biodiesel production.

REFERENCES

1. Richmond, A. Handbook of Microalgal Culture: Biotechnology and Applied Phycology; Blackwell Science Ltd.: Hudson County, NJ, USA, 2004.
2. Mata, T.M.; Martins, A.A.; Caetano, N.S. Microalgae for biodiesel production and other applications: A review. Renew. Sustain. Energy Rev. 2010, 14, 217–232.
3. CSIRO. Australian national algae culture collection. Available online: http://www.csiro.au/ Organisation-Structure/National-Facilities/Australian-National-Algae-Culture-Collection.aspx (accessed on 19 January 2012).
4. Chisti, Y. Biodiesel from microalgae. Biotechnol. Adv. 2007, 25, 294–306.
5. Li, Q.; Du, W.; Liu, D. Perspectives of microbial oils for biodiesel production. Appl. Microbiol. Biotechnol. 2008, 80, 749–756.
6. European Biodiesel Board. The EU biodiesel industry. Available online: http://www.ebb-eu.org/stats.php (accessed on 18 January 2012).
7. Carriquiry, M. U.S. Bidiesel production: Recent developments and prospects. Iowa Agric. Rev. Online 2007, 13, 8–9.
8. TUSNBB. Production statistics. Available online: http://www.biodiesel.org/production/production-statistics (accesssed on 18 January 2012).
9. Wang, B.; Li, Y.; Wu, N.; Lan, C.Q. CO2 bio-mitigation using microalgae. Appl. Microbiol. Biotechnol. 2008, 79, 707–718.
10. Sheehan, J.; Dunahay, T.; Benemann, J.; Roessler, P. A Look Back at the U.S. Department of Energy's Aquatic Species Program: Biodiesel from Algae; National Renewable Energy Laboratory: Golden, Colorado, USA, 1998.
11. Delucchi, M.A. A Lifecycle Emission Model (LEM): Lifecycle Emissions from Transportation Fuels; Motor Vehicles, Transportation Modes, Electricity Use, Heating and Cooking Fuels; Institute of Transport Studies, University of California: Davis, CA, USA, 2003.
12. Paulson, N.D.; Ginder, R.D. The Growth and Direction of Biodiesel Industry in the United States; Center for Agricultural and Rural Development: Iowa State University, IA, USA, 2007.
13. Laboratory, N.R.E. Biodiesel Handling and Use Guide; The U.S. Department of Energy: Golden, Colorado, USA, 2009.
14. Fukuda, H.; Kondo, A.; Noda, H. Biodiesel fuel production by transesterification of oils. J. Biosci. Bioeng. 2001, 92, 405–416.
15. Bahadur, N.P.; Boocock, D.G.B.; Konar, S.K. Liquid hydrocarbons from catalytic pyrolysis of sewage sludge lipid and canola oil: Evaluation of fuel properties. Energy Fuels 1995, 9, 248–256.
16. Boateng, A.A.; Mullen, C.A.; Goldberg, N.; Hicks, K.B.; Jung, H.-J.G.; Lamb, J.F.S. Production of bio-oil from alfalfa stems by fluidized-Bed fast pyrolysis. Ind. Eng. Chem. Res. 2008, 47, 4115–4122.
17. Davey, H.M.; Kell, D.B. Flow cytometry and cell sorting of heterogeneous microbial populations: The importance of single-cell analyses. Microbiol. Rev. 1996, 60, 641–696.

18. Reckermann, M. Flow sorting in aquatic ecology. Sci. Mar. 2000, 64, 235–246.
19. Dinh, L.T.T.; Guo, Y.; Mannan, M.S. Sustainability evaluation of biodiesel production using multicriteria decision-making. Environ. Prog. Sustain. Energy 2009, 28, 38–46.
20. Rismani-Yazdi, H.; Haznedaroglu, B.Z.; Bibby, K.; Peccia, J. Transcriptome sequencing and annotation of the microalgae Dunaliella tertiolecta: Pathway description and gene discovery for production of next-generation biofuels. BMC Genomics 2011, 12, 148.
21. Tonon, T.; Harvey, D.; Qing, R.; Li, Y.; Larson, T.R.; Graham, I.A. Identification of a fatty acid Δ11-desaturase from the microalga Thalassiosira pseudonana. FEBS Lett. 2004, 563, 28–34.
22. Berard, A.; Dorigo, U.; Humbert, J.F.; Martin-Laurent, F. Microalgae community structure analysis based on 18S rDNA amplification from DNA extracted directly from soil as a potential soil bioindicator. Agronomie 2005, 25, 1–7.
23. Rasoul-Amini, S.; Ghasemi, Y.; Morowvat, M.H.; Mohagheghzadeh, A. PCR amplification of 18S rRNA, single cell protein production and fatty acid evaluation of some naturally isolated microalgae. Food Chem. 2009, 116, 129–136.
24. Matsumoto, M.; Sugiyama, H.; Maeda, Y.; Sato, R.; Tanaka, T.; Matsunaga, T. Marine diatom, Navicula sp. strain JPCC DA0580 and marine green alga, Chlorella sp. strain NKG400014 as potential sources for biodiesel production. Appl. Biochem. Biotechnol. 2010, 161, 483–490.
25. Timmins, M.; Thomas-Hall, S.R.; Darling, A.; Zhang, E.; Hankamer, B.; Marx, U.C.; Schenk, P.M. Phylogenetic and molecular analysis of hydrogen-producing green algae. J. Exp. Bot. 2009, 60, 1691–1702.
26. Yu, Y.; Chen, B.; You, W. Identification of the alga known as Nannochloropsis Z-1 isolated from a prawn farm in Hainan, China as Chlorella. World J. Microbiol. Biotechnol. 2007, 23, 207–210.
27. Radakovits, R.; Jinkerson, R.E.; Darzins, A.; Posewitz, M.C. Genetic engineering of algae for enhanced biofuel production. Eukaryot. Cell 2010, 9, 486–501.
28. Schuhmann, H.; Lim, D.K.Y.; Schenk, P.M. Perspectives on metabolic engineering for increased lipid contents in microalgae. Biofuels 2012, 3, 71–86.
29. Radakovits, R.; Jinkerson, R.E.; Fuerstenberg, S.I.; Tae, H.; Settlage, R.E.; Boore, J.L.; Posewitz, M.C. Draft genome sequence and genetic transformation of the oleaginous alga Nannochloropis gaditana. Nat. Commun. 2012, 3, 686.
30. Bligh, E.G.; Dyer, W.J. A rapid method for total lipid extraction and purification. Can. J. Biochem. Phys. 1959, 37, 911–917.
31. Eltgroth, M.L.; Watwood, R.L.; Wolfe, G.V. Production and cellular localization of neutral long-chain lipids in the haptophyte algae Isochrysis galbana and Emiliania huxleyi. J. Phycol. 2005, 41, 1000–1009.
32. Greenspan, P.; Mayer, E.P.; Fowler, S.D. Nile red—A selective fluorescent stain for intracellular lipid droplets. J. Cell Biol. 1985, 100, 965–973.
33. Chen, W.; Zhang, C.; Song, L.; Sommerfeld, M.; Hu, Q. A high throughput Nile red method for quantitative measurement of neutral lipids in microalgae. J. Microbiol. Methods 2009, 77, 41–47.

34. Chen, W.; Sommerfeld, M.; Hu, Q. Microwave-assisted Nile red method for in vivo quantification of neutral lipids in microalgae. Bioresour. Technol. 2011, 102, 135–141.
35. Chiu, K.Y.; Chan, K.W. Rapid immunofluorescence staining of human renal biopsy speciments using microwave irradiation. J. Clin. Pathol. 1987, 40, 689–692.
36. Muñoz, T.E.; Giberson, R.T.; Demaree, R.; Day, J.R. Microwave-assisted immunostaining: A new approach yields fast and consistent results. J. Neurosci. Methods 2004, 137, 133–139.
37. Spaulding, B.W. A Nile red staining method for the fluorescence detection of lipid in algae utilizing a FlowCAM: Biofuels digest. Available online: http://www.biofuelsdigest.com/bdigest/2010/05/05/a-nile-red-staining-method-for-the-fluorescence-detection-of-lipid-in-algae-utilizing-a-flowcam/ (accessed on 31 January 2012).
38. Schenk, P.; Thomas-Hall, S.; Stephens, E.; Marx, U.; Mussgnug, J.; Posten, C.; Kruse, O.; Hankamer, B. Second generation biofuels: High-efficiency microalgae for biodiesel production. Bioenerg. Res. 2008, 1, 20–43.
39. Rodolfi, L.; Chini Zittelli, G.; Bassi, N.; Padovani, G.; Biondi, N.; Bonini, G.; Tredici, M.R. Microalgae for oil: Strain selection, induction of lipid synthesis and outdoor mass cultivation in a low-cost photobioreactor. Biotechnol. Bioeng. 2009, 102, 100–112.
40. Li, Y.; Qin, J.G. Comparison of growth and lipid content in three Botryococcus braunii strains. J. Appl. Phycol. 2005, 17, 6.
41. Thomas, W.H.; Tornabene, T.G.; Weissman, J. Screening for Lipid Yielding Microalgae: Activities for 1983; Solar Energy Research Institute: Golden, Colorado, USA, 1984.
42. Al-Hasan, R.; Ali, A.; Ka'wash, H.; Radwan, S. Effect of salinity on the lipid and fatty acid composition of the halophyte Navicula sp.: Potential in mariculture. J. Appl. Phycol. 1990, 2, 215–222.
43. Pulz, O.P. Photobioreactors: Production systems for phototrophic microorganisms. Appl. Microbiol. Biotechnol. 2001, 57, 287–293.
44. Brown, A.C.; Knights, B.A.; Conway, E. Hydrocarbon content and its relationship to physiological state in the green alga Botryococcus braunii. Phytochemistry 1969, 8, 5.
45. Huerlimann, R.; de Nys, R.; Heimann, K. Growth, lipid content, productivity, and fatty acid composition of tropical microalgae for scale-up production. Biotechnol. Bioeng. 2010, 107, 245–257.
46. Miao, X.; Wu, Q. Biodiesel production from heterotrophic microalgal oil. Bioresour. Technol. 2006, 97, 841–846.
47. Knothe, G. Analyzing biodiesel: Standards and other methods. J. Am. Oil Chem. Soc. 2006, 83, 823–833.
48. Yan, L.; Schenk, P.M. Selection of Cultured Microalgae for Producing Omega-3 Bio-Lipid Oil; Report for Queensland Sea Scallop Trading Pty Ltd.; The University of Queensland: Queensland, Austrilia, 2011; p. 36.
49. Scholz, M.; Hoshino, T.; Johnson, D.; Riley, M.R.; Cuello, J. Flocculation of wall-deficient cells of Chlamydomonas reinhardtii mutant cw15 by calcium and methanol. Biomass Bioenerg. 2011, 35, 4835–4840.

50. Christenson, L.; Sims, R. Production and harvesting of microalgae for wastewater treatment, biofuels, and bioproducts. Biotechnol. Adv. 2011, 29, 686–702.
51. Park, J.B.K.; Craggs, R.J.; Shilton, A.N. Recycling algae to improve species control and harvest efficiency from a high rate algal pond. Water Res. 2011, 45, 6637–6649.

This chapter was originally published under the Creative Commons Attribution License. Duong, V. T., Li, Y., Nowak, E., and Schenk, P. M. Microalgae Isolation and Selection for Prospective Biodiesel Production. Energies 2012: 5; 1835–1849.

PART III

NEXT GENERATION RESEARCH

CHAPTER 10

COMPARISON OF NEXT-GENERATION SEQUENCING SYSTEMS

LIN LIU, YINHU LI, SILIANG LI, NI HU, YIMIN HE, RAY PONG, DANNI LIN, LIHUA LU, and MAGGIE LAW

10.1 INTRODUCTION

(Deoxyribonucleic acid) DNA was demonstrated as the genetic material by Oswald Theodore Avery in 1944. Its double helical strand structure composed of four bases was determined by James D. Watson and Francis Crick in 1953, leading to the central dogma of molecular biology. In most cases, genomic DNA defined the species and individuals, which makes the DNA sequence fundamental to the research on the structures and functions of cells and the decoding of life mysteries [1]. DNA sequencing technologies could help biologists and health care providers in a broad range of applications such as molecular cloning, breeding, finding pathogenic genes, and comparative and evolution studies. DNA sequencing technologies ideally should be fast, accurate, easy-to-operate, and cheap. In the past thirty years, DNA sequencing technologies and applications have undergone tremendous development and act as the engine of the genome era which is characterized by vast amount of genome data and subsequently broad range of research areas and multiple applications. It is necessary to look back on the history of sequencing technology development to review the NGS systems (454, GA/HiSeq, and SOLiD), to compare their advantages and disadvantages, to discuss the various applications, and to evaluate the recently introduced PGM (personal genome machines) and third-generation sequencing technologies and applications. All of these aspects will be described in this paper. Most data and conclusions are from independent users who have extensive first-hand experience in these typical NGS systems in BGI (Beijing Genomics Institute).

Before talking about the NGS systems, we would like to review the history of DNA sequencing briefly. In 1977, Frederick Sanger developed DNA sequencing technology which was based on chain-termination method (also known as Sanger sequencing), and Walter Gilbert developed another sequencing technology based on chemical modification of DNA and subsequent cleavage at specific bases. Because of its high efficiency and low radioactivity, Sanger sequencing was adopted as the primary technology in the "first generation" of laboratory and commercial sequencing applications [2]. At that time, DNA sequencing was laborious and radioactive materials were required. After years of improvement, Applied Biosystems introduced the first automatic sequencing machine (namely AB370) in 1987, adopting capillary electrophoresis which made the sequencing faster and more accurate. AB370 could detect 96 bases one time, 500 K bases a day, and the read length could reach 600 bases. The current model AB3730xl can output 2.88 M bases per day and read length could reach 900 bases since 1995. Emerged in 1998, the automatic sequencing instruments and associated software using the capillary sequencing machines and Sanger sequencing technology became the main tools for the completion of human genome project in 2001 [3]. This project greatly stimulated the development of powerful novel sequencing instrument to increase speed and accuracy, while simultaneously reducing cost and manpower. Not only this, X-prize also accelerated the development of next-generation sequencing (NGS) [4]. The NGS technologies are different from the Sanger method in aspects of massively parallel analysis, high throughput, and reduced cost. Although NGS makes genome sequences handy, the followed data analysis and biological explanations are still the bottle-neck in understanding genomes.

Following the human genome project, 454 was launched by 454 in 2005, and Solexa released Genome Analyzer the next year, followed by

(Sequencing by Oligo Ligation Detection) SOLiD provided from Agencourt, which are three most typical massively parallel sequencing systems in the next-generation sequencing (NGS) that shared good performance on throughput, accuracy, and cost compared with Sanger sequencing (shown in Table 1(a)). These founder companies were then purchased by other companies: in 2006 Agencourt was purchased by Applied Biosystems, and in 2007, 454 was purchased by Roche, while Solexa was purchased by Illumina. After years of evolution, these three systems exhibit better performance and their own advantages in terms of read length, accuracy, applications, consumables, man power requirement and informatics infrastructure, and so forth. The comparison of these three systems will be focused and discussed in the later part of this paper (also see Tables 1(a), 1(b), and 1(c)).

TABLE 1: (a) Advantage and mechanism of sequencers. (b) Components and cost of sequencers. (c) Application of sequencers.

(A)

Sequencer	454 GS FLX	HiSeq 2000	SOLiDv4	Sanger 3730xl
Sequencing mechanism	Pyrosequencing	Sequencing by synthesis	Ligation and two-base coding	Dideoxy chain termination
Read length	700 bp	50SE, 50PE, 101PE	50 + 35 bp or 50 + 50 bp	400 ~ 900 bp
Accuracy	99.9%*	98%, (100PE)	99.94% *raw data	99.999%
Reads	1 M	3 G	1200~1400 M	—
Output data/run	0.7 Gb	600 Gb	120 Gb	1.9~84 Kb
Time/run	24 Hours	3~10 Days	7 Days for SE 14 Days for PE	20 Mins~3 Hours
Advantage	Read length, fast	High throughput	Accuracy	High quality, long read length
Disadvantage	Error rate with polybase more than 6, high cost, low throughput	Short read assembly	Short read assembly	High cost low throughput

TABLE 1: *Cont.*

(B)

Sequencers	454 GS FLX	HiSeq 2000	SOLiDv4	3730xl
Instrument price	Instrument $500,000, $7000 per run	Instrument $690,000, $6000/(30x) human genome	Instrument $495,000, $15,000/100 Gb	Instrument $95,000, about $4 per 800 bp reaction
CPU	2* Intel Xeon X5675	2* Intel Xeon X5560	8* processor 2.0 GHz	Pentium IV 3.0 GHz
Memory	48 GB	48 GB	16 GB	1 GB
Hard disk	1.1 TB	3 TB	10 TB	280 GB
Automation in library preparation	Yes	Yes	Yes	No
Other required device	REM e system	cBot system	EZ beads system	No
Cost/million bases	$10	$0.07	$0.13	$2400

(C)

Sequencers	454 GS FLX	HiSeq 2000	SOLiDv4	3730xl
Resequencing		Yes	Yes	
De novo	Yes	Yes		Yes
Cancer	Yes	Yes	Yes	
Array	Yes	Yes	Yes	Yes
High GC sample	Yes	Yes	Yes	
Bacterial	Yes	Yes	Yes	
Large genome	Yes	Yes		
Mutation detection	Yes	Yes	Yes	Yes

(1) All the data is taken from daily average performance runs in BGI. The average daily sequence data output is about 8 Tb in BGI when about 80% sequencers (mainly HiSeq 2000) are running.
(2) The reagent cost of 454 GS FLX Titanium is calculated based on the sequencing of 400 bp; the reagent cost of HiSeq 2000 is calculated based on the sequencing of 200 bp; the reagent cost of SOLiDv4 is calculated based on the sequencing of 85 bp.
(3) HiSeq 2000 is more flexible in sequencing types like 50SE, 50PE, or 101PE.
(4) SOLiD has high accuracy especially when coverage is more than 30x, so it is widely used in detecting variations in resequencing, targeted resequencing, and transcriptome sequencing. Lanes can be independently run to reduce cost.

10.2 ROCHE 454 SYSTEM

Roche 454 was the first commercially successful next generation system. This sequencer uses pyrosequencing technology [5]. Instead of using dideoxynucleotides to terminate the chain amplification, pyrosequencing technology relies on the detection of pyrophosphate released during nucleotide incorporation. The library DNAs with 454-specific adaptors are denatured into single strand and captured by amplification beads followed by emulsion PCR [6]. Then on a picotiter plate, one of dNTP (dATP, dGTP, dCTP, dTTP) will complement to the bases of the template strand with the help of ATP sulfurylase, luciferase, luciferin, DNA polymerase, and adenosine 5′ phosphosulfate (APS) and release pyrophosphate (PPi) which equals the amount of incorporated nucleotide. The ATP transformed from PPi drives the luciferin into oxyluciferin and generates visible light [7]. At the same time, the unmatched bases are degraded by apyrase [8]. Then another dNTP is added into the reaction system and the pyrosequencing reaction is repeated.

The read length of Roche 454 was initially 100–150 bp in 2005, 200000+ reads, and could output 20 Mb per run [9, 10]. In 2008 454 GS FLX Titanium system was launched; through upgrading, its read length could reach 700 bp with accuracy 99.9% after filter and output 0.7 G data per run within 24 hours. In late 2009 Roche combined the GS Junior a bench top system into the 454 sequencing system which simplified the library preparation and data processing, and output was also upgraded to 14 G per run [11, 12]. The most outstanding advantage of Roche is its speed: it takes only 10 hours from sequencing start till completion. The read length is also a distinguished character compared with other NGS systems (described in the later part of this paper). But the high cost of reagents remains a challenge for Roche 454. It is about $\$12.56 \times 10^{-6}$ per base (counting reagent use only). One of the shortcomings is that it has relatively high error rate in terms of poly-bases longer than 6 bp. But its library construction can be automated, and the emulsion PCR can be semiautomated which could reduce the manpower in a great extent. Other informatics infrastructure and sequencing advantages are listed and compared with HiSeq 2000 and SOLiD systems in Tables 1(a), 1(b), and 1(c).

10.2.1 454 GS FLX TITANIUM SOFTWARE

GS RunProcessor is the main part of the GS FLX Titanium system. The software is in charge of picture background normalization, signal location correction, cross-talk correction, signals conversion, and sequencing data generation. GS RunProcessor would produce a series of files including SFF (standard flowgram format) files each time after run. SFF files contain the basecalled sequences and corresponding quality scores for all individual, high-quality reads (filtered reads). And it could be viewed directly from the screen of GS FLX Titanium system. Using GS De Novo Assembler, GS Reference Mapper and GS Amplicon Variant Analyzer provided by GS FLX Titanium system, SFF files can be applied in multiaspects and converted into fastq format for further data analyzing.

10.3 AB SOLID SYSTEM

(Sequencing by Oligo Ligation Detection) SOLiD was purchased by Applied Biosystems in 2006. The sequencer adopts the technology of two-base sequencing based on ligation sequencing. On a SOLiD flowcell, the libraries can be sequenced by 8 base-probe ligation which contains ligation site (the first base), cleavage site (the fifth base), and 4 different fluorescent dyes (linked to the last base) [10]. The fluorescent signal will be recorded during the probes complementary to the template strand and vanished by the cleavage of probes' last 3 bases. And the sequence of the fragment can be deduced after 5 round of sequencing using ladder primer sets.

The read length of SOLiD was initially 35 bp reads and the output was 3 G data per run. Owing to two-base sequencing method, SOLiD could reach a high accuracy of 99.85% after filtering. At the end of 2007, ABI released the first SOLiD system. In late 2010, the SOLiD 5500xl sequencing system was released. From SOLiD to SOLiD 5500xl, five upgrades were released by ABI in just three years. The SOLiD 5500xl realized improved read length, accuracy, and data output of 85 bp, 99.99%, and 30 G per run, respectively. A complete run could be finished within 7 days. The sequencing cost is about $40 × 10^{-9}$ per base estimated from reagent use only by BGI users. But the short read length and resequencing only in applications is still its major shortcoming [13]. Application of SOLiD includes whole genome resequencing, targeted resequencing, transcriptome research (including gene expression profiling, small RNA analysis, and whole transcriptome analysis), and epigenome (like ChIP-Seq and methylation). Like other NGS systems, SOLiD's computational infrastructure is expensive and not trivial to use; it requires an air-conditioned data center, computing cluster, skilled personnel in computing, distributed memory cluster, fast networks, and batch queue system. Operating system used by most researchers is GNU/LINUX. Each solid sequencer run takes 7 days and generates around 4TB of raw data. More data will be generated after bioinformatics analysis. This information is listed and compared with other NGS systems in Tables 1(a), 1(b), and 1(c). Automation can be used in library preparations, for example, Tecan system which integrated a Covaris A and Roche 454 REM e system [14].

10.3.1 SOLID SOFTWARE

After the sequencing with SOLiD, the original sequence of color coding will be accumulated. According to double-base coding matrix, the original color sequence can be decoded to get the base sequence if we knew the base types for one of any position in the sequence. Because of a kind of color corresponding four base pair, the color coding of the base will directly influence the decoding of its following base. It said that a wrong color coding will cause a chain decoding mistakes. BioScope is SOLiD data analysis package which provides a validated, single framework for resequencing, ChIP-Seq, and whole transcriptome analysis. It depends on reference for the follow-up data analysis. First, the software converts the base sequences of references into color coding sequence. Second, the color-coding sequence of references is compared with the original sequence of color-coding to get the information of mapping with newly developed mapping algorithm MaxMapper.

10.4 ILLUMINA GA/HISEQ SYSTEM

In 2006, Solexa released the Genome Analyzer (GA), and in 2007 the company was purchased by Illumina. The sequencer adopts the technology of sequencing by synthesis (SBS). The library with fixed adaptors is denatured to single strands and grafted to the flowcell, followed by bridge amplification to form clusters which contains clonal DNA fragments. Before sequencing, the library splices into single strands with the help of linearization enzyme [10], and then four kinds of nucleotides (ddATP, ddGTP, ddCTP, ddTTP) which contain different cleavable fluorescent dye and a removable blocking group would complement the template one base at a time, and the signal could be captured by a (charge-coupled device) CCD.

At first, solexa GA output was 1 G/run. Through improvements in polymerase, buffer, flowcell, and software, in 2009 the output of GA increased to 20 G/run in August (75PE), 30 G/run in October (100PE), and 50 G/run in December (Truseq V3, 150PE), and the latest GAIIx series can attain 85 G/run. In early 2010, Illumina launched HiSeq 2000, which adopts the same sequencing strategy with GA, and BGI was among the first globally to adopt the HiSeq system. Its output was 200 G per run initially, improved to 600 G per run currently which could be finished in 8 days. In the foreseeable future, it could reach 1 T/run when a personal genome cost could drop below $1 K. The error rate of 100PE could be below 2% in average after filtering (BGI's data). Compared with 454 and SOLiD, HiSeq 2000 is the cheapest in sequencing with $0.02/million bases (reagent counted only by BGI). With multiplexing incorporated in P5/P7 primers and adapters, it could handle thousands of samples simultaneously. HiSeq 2000 needs (HiSeq control software) HCS for program control, (real-time analyzer software) RTA to do on-instrument base-calling, and CASAVA for secondary analysis. There is a 3 TB hard disk in HiSeq 2000. With the aid of Truseq v3 reagents and associated softwares, HiSeq 2000 has improved much on high GC sequencing. MiSeq, a bench top sequencer launched in 2011 which shared most technologies with HiSeq, is especially convenient for amplicon and bacterial sample sequencing. It could sequence 150PE and generate 1.5 G/run in about 10 hrs including sample and library preparation time. Library preparation and their concentration measurement can both be automated with compatible systems like Agilent Bravo, Hamilton Banadu, Tecan, and Apricot Designs.

10.4.1 HISEQ SOFTWARE

HiSeq control system (HCS) and real-time analyzer (RTA) are adopted by HiSeq 2000. These two softwares could calculate the number and position of clusters based on their first 20 bases, so the first 20 bases of each sequencing would decide each sequencing's output and quality. HiSeq 2000 uses two lasers and four filters to detect four types of nucleotide (A, T, G, and C). The emission spectra of these four kinds of nucleotides have cross-talk, so the images of four nucleotides are not independent and the distribution of bases would affect the quality of sequencing. The standard sequencing output files of the HiSeq 2000 consist of *bcl files, which contain the base calls and quality scores in each cycle. And then it is converted into *_qseq.txt files by BCL Converter. The ELAND program of CASAVA (offline software provided by Illumina) is used to match a large number of reads against a genome.

In conclusion, of the three NGS systems described before, the Illumina HiSeq 2000 features the biggest output and lowest reagent cost, the SOLiD system has the highest accuracy [11], and the Roche 454 system has the longest read length. Details of three sequencing system are list in Tables 1(a), 1(b), and 1(c).

10.5 COMPACT PGM SEQUENCERS

Ion Personal Genome Machine (PGM) and MiSeq were launched by Ion Torrent and Illumina. They are both small in size and feature fast turnover rates but limited data throughput. They are targeted to clinical applications and small labs.

10.5.1 ION PGM FROM ION TORRENT

Ion PGM was released by Ion Torrent at the end of 2010. PGM uses semiconductor sequencing technology. When a nucleotide is incorporated into the DNA molecules by the polymerase, a proton is released. By detecting the change in pH, PGM recognized whether the nucleotide is added or not. Each time the chip was flooded with one nucleotide after another, if it is not the correct nucleotide, no voltage will be found; if there is 2 nucleotides added, there is double voltage detected [15]. PGM is the first commercial sequencing machine that does not require fluorescence and camera scanning, resulting in higher speed, lower cost, and smaller instrument size. Currently, it enables 200 bp reads in 2 hours and the sample preparation time is less than 6 hours for 8 samples in parallel.

An exemplary application of the Ion Torrent PGM sequencer is the identification of microbial pathogens. In May and June of 2011, an ongoing outbreak of exceptionally virulent Shiga-toxin- (Stx) producing *Escherichia coli* O104:H4 centered in Germany [16, 17], there were more than 3000 people infected. The whole genome sequencing on Ion Torrent PGM sequencer and HiSeq 2000 helped the scientists to identify the type of *E. coli* which would directly apply the clue to find the antibiotic resistance. The strain appeared to be a hybrid of two *E. coli* strains—entero aggregative *E. coli* and entero hemorrhagic *E. coli*—which may help explain why it has been particularly pathogenic. From the sequencing result of *E. coli* TY2482 [18], PGM shows the potential of having a fast, but limited throughput sequencer when there is an outbreak of new disease.

In order to study the sequencing quality, mapping rate, and GC depth distribution of Ion Torrent and compare with HiSeq 2000, a high GC Rhodobacter sample with high GC content (66%) and 4.2 Mb genome was sequenced in these two different sequencers (Table 2). In another experiment, *E. coli* K12 DH10B (NC_010473.1) with GC 50.78% was sequenced by Ion Torrent for analysis of quality value, read length, position accuracies, and GC distribution (Figure 1).

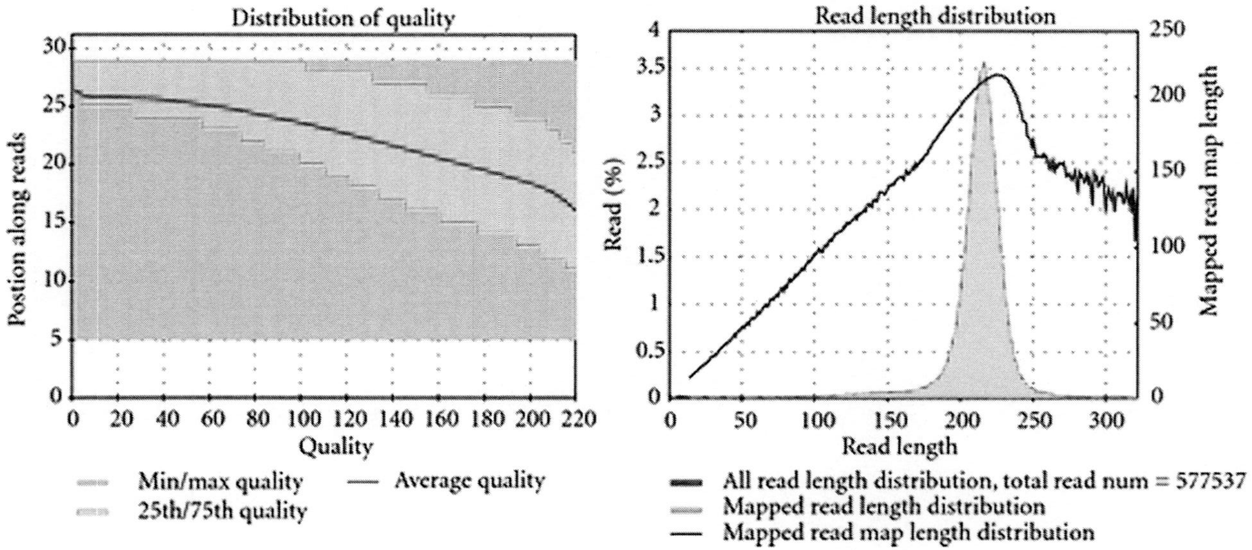

FIGURE 1: Ion Torrent sequencing quality. *E. coli* K12 DH10B (NC_010473.1) with GC 50.78% was used for this experiment. (a) is 314–200 bp from Ion Torrent. The left figure is quality value: pink range represents quality minimum and maximum values each position has. Green area represents the top and bottom quarter (1/4) reads of quality. Red line represents the average quality value in the position. The right figure is read length analysis: colored histogram represents the real read length. The black line represents the mapped length, and because it allows 3′ soft clipping, the length is different from the real read length. (b) is accuracy analysis. In each position, accuracy type including mismatch, insertion, and deletion is shown on the left y-axis. The average accuracy is shown the right y-axis. Accuracy of 200 bp sequencing could reach 99%. (c) is base composition along reads (left) and GC distribution analysis (right). The left figure is base composition in each position of reads. Base line splits after about 95 cycles indicating an inaccurate sequencing. The right one uses 500 bp window and the GC distribution is quite even. The data using high GC samples also indicates a good performance in Ion Torrent (data not shown).

TABLE 2: Comparison in alignment between Ion Torrent and HiSeq 2000.

	Ion Torrenta	HiSeq 2000b
Total reads num	165518	205683
Total bases num	18574086	18511470
Max read length	201	90
Min read length	15	90
Map reads num	157258	157511
Map rate	95%	76.57%
Covered rate	96.50%	93.11%
Total map length	15800258	14176420
Total mismatch base	53475	142425
Total insertion base	109550	1397
Total insertion num	95740	1332
Total deletion base	152495	431
Total deletion num	139264	238
Ave mismatch rate	0.338%	1.004%
Ave insertion rate	0.693%	0.009%
Ave deletion rate	0.965%	0.003%

a: use TMAP to align; b: use SOAP2 to align.

10.5.1.1 SEQUENCING QUALITY

The quality of Ion Torrent is more stable, while the quality of HiSeq 2000 decreases noticeably after 50 cycles, which may be caused by the decay of fluorescent signal with increasing the read length (shown in Figure 1).

10.5.1.2 MAPPING

The insert size of library of Rhodobacter was 350 bp, and 0.5 Gb data was obtained from HiSeq. The sequencing depth was over 100x, and the contig and scaffold N50 were 39530 bp and 194344 bp, respectively. Based on the assembly result, we used 33 Mb which is obtained from ion torrent with 314 chip to analyze the map rate. The alignment comparison is Table 2.

The map rate of Ion Torrent is higher than HiSeq 2000, but it is incomparable because of the different alignment methods used in different sequencers. Besides the significant difference on data including mismatch rate, insertion rate, and deletion rate, HiSeq 2000 and Ion Torrent were still incomparable because of the different sequencing principles. For example, the polynucleotide site could not be indentified easily in Ion Torrent. But it is shown that Ion Torrent has a stable quality along sequencing reads and a good performance on mismatch accuracies, but rather a bias in detection of indels. Different types of accuracy are analyzed and shown in Figure 1.

10.5.1.3 GC DEPTH DISTRIBUTION

The GC depth distribution is better in Ion Torrent from Figure 1. In Ion Torrent, the sequencing depth is similar while the GC content is from 63% to 73%. However in HiSeq 2000, the average sequencing depth is 4x when the GC content is 60%, while it is 3x with 70% GC content.

Ion Torrent has already released Ion 314 and 316 and planned to launch Ion 318 chips in late 2011. The chips are different in the number of wells resulting in higher production within the same sequencing time. The Ion 318 chip enables the production of >1 Gb data in 2 hours. Read length is expected to increase to >400 bp in 2012.

10.5.2 MISEQ FROM ILLUMINA

MiSeq which still uses SBS technology was launched by Illumina. It integrates the functions of cluster generation, SBS, and data analysis in a single instrument and can go from sample to answer (analyzed data) within a single day (as few as 8 hours). The Nextera, TruSeq, and Illumina's reversible terminator-based sequencing by synthesis chemistry was used in this innovative engineering. The highest integrity data and broader range of application, including amplicon sequencing, clone checking, ChIP-Seq, and small genome sequencing, are the outstanding parts of MiSeq. It is also flexible to perform single 36 bp reads (120 MB output) up to 2 × 150 paired-end reads (1–1.5 GB output) in MiSeq. Due to its significant improvement in read length, the resulting data performs better in contig assembly compared with HiSeq (data not shown). The related sequencing result of MiSeq is shown in Table 3. We also compared PGM with MiSeq in Table 4.

TABLE 3: MiSeq 150PE data.

Sample	GC	Q20	Q30
Human HPV	33.57; 33.62	98.26; 95.52	93.64; 88.52
Bacteria	61.33; 61.43	90.84; 83.86	78.46; 69.04

(1) The data in the table includes both read 1 and read 2 from paired-end sequencing.
(2) GC represents the GC content of libraries.
(3) Q20 value is the average Q20 of all bases in a read, which represents the ratio of bases with probability of containing no more than one error in 100 bases. Q30 value is the average Q30 of all bases in a read, which represents the ratio of bases with probability of containing no more than one error in 1,000 bases.

TABLE 4: The comparison between PGM and MiSeq.

	PGM	MiSeq
Output	10 MB–100 MB	120 MB–1.5 GB
Read length	~200 bp	Up to 2 × 150 bp
Sequencing time	2 hours for 1 × 200 bp	3 hours for 1 × 36 single read
		27 hours for 2 × 150 bp pair end read
Sample preparation time	8 samples in parallel, less than 6 hrs	As fast as 2 hrs, with 15 minutes hand on time
Sequencing method	semiconductor technology with a simple	
sequencing chemistry	Sequencing by synthesis (SBS)	
Potential for development	Various parameters	
(read length, cycle time, accuracy, etc.)	Limited factors, major concentrate in flowcell surface size, insert sizes, and how to pack cluster in tighter	
Input amount	μg	Ng (Nextera)
Data analysis	Off instrument	On instrument

10.5.3 COMPLETE GENOMICS

Complete genomics has its own sequencer based on Polonator G.007, which is ligation-based sequencer. The owner of Polonator G.007, Dover, collaborated with the Church Laboratory of Harvard Medical School, which is the same team as SOLiD system, and introduced this cheap open system. The Polonator could combine a high-performance instrument at very low price and the freely downloadable, open-source software and protocols in this sequencing system. The Polonator G.007 is ligation detection sequencing, which decodes the base by the single-base probe in nonanucleotides (nonamers), not by dual-base coding [19]. The fluorophore-tagged nonamers will be degenerated by selectively ligate onto a series of anchor primers, whose four components are labeled with one of four fluorophores with the help of T4 DNA ligase, which correspond to the base type at the query position. In the ligation progress, T4 DNA ligase is particularly sensitive to mismatches on 3′-side of the gap which is benefit to improve the accuracy of sequencing. After imaging, the Polonator chemically strips the array of annealed primer-fluorescent probe complex; the anchor primer is replaced and the new mixture are fluorescently tagged nonamers is introduced to sequence the adjacent base [20]. There are two updates compared with Polonator G.007, DNA nanoball (DNB) arrays, and combinatorial probe-anchor ligation (cPAL). Compared with DNA cluster or microsphere, DNA nanoball arrays obtain higher density of DNA cluster on the surface of a silicon chip. As the seven 5-base segments are discontinuous, so the system of hybridization-ligation-detection cycle has higher fault-tolerant ability compared with SOLiD. Complete genomics claim to have 99.999% accuracy with 40x depth and could analyze SNP, indel, and CNV with price 5500$–9500$. But Illumina reported a better performance of HiSeq 2000 use only 30x data (Illumina Genome Network). Recently some researchers compared CG's human genome sequencing data with Illumina system [21], and there are notable differences in detecting SNVs, indels, and system-specific detections in variants.

10.5.4 THE THIRD GENERATION SEQUENCER

While the increasing usage and new modification in next generation sequencing, the third generation sequencing is coming out with new insight in the sequencing. Third-generation sequencing has two main characteristics. First, PCR is not needed before sequencing, which shortens DNA preparation time for sequencing. Second, the signal is captured in real time, which means that the signal, no matter whether it is fluorescent (Pacbio) or electric current (Nanopore), is monitored during the enzymatic reaction of adding nucleotide in the complementary strand.

Single-molecule real-time (SMRT) is the third-generation sequencing method developed by Pacific Bioscience (Menlo Park, CA, USA), which made use of modified enzyme and direct observation of the enzymatic reaction in real time. SMRT cell consists of millions of zero-mode waveguides (ZMWs), embedded with only one set of enzymes and DNA template that can be detected during the whole process. During the reaction, the enzyme will incorporate the nucleotide into the complementary strand and cleave off the fluorescent dye previously linked with the nucleotide. Then the camera inside the machine will capture signal in a movie format in real-time observation [19]. This will give out not only the fluorescent signal but also the signal difference along time, which may be useful for the prediction of structural variance in the sequence, especially useful in epigenetic studies such as DNA methlyation [22].

Comparing to second generation, PacBio RS (the first sequencer launched by PacBio) has several advantages. First the sample preparation is very fast; it takes 4 to 6 hours instead of days. Also it does not need PCR step in the preparation step, which reduces bias and error caused by PCR. Second, the turnover rate is quite fast; runs are finished within a day. Third, the average read length is 1300 bp, which is longer than that of any second-generation sequencing technology. Although the throughput of the PacBioRS is lower than second-generation sequencer, this technology is quite useful for clinical laboratories, especially for microbiology research. A paper has been published using PacBio RS on the Haitian cholera outbreak [19].

	Prefilter	Post-QC filter*
Number of bases	84, 110, 272 bp	22, 373, 400 bp
Number of reads	46, 861	6, 754
Mean read length	513 bp	2, 566 bp
Mean read score	0.144	0.819

* MinRL = 50, MinRS = 0.75

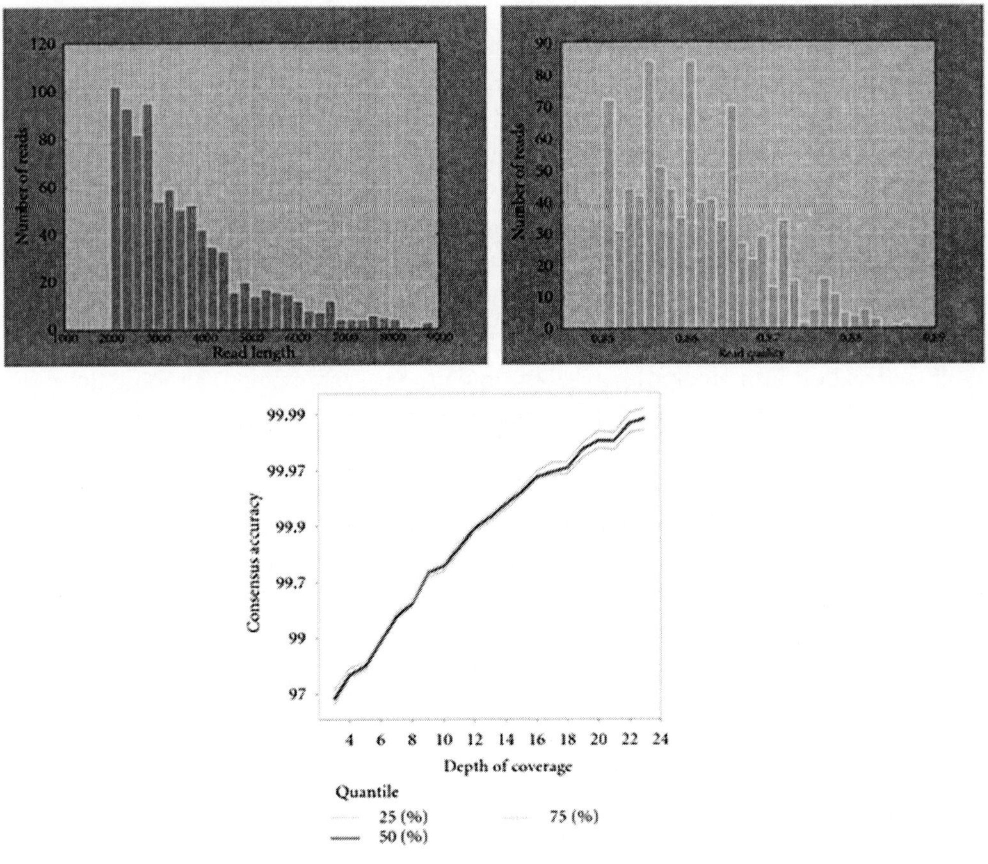

FIGURE 2: Sequencing of a fosmid DNA using Pacific Biosciences sequencer. With coverage, the accuracy could be above 97%. The figure was constructed by BGI's own data.

We have run a de novo assembly of DNA fosmid sample from Oyster with PacBio RS in standard sequencing mode (using LPR chemistry and SMRTcells instead of the new version FCR chemistry and SMRTcells). An SMRT belt template with mean insert size of 7500 kb is made and run in one SMRT cell and a 120-minute movie is taken. After Post-QC filter, 22,373,400 bp reads in 6754 reads (average 2,566 bp) were sequenced with the average Read Score of 0.819. The Coverage is 324x with mean read score of 0.861 and high accuracy (~99.95). The result is exhibited in Figure 2.

Nanopore sequencing is another method of the third generation sequencing. Nanopore is a tiny biopore with diameter in nanoscale [23], which can be found in protein channel embedded on lipid bilayer which facilitates ion exchange. Because of the biological role of nanopore, any particle movement can disrupt the voltage across the channel. The core concept of nanopore sequencing involves putting a thread of single-stranded DNA across α-haemolysin (αHL) pore. αHL, a 33 kD protein isolated from *Staphylococcus aureus* [20], undergoes self-assembly to form a heptameric transmembrane channel [23]. It can tolerate extraordinary voltage up to 100 mV with current 100 pA [20]. This unique property supports its role as building block of nanopore. In nanopore sequencing, an ionic flow is applied continuously. Current disruption is simply detected by standard electrophysiological technique. Readout is relied on the size difference between all deoxyribonucleoside monophosphate (dNMP). Thus, for given dNMP, characteristic current modulation is shown for discrimination. Ionic current is resumed after trapped nucleotide entirely squeezing out.

Nanopore sequencing possesses a number of fruitful advantages over existing commercialized next-generation sequencing technologies. Firstly, it potentially reaches long read length >5 kbp with speed 1 bp/ns [19]. Moreover, detection of bases is fluorescent tag-free. Thirdly, except the use of exonuclease for holding up ssDNA and nucleotide cleavage [24], involvement of enzyme is remarkably obviated in nanopore sequencing [22]. This implies that nanopore sequencing is less sensitive to temperature throughout the sequencing reaction and reliable outcome can be maintained. Fourthly, instead of sequencing DNA during polymerization, single DNA strands are sequenced through nanopore by means of DNA strand depolymerization. Hence, hand-on time for sample preparation such as cloning and amplification steps can be shortened significantly.

10.6 DISCUSSION OF NGS APPLICATIONS

Fast progress in DNA sequencing technology has made for a substantial reduction in costs and a substantial increase in throughput and accuracy. With more and more organisms being sequenced, a flood of genetic data is inundating the world every day. Progress in genomics has been moving steadily forward due to a revolution in sequencing technology. Additionally, other of types-large scale studies in exomics, metagenomics, epigenomics, and transcriptomics all become reality. Not only do these studies provide the knowledge for basic research, but also they afford immediate application benefits. Scientists across many fields are utilizing these data for the development of better-thriving crops and crop yields and livestock and improved diagnostics, prognostics, and therapies for cancer and other complex diseases.

BGI is on the cutting edge of translating genomics research into molecular breeding and disease association studies with belief that agriculture, medicine, drug development, and clinical treatment will eventually enter a new stage for more detailed understanding of the genetic components of all the organisms. BGI is primarily focused on three projects. (1) The Million Species/Varieties Genomes Project, aims to sequence a million economically and scientifically important plants, animals, and model organisms, including different breeds, varieties, and strains. This project is best represented by our sequencing of the genomes of the Giant panda, potato, macaca, and others, along with multiple resequencing projects. (2) The Million Human Genomes Project focuses on large-scale population and association studies that use whole-genome or whole-exome sequencing strategies. (3) The Million Eco-System Genomes Project has the objective of sequencing the metagenome and cultured microbiome of several different environments, including microenvironments within the human body [25]. Together they are called 3 M project.

In the following part, each of the following aspects of applications including de novo sequencing, mate-pair, whole genome or target-region resequencing, small RNA, transcriptome, RNA seq, epigenomics, and metagenomics, is briefly summarized.

In DNA de novo sequencing, the library with insert size below 800 bp is defined as DNA short fragment library, and it is usually applied in de novo and resequencing research. Skovgaard et al. [26] have applied a combination method of WGS (whole-genome sequencing) and genome copy number analysis to identify the mutations which could suppress the growth deficiency imposed by excessive initiations from the *E. coli* origin of replication, oriC.

Mate-pair library sequencing is significant beneficial for de novo sequencing, because the method could decrease gap region and extend scaffold length. Reinhardt et al. [27] developed a novel method for de novo genome assembly by analyzing sequencing data from high-throughput short read sequencing technology. They assembled genomes into large scaffolds at a fraction of the traditional cost and without using reference sequence. The assembly of one sample yielded an N50 scaffold size of 531,821 bp with >75% of the predicted genome covered by scaffolds over 100,000 bp.

Whole genome resequencing sequenced the complete DNA sequence of an organism's genome including the whole chromosomal DNA at a single time and alignment with the reference sequence. Mills et al. [28] constructed a map of unbalanced SVs (genomic structural variants) based on whole genome DNA sequencing data from 185 human genomes with SOLiD platform; the map encompassed 22,025 deletions and 6,000 additional SVs, including insertions and tandem duplications [28]. Most SVs (53%) were mapped to nucleotide resolution, which facilitated analyzing their origin and functional impact [28].

The whole genome resequencing is an effective way to study the functional gene, but the high cost and massive data are the main problem for most researchers. Target region sequencing is a solution to solve it. Microarray capture is a popular way of target region sequencing, which uses hybridization to arrays containing synthetic oligo-nucleotides matching the target DNA sequencing. Gnirke et al. [29] developed a captured method that uses an RNA "baits" to capture target DNA fragments from the "pond" and then uses the Illumina platform to read out the sequence. About 90% of uniquely aligning bases fell on or near bait sequence; up to 50% lay on exons proper [29].

Fehniger et al. used two platforms, Illumina GA and ABI SOLiD, to define the miRNA transcriptomes of resting and cytokine-activated primary murine NK (natural killer) cells [30]. The identified 302 known and 21 novel mature miRNAs were analyzed by unique bioinformatics pipeline from small RNA libraries of NK cell. These miRNAs are overexpressed in broad range and exhibit isomiR complexity, and a subset is differentially expressed following cytokine activation, which were the clue to identify the identification of miRNAs by the Illumina GA and SOLiD instruments [30].

The transcriptome is the set of all RNA molecules, including mRNA, rRNA, tRNA, and other noncoding RNA produced in one or a population of cells. In these years, next-generation sequencing technology is used to study the transcriptome compares with DNA microarray technology in the past. The S. mediterranea transcriptome could be sequenced by an efficient sequencing strategy which designed by Adamidi et al. [31]. The catalog of assembled transcripts and the identified peptides in this study dramatically expand and refine planarian gene annotation, which is demonstrated by validation of several previously unknown transcripts with stem cell-dependent expression patterns.

RNA-seq is a new method in RNA sequencing to study mRNA expression. It is similar to transcriptome sequencing in sample preparation, except the enzyme. In order to estimate the technical variance, Marioni et al. [32] analyzed a kidney RNA samples on both Illumina platform and Affymetrix arrays. The additional analyses such as low-expressed genes, alternative splice variants, and novel transcripts were found on Illumina platform. Bradford et al. [33] compared the data of RNA-seq library on the SOLiD platform and Affymetrix Exon 1.0ST arrays and found a high degree of correspondence between the two platforms in terms of exon-level fold changes and detection. And the greatest detection correspondence was seen when the background error rate is extremely low in RNA-seq. The difference between RNA-seq and transcriptome on SOLiD is not so obvious as Illumina.

There are two kinds of application of epigenetic, Chromatin immunoprecipitation and methylation analysis. Chromatin immunoprecipitation (ChIP) is an immunoprecipitation technique which is used to study the interaction between protein and DNA in a cell, and the histone modifies would be found by the specific location in genome. Based on next-generation sequencing technology, Johnson et al. [34] developed a large-scale chromatin immunoprecipitation assay to identify motif, especially noncanonical NRSF-binding motif. The data displays sharp resolution of binding position (±50 bp), which is important to infer new candidate interaction for the high sensitivity and specificity (ROC (receiver operator characteristic) area ≥0.96) and statistical confidence (P < 10–4). Another important application in epigenetic is DNA methylation analysis. DNA methylation exists typically in vertebrates at CpG sites; the methylation caused the conversion of the cytosine to 5-methylcytosine. Chung presented a whole methylome sequencing to study the difference between two kinds of bisulfite conversion methods (in solution versus in gel) by SOLiD platform [35].

The world class genome projects include the 1000 genome project, and the human ENCODE project, the human Microbiome (HMP) project, to name a few. BGI takes an active role in these and many more ongoing projects like 1000 Animal and Plant Genome project, the MetaHIT project, Yanhuang project, LUCAMP (Diabetes-associated Genes and Variations Study), ICGC (international cancer genome project), Ancient human genome, 1000 Mendelian Disorders Project, Genome 10K Project, and so forth [25]. These internationally collaborated genome projects greatly enhanced genomics study and applications in healthcare and other fields.

To manage multiple projects including large and complex ones with up to tens of thousands of samples, a superior and sophisticated project management system is required handling information processing from the very beginning of sample labeling and storage to library construction, multiplexing, sequencing, and informatics analysis. Research-oriented bioinformatics analysis and followup experiment processed are not included. Although automation techniques' adoption has greatly simplified bioexperiment human interferences, all other procedures carried out by human power have to be managed. BGI has developed BMS system and Cloud service for efficient information exchange and project management. The behavior management mainly follows Japan 5S onsite model. Additionally, BGI has passed ISO9001 and CSPro (authorized by Illumina) QC system and is currently taking (Clinical Laboratory Improvement Amendments) CLIA and (American Society for Histocompatibility and Immunogenetics) AShI tests. Quick, standard, and open reflection system guarantees an efficient troubleshooting pathway and high performance, for example, instrument design failure of Truseq v3 flowcell resulting in bubble appearance (which is defined as "bottom-middle-swatch" phenomenon by Illumina) and random N in reads. This potentially hazards sequencing quality, GC composition as well as throughput. It not only effects a small area where the bubble locates resulting in reading N but also effects the focus of the place nearby, including the whole swatch, and the adjacent swatch. Filtering parameters have to be determined to ensure quality raw data for bioinformatics processing. Lead by the NGS tech group, joint meetings were called for analyzing and troubleshooting this problem, to discuss strategies to best minimize effect in terms of cost and project time, to construct communication channel, to statistically summarize compensation, in order to provide best project management strategies in this time. Some reagent QC examples are summaried in Liu et al. [36].

BGI is establishing their cloud services. Combined with advanced NGS technologies with multiple choices, a plug-and-run informatics service is handy and affordable. A series of softwares are available including BLAST, SOAP, and SOAP SNP for sequence alignment and pipelines for RNAseq data. Also SNP calling programs such as Hecate and Gaea are about to be released. Big-data studies from the whole spectrum of life and biomedical sciences now can be shared and published on a new journal GigaSicence cofounded by BGI and Biomed Central. It has a novel publication format: each piece of data links to a standard manuscript publication with an extensive database which hosts all associated data, data analysis tools, and cloud-computing resources. The scope covers not just omic type data and the fields of high-throughput biology currently serviced by large public repositories but also the growing range of more difficult-to-access data, such as imaging, neuroscience, ecology, cohort data, systems biology, and other new types of large-scale sharable data.

REFERENCES

1. G. M. Church and W. Gilbert, "Genomic sequencing," Proceedings of the National Academy of Sciences of the United States of America, vol. 81, no. 7, pp. 1991–1995, 1984.
2. http://en.wikipedia.org/wiki/DNA_sequencing/.
3. F. S. Collins, M. Morgan, and A. Patrinos, "The Human Genome Project: lessons from large-scale biology," Science, vol. 300, no. 5617, pp. 286–290, 2003.
4. http://genomics.xprize.org/.

5. http://my454.com/products/technology.asp.
6. J. Berka, Y. J. Chen, J. H. Leamon, et al., "Bead emulsion nucleic acid amplification," U.S. Patent Application, 2005.
7. T. Foehlich, et al., "High-throughput nucleic acid analysis," U.S. Patent, 2010.
8. http://www.pyrosequencing.com/DynPage.aspx.
9. http://www.roche-applied-science.com/.
10. E. R. Mardis, "The impact of next-generation sequencing technology on genetics," Trends in Genetics, vol. 24, no. 3, pp. 133–141, 2008.
11. S. M. Huse, J. A. Huber, H. G. Morrison, M. L. Sogin, and D. M. Welch, "Accuracy and quality of massively parallel DNA pyrosequencing," Genome Biology, vol. 8, no. 7, article R143, 2007.
12. "The new GS junior sequencer," http://www.gsjunior.com/instrument-workflow.php.
13. "SOLiD system accuray," http://www.appliedbiosystems.com/absite/us/en/home/applications-technologies/solid-next-generation-sequencing.html.
14. http://www.tecan.com/platform/apps/product/index.asp?MenuID=3465&ID=7191&Menu=1&Item=33.52.2.
15. B. A. Flusberg, D. R. Webster, J. H. Lee et al., "Direct detection of DNA methylation during single-molecule, real-time sequencing," Nature Methods, vol. 7, no. 6, pp. 461–465, 2010.
16. A. Mellmann, D. Harmsen, C. A. Cummings et al., "Prospective genomic characterization of the german enterohemorrhagic Escherichia coli O104:H4 outbreak by rapid next generation sequencing technology," PLoS ONE, vol. 6, no. 7, Article ID e22751, 2011.
17. H. Rohde, J. Qin, Y. Cui, et al., "Open-source genomic analysis of Shiga-toxin-producing *E. coli* O104:H4," New England Journal of Medicine, vol. 365, no. 8, pp. 718–724, 2011.
18. C. S. Chin, J. Sorenson, J. B. Harris et al., "The origin of the Haitian cholera outbreak strain," New England Journal of Medicine, vol. 364, no. 1, pp. 33–42, 2011.
19. W. Timp, U. M. Mirsaidov, D. Wang, J. Comer, A. Aksimentiev, and G. Timp, "Nanopore sequencing: electrical measurements of the code of life," IEEE Transactions on Nanotechnology, vol. 9, no. 3, pp. 281–294, 2010.
20. D. W. Deamer and M. Akeson, "Nanopores and nucleic acids: prospects for ultrarapid sequencing," Trends in Biotechnology, vol. 18, no. 4, pp. 147–151, 2000.
21. "Performance comparison of whole-genome sequencing systems," Nature Biotechnology, vol. 30, pp. 78–82, 2012.
22. D. Branton, D. W. Deamer, A. Marziali, et al., "The potential and challenges of nanopore sequencing," Nature Biotechnology, vol. 26, no. 10, pp. 1146–1153, 2008.
23. L. Song, M. R. Hobaugh, C. Shustak, S. Cheley, H. Bayley, and J. E. Gouaux, "Structure of staphylococcal α-hemolysin, a heptameric transmembrane pore," Science, vol. 274, no. 5294, pp. 1859–1866, 1996.
24. J. Clarke, H. C. Wu, L. Jayasinghe, A. Patel, S. Reid, and H. Bayley, "Continuous base identification for single-molecule nanopore DNA sequencing," Nature Nanotechnology, vol. 4, no. 4, pp. 265–270, 2009.
25. Website of BGI, http://www.genomics.org.cn.

26. O. Skovgaard, M. Bak, A. Løbner-Olesen, et al., "Genome-wide detection of chromosomal rearrangements, indels, and mutations in circular chromosomes by short read sequencing," Genome Research, vol. 21, no. 8, pp. 1388–1393, 2011.
27. J. A. Reinhardt, D. A. Baltrus, M. T. Nishimura, W. R. Jeck, C. D. Jones, and J. L. Dangl, "De novo assembly using low-coverage short read sequence data from the rice pathogen Pseudomonas syringae pv. oryzae," Genome Research, vol. 19, no. 2, pp. 294–305, 2009.
28. R. E. Mills, K. Walter, C. Stewart et al., "Mapping copy number variation by population-scale genome sequencing," Nature, vol. 470, no. 7332, pp. 59–65, 2011.
29. A. Gnirke, A. Melnikov, J. Maguire et al., "Solution hybrid selection with ultra-long oligonucleotides for massively parallel targeted sequencing," Nature Biotechnology, vol. 27, no. 2, pp. 182–189, 2009.
30. T. A. Fehniger, T. Wylie, E. Germino et al., "Next-generation sequencing identifies the natural killer cell microRNA transcriptome," Genome Research, vol. 20, no. 11, pp. 1590–1604, 2010.
31. C. Adamidi, Y. Wang, D. Gruen et al., "De novo assembly and validation of planaria transcriptome by massive parallel sequencing and shotgun proteomics," Genome Research, vol. 21, no. 7, pp. 1193–1200, 2011.
32. J. C. Marioni, C. E. Mason, S. M. Mane, M. Stephens, and Y. Gilad, "RNA-seq: an assessment of technical reproducibility and comparison with gene expression arrays," Genome Research, vol. 18, no. 9, pp. 1509–1517, 2008.
33. J. R. Bradford, Y. Hey, T. Yates, Y. Li, S. D. Pepper, and C. J. Miller, "A comparison of massively parallel nucleotide sequencing with oligonucleotide microarrays for global transcription profiling," BMC Genomics, vol. 11, no. 1, article 282, 2010.
34. D. S. Johnson, A. Mortazavi, R. M. Myers, and B. Wold, "Genome-wide mapping of in vivo protein-DNA interactions," Science, vol. 316, no. 5830, pp. 1497–1502, 2007.
35. H. Gu, Z. D. Smith, C. Bock, P. Boyle, A. Gnirke, and A. Meissner, "Preparation of reduced representation bisulfite sequencing libraries for genome-scale DNA methylation profiling," Nature Protocols, vol. 6, no. 4, pp. 468–481, 2011.
36. L. Liu, N. Hu, B. Wang, et al., "A brief utilization report on the Illumina HiSeq 2000 sequencer," Mycology, vol. 2, no. 3, pp. 169–191, 2011.

This chapter was originally published under the Creative Commons Attribution License. Liu, L., Li, Y., Li, S., Hu, N., He, Y., Pong, R., Lin, D., Lu L., and Law, M. Comparison of Next-Generation Sequencing Systems. Journal of Biomedicine and Biotechnology 2012: 2012.

CHAPTER 11

ALGAL FUNCTIONAL ANNOTATION TOOL: A WEB-BASED ANALYSIS SUITE TO FUNCTIONALLY INTERPRET LARGE GENE LISTS USING INTEGRATED ANNOTATION AND EXPRESSION DATA

DAVID LOPEZ, DAVID CASERO, SHAWN J. COKUS, SABEEHA S. MERCHANT, and MATTEO PELLEGRINI

11.1 BACKGROUND

Next-generation sequencers are revolutionizing our ability to sequence the genomes of new algae efficiently and in a cost effective manner. Several assembly tools have been developed that take short read data and assemble it into large continuous fragments of DNA. Gene prediction tools are also available which identify coding structures within these fragments. The resulting transcripts can then be analyzed to generate predicted protein sequences. The function of these protein sequences are subsequently determined by searching for close homologs in protein databases and transferring the annotation between the two proteins. While some versions of the previously described data processing pipeline have become commonplace in genome projects, the resulting functional annotation is typically fairly minimal and includes only limited biological pathway information and protein structure annotation. In contrast, the integration of a variety of pathway, function and protein databases allows for the generation of much richer and more valuable annotations for each protein.

A second challenge is the use of these protein-level annotations to interpret the output of genome-scale profiling experiments. High-throughput genomic techniques, such as RNA-seq experiments, produce measurements of large numbers of genes relevant to the biological processes being studied. In order to interpret the biological relevance of these gene lists, which commonly range in size from hundreds to thousands of genes, the members must be functionally classified into biological pathways and cellular mechanisms. Traditionally, the genes within these lists are examined using independent annotation databases to assign functions and pathways. Several of these annotation databases, such as the Kyoto Encyclopedia of Genes and Genomes (KEGG) [1], MetaCyc [2], and Pfam [3], include a rich set of functional data useful for these purposes.

However, presently researchers must explore these different knowledge bases separately, which requires a substantial amount of time and effort. Furthermore, without systematic integration of annotation data, it may be difficult to arrive at a cohesive biological picture. In addition, many of these annotation databases were designed to accommodate a single gene search, a methodology not optimal for functionally interpreting the large lists of genes derived from high-throughput genomic techniques. Thus, while modern genomic experiments generate data for many genes in parallel, their output must often still be analyzed on a gene-by-gene basis across different databases. This fragmented analysis approach presents a significant bottleneck in the pipeline of biological discovery.

One approach to solving this problem is integrating information from multiple annotation databases and providing access to the combined biological data from a single comprehensive portal that is equipped with the proper statistical foundations to effectively analyze large gene lists. For example, the DAVID database integrates information from several pathway, ontology, and protein family databases [4]. Similarly, Ingenuity Pathway Analysis (IPA) provides an integrated knowledge base derived from published literature for the human genome [5]. The integrated functional information and annotation terms are then assigned to lists of genes and for some analyses, enrichment tests are performed to determine which biological terms are overrepresented within the group of genes. By combining the information found in a number of knowledge bases and performing the analysis of lists of genes, these tools permit the efficient processing of high-throughput genomic experiments and thus expedite the process of biological discovery. However, most of these integrated databases have been developed for the analysis of well-annotated and thoroughly studied organisms, and are lacking for many newly genome-enabled organisms.

One large group of organisms for which integrated functional databases are lacking are the algae. The algae constitute a branch in the plant kingdom, although they form a polyphyletic group as they do not include all the descendants of their last common ancestor. As many as 10 algal genomes have been sequenced, including those of a red alga and several chlorophyte algae, with several more in the pipeline [6-11]. Algal genomic studies have provided insights into photosymbiosis, evolutionary relationships between the different species of algae, as well as their unique properties and adaptations. Recently, there has been a renewed interest in the study of algal biochemistry and biology for their potential use in the development of renewable biofuels [reviewed in [12]]. This has promoted the study of varied biochemical processes in diverse algae, such as hydrogen metabolism, fermentation, lipid biosynthesis, photosynthesis and nutrient assimilation [13-20]. One of the most studied algae is *Chlamydomonas reinhardtii*. It has a sequenced genome that has been assembled into large scaffolds that are placed on to chromosomes [6]. For many years, *Chlamydomonas* has served as a reference organism for the study of photosynthesis, photoreceptors, chloroplast biology and diseases involving flagellar dysfunction [21-25]. Its transcriptome has recently been profiled by RNA-seq experiments under various conditions of nutrient deprivation [[26,27], unpublished data (Castruita M., et al.)].

While *Chlamydomonas* has been extensively characterized experimentally, annotation of its genome is still approximate. Although KEGG categorizes some *C. reinhardtii* gene models into biological pathways, other databases - such as Reactome [28] - do not directly provide information for proteins of this green alga. Complicating the analysis of *Chlamydomonas* genes is the fact that there are two assemblies of the genome in use (version 3 and version 4) and multiple sets of gene models have been developed that are catalogued under diverse identifiers: Joint Genome Institute (JGI) FM3.1 protein IDs for the version 3 assembly, and JGI version FM4 protein IDs and Augustus version 5 IDs for the version 4 assembly [11,29]. The differences between these assemblies are significant; for example, the version 3 assembly contains 1,557 continuous segments of sequence while the fourth version contains 88. Although the version 3 assembly is superseded by version 4, users presently access version 3 because of the richer user-based functional annotations. In addition, other sets of gene predictions have been generated using a variety of additional data, including ESTs and RNA-seq data, to more accurately delineate start and stop positions and improve upon existing gene models. One such gene prediction set is Augustus u10.2. As such, there are a variety of gene models between different assemblies being simultaneously used by researchers, presenting complications in genomics studies. To facilitate the analysis of *Chlamydomonas* genome-scale data, we developed the Algal Functional Annotation Tool, which provides a comprehensive analysis suite for functionally interpreting *C. reinhardtii* genes across all available protein identifiers. This web-based tool provides an integrative data-mining environment that assigns pathway, ontology, and protein family terms to proteins of *C. reinhardtii* and enables term enrichment analysis for lists of genes. Expression data for several experimental conditions are also integrated into the tool, allowing the determination of overrepresented differentially expressed conditions. Additionally, a gene similarity search tool allows for genes with similar expression patterns to be identified based on expression levels across these conditions.

11.2 CONSTRUCTION AND CONTENT

11.2.1 INTEGRATION OF MULTIPLE ANNOTATION DATABASES

The Algal Functional Annotation Tool integrates annotation data from the biological knowledge bases listed in Table 1. Publically available flat files containing annotation data were downloaded and parsed for each individual resource. *Chlamydomonas* reinhardtii proteins were assigned KEGG pathway annotations by means of sequence similarity to proteins within the KEGG genes database [1]. MetaCyc [2], Reactome [28], and Panther [30] pathway annotations were assigned to *C. reinhardtii* proteins by sequence similarity to subsets of UniProt IDs annotated in each corresponding database. In all cases, sequence similarity was determined by BLAST. BLAST results were filtered to contain only best hits with an E-value < 1e-05.

TABLE 1: List of annotation resources integrated into the Algal Functional Annotation Tool

Resource	URL	Reference
KEGG	http://www.genome.jp/kegg/	[1]
MetaCyc	http://www.metacyc.org/	[2]
Pfam	http://pfam.sanger.ac.uk/	[3]
Reactome	http://www.reactome.org/	[28]
Panther	http://www.pantherdb.org/pathway	[30]
Gene Ontology	http://www.geneontology.org/	[31]
InterPro	http://www.ebi.ac.uk/interpro	[32]
MapMan Ontology	http://mapman.gabipd.org/	[33]
KOG	http://www.ncbi.nlm.nih.gov/COG/grace/shokog.cgi	[35]

Primary databases used to functionally annotate gene models and integrated into the Algal Functional Annotation Tool.

Gene Ontology (GO) [31] terms were downloaded from the *Chlamydomonas* reinhardtii annotation provided by JGI. These GO terms were associated with their respective ancestors in the hierarchical ontology structure to include broader functional terms and provide a complete annotation set. Pfam domain annotations were assigned by direct search against protein domain signatures provided by Pfam. InterPro [32] and user-submitted manual annotations are based on those contained within JGI's annotation of the *C. reinhardtii* genome [11]. These methods were applied to four types of gene identifiers commonly used for *C. reinhardtii* proteins: JGI protein identifiers (versions 3 and 4) and Augustus gene models (versions 5 and 10.2). In total, over 12,600 unique functional annotation terms were assigned to 65,494 *C. reinhardtii* gene models spanning four different gene identifier types by these methods (Table 2). These assigned annotations may be explored for single genes using a built-in keyword search tool as well as an integrated annotation lookup tool which displays all annotations for a particular identifier.

11.2.2 ASSIGNMENT OF ANNOTATION FROM ARABIDOPSIS THALIANA

To extend the terms associated with *C. reinhartdii* genes, functional terms were inferred by homology to the annotation set of the plant Arabidopsis thaliana (thale cress). Identification of orthologous proteins was based on sequence similarity and subsequent filtering of the results by retaining only mutual best hits between the two sets of protein sequences. The corresponding Arabidopsis thaliana annotation was used to supplement GO terms and was similarly expanded to contain term ancestry. The A. thaliana annotations of the MapMan Ontology [33] and MetaCyc Pathway database [2] were also used to provide more complete annotation coverage of the C. reinhardii genome.

TABLE 2: Number of gene identifiers associated with annotation databases

Identifier Type	Total Gene IDs	KEGG	Reactome	Panther	Gene Ontology	MapMan	KOG	Pfam	InterPro
JGI v3.0	14598	5348	2740	1147	6563	5214	9139	7166	7532
JGI v4.0	16706	4232	1949	1085	7568	3171	9973	7305	8151
Augustus v5.0	16888	4686	2983	1673	4334	3160	5123	8202	5202
Augustus u10.2	17302	4583	3326	1913	6956	3892	8977	8691	7464

Number of Chlamydomonas reinhardtii identifiers with at least one functional annotation for each primary database, shown per identifier type.

11.2.3 FUNCTIONAL TERM ENRICHMENT TESTING

The hypergeometric distribution is commonly used to determine the significance of functional term enrichment within a list of genes. In this test, the occurrence of a functional term within a gene list is compared to the background level of occurrence across all genes in the genome to determine the degree of enrichment. A p-value based on this test can be calculated from four parameters: (1) the number of genes within the list, (2) the frequency of a term within the gene list, (3) the total number of genes within the genome, and (4) the frequency of a term across all genes in the genome. This test effectively distinguishes truly overrepresented terms from those occurring at a high frequency across all genes in the genome and therefore within the gene list as well. The cumulative hypergeometric test assigns a p-value to each functional term associated with genes within a given list, and all functional terms are ranked by ascending p-value (i.e. by descending levels of enrichment). Huang et al. reviews the use of the hypergeometric test for functional term enrichment [34]. The Algal Functional Annotation Tool computes hypergeometric p-values using a Perl wrapper for the GNU Scientific Library cumulative hypergeometric function written in C to provide a quick and accurate implementation of this statistical test.

11.2.4 DYNAMIC VISUALIZATION OF KEGG PATHWAY MAPS

Individual pathway maps from KEGG provide information on protein localization within the cell, compartmentalization into different cellular components, or of reactions within a larger metabolic process. Visualization of proteins from gene lists onto pathway maps is useful for their interpretation. The Algal Functional Annotation Tool utilizes the publicly available KEGG application programming interface (API) for pathway highlighting. The information linking *C. reinhardtii* proteins to identifiers within the KEGG database is used to determine the subset of KEGG IDs within the supplied gene list associated with a particular pathway. The Algal Functional Annotation Tool also deduces which proteins within the pathway are located within the genome of *C. reinhardtii* but not found in the gene list and sends the corresponding identifiers to the KEGG API to be highlighted in a different background color. This API interface is implemented using the SOAP architecture for web applications.

11.2.5 INTEGRATION OF EXPRESSION DATA

The expression levels of *C. reinhardtii* genes have been experimentally characterized under numerous conditions using high-throughput methods such as RNA-seq [[26,27], unpublished data (Castruita M., et al.)]. These expression data were compiled and analyzed to determine which genes are over- and under-expressed in each experimental condition. The expression data was preprocessed to normalize the counts for uniquely mappable reads in any experiment. Genes exhibiting greater than a two-fold change in expression compared to average expression across all conditions with a Poisson cumulative p-value of less than 0.05 were considered differentially expressed. Using this data, *C. reinhardtii* genes were associated with conditions in which they were over- and under-expressed.

The compiled expression data was also analyzed to find functionally related genes based on their expression levels across the different experimental conditions [[26,27], unpublished data (Castruita M., et al.)]. Genes demonstrating low variance of expression across all samples were not considered. This analysis was performed for three representations of the expression data: absolute counts, log counts, and log ratios of expression. By this method, *C. reinhardtii* genes are each associated with 100 genes with the most similar expression patterns to determine potentially functionally related genes.

11.2.6 GENE IDENTIFIER CONVERSION

Due to the existence of several protein identifier types (FM3.1, FM4, Au5, Au10.2), different identifiers are associated with an individual protein within the *Chlamydomonas* genome. In order to extend annotations from one identifier type to another, matching protein identifiers are deduced by sequence similarity filtering for mutual best hits between identifiers using BLAST. Matching identifiers with 100% sequence coverage are kept, and the rest of the mutual best hits are filtered to include only those proteins with matches with at least 75% coverage. Potential ambiguities involving proteins similar to multiple other proteins are resolved by considering only the reciprocal best hit from the BLAST query in the opposite direction. The information derived by this analysis is used to convert gene identifiers between different types, which allows the Algal Annotation Tool to work with multiple protein identifier types.

11.2.7 WEB-BASED INTERFACE AND UPDATES

The web interface of the Algal Functional Annotation Tool consists of a set of portals that give access to the different types of analyses available. Results are shown within expandable/collapsible HTML tables that display annotation information along with the statistical results of the analysis. When expanded, the results table shows which gene identifiers contain a specific annotation along with further information regarding matching gene identifiers and BLAST E-values. Updates to the Algal Functional Annotation Tool are semi-automated using a set of Perl scripts that parse and process updated flat files from the various integrated annotation databases at regular intervals. Currently, functional data from the primary annotation databases is set to be updated every 4 months.

11.3 UTILITY AND DISCUSSION

11.3.1 COMPREHENSIVE, INTEGRATED DATA-MINING ENVIRONMENT

The Algal Functional Annotation Tool is composed of three main components - functional term enrichment tests (which are separated by type), a batch gene identifier conversion tool, and a gene similarity search tool. A 'Quick Start' analysis is provided from the front page, featuring enrichment analysis using a sample set of databases containing the richest set of annotations (Figure 1). From any page, the sidebar provides access to the 'Quick Start' function of the tool.

Numerous other enrichment analyses - including enrichment using pathway, ontology, protein family, or differential expression data - are available within the Algal Functional Annotation Tool. Enrichment results

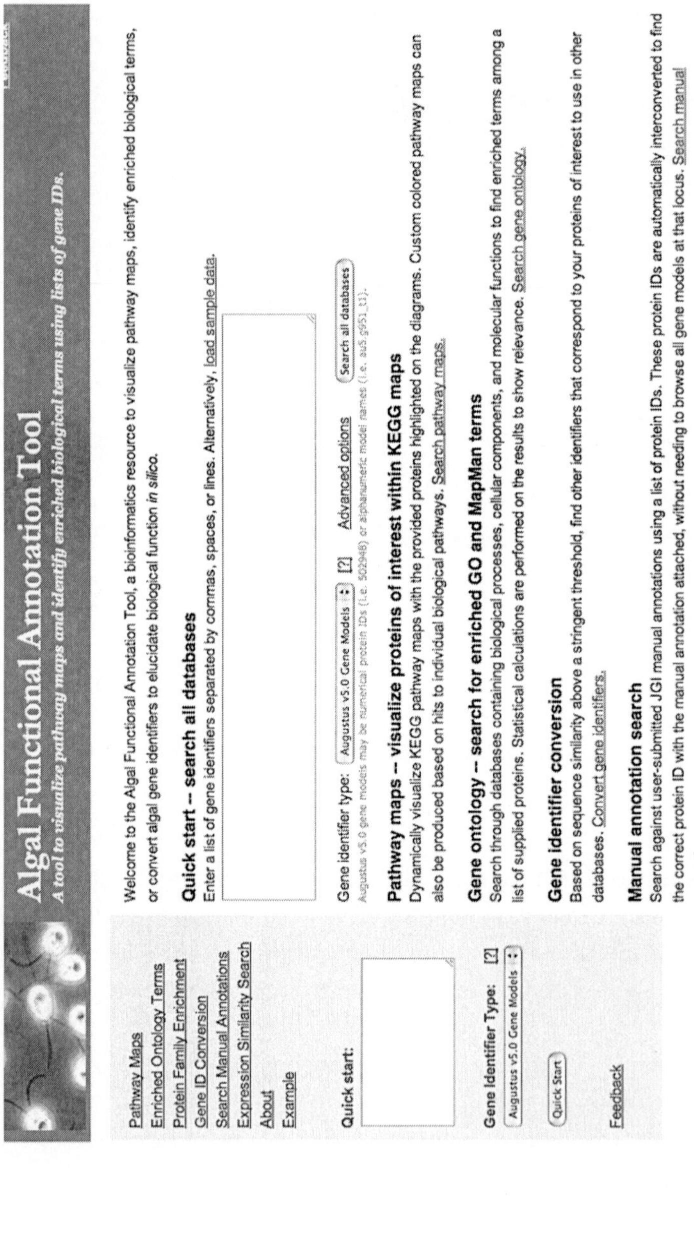

FIGURE 1: Algal Functional Annotation Tool. The front page of the Algal Functional Annotation Tool. A 'Quick Start' analysis is available to test for enrichment using the richest annotation databases included in the tool. Other features accessible from the sidebar include more specific enrichment tests (based on biological pathways, ontology terms, or protein families), a gene identifier conversion tool, a manual annotation search tool, and an expression similarity search tool.

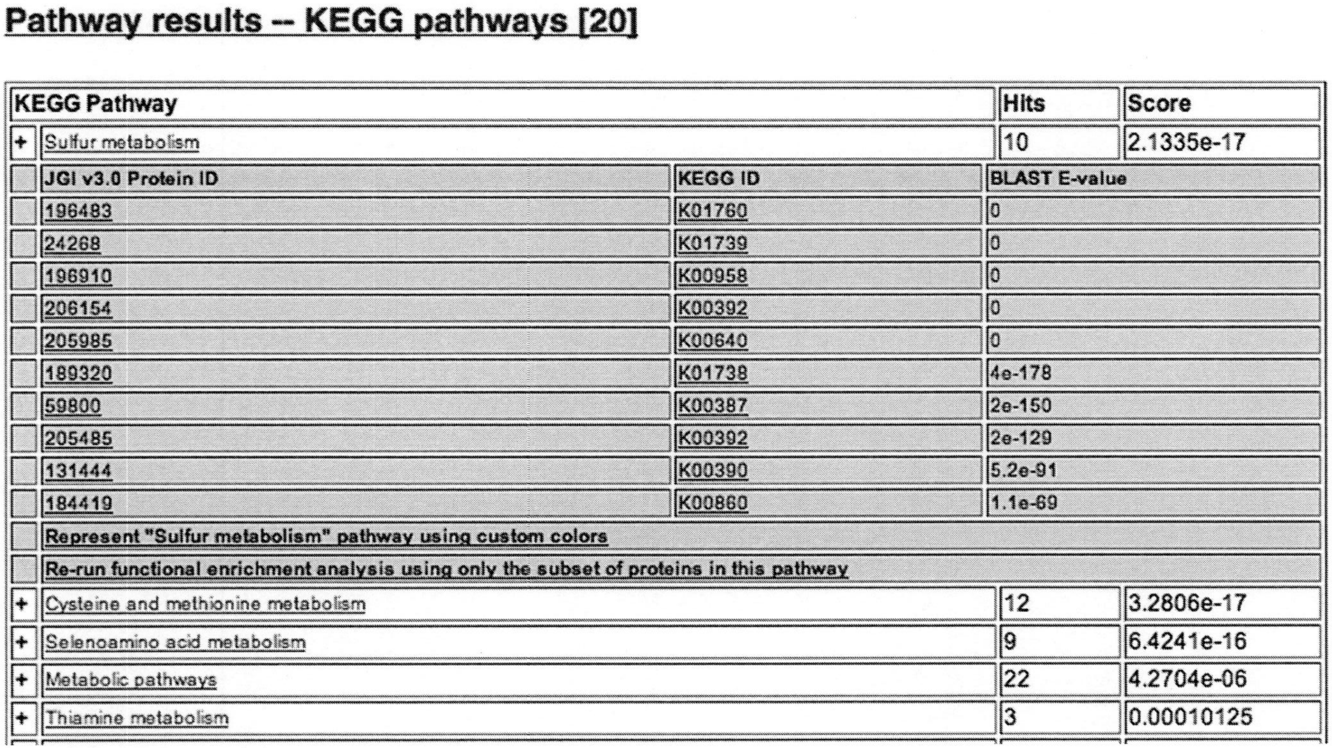

Pathway results -- KEGG pathways [20]

	KEGG Pathway	Hits	Score
+	Sulfur metabolism	10	2.1335e-17
	JGI v3.0 Protein ID	KEGG ID	BLAST E-value
	196483	K01760	0
	24268	K01739	0
	196910	K00958	0
	206154	K00392	0
	205985	K00640	0
	189320	K01738	4e-178
	59800	K00387	2e-150
	205485	K00392	2e-129
	131444	K00390	5.2e-91
	184419	K00860	1.1e-69
	Represent "Sulfur metabolism" pathway using custom colors		
	Re-run functional enrichment analysis using only the subset of proteins in this pathway		
+	Cysteine and methionine metabolism	12	3.2806e-17
+	Selenoamino acid metabolism	9	6.4241e-16
+	Metabolic pathways	22	4.2704e-06
+	Thiamine metabolism	3	0.00010125

FIGURE 2: Annotation Enrichment Results. Annotation enrichment results, sorted by ascending hypergeometric p-values, are shown in expandible/collapsible HTML tables such as the one shown. When expanded, the genes within the user-submitted list containing the expanded annotation are shown alongside additional statistical information. All results are downloadable as tab-delimited text files.

are always sorted by hypergeometric p-value and whenever possible contain links to the primary database's entry for that annotation or to the protein page of the gene identifier. The number of hits to a certain annotation term are also displayed alongside the p-value, and results may always be expanded to show additional details, such as the specific gene IDs within the list matching a certain annotation (Figure 2). These results are downloadable as tab-delimited text files which may then be further analyzed or used in conjunction with other databases.

Dynamic visualization of KEGG pathway maps may be accessed from the results table for KEGG pathway enrichment by clicking on any pathway name. The proteins in the list that are members of the particular biological pathway will appear in red, while those proteins existing in Chlamyomonas reinhardtii but not in the list appear in green (Figure 3). Alternatively, by expanding the pathway results and following the link at the bottom, the user may select a custom color scheme for visualizing the proteins on pathway maps. These custom color schemes may be designed on a gene-by-gene basis (choosing colors individually for genes) or in a group-by-group fashion (such as choosing a color for those proteins found within the organism but not in the gene list).

A list of genes may also be converted into a list of gene identifiers of another type. This feature allows easy transformation of gene IDs into corresponding models for use in other databases that may have additional annotation information. Additionally, the resulting list of gene identifiers may be used as a new starting point for enrichment analysis. Because of the different annotations associated with other gene identifier types (albeit of the same proteins), enrichment results using a converted set of gene IDs may yield new biological information.

The gene similarity search tool, the third component of the Algal Functional Annotation Tool, accepts single genes and returns functionally related genes (based on gene expression across different experimental conditions) using user-specified distance metrics and thresholds. Presently, functionally related genes may be determined using correlation distance based on absolute counts, log counts, or log ratios of expression. The results page shows the original query gene at the top in gray and any resulting genes, sorted by similarity, are shown below the query gene (Figure 4). A colormap based on gene expression is generated for the different genes

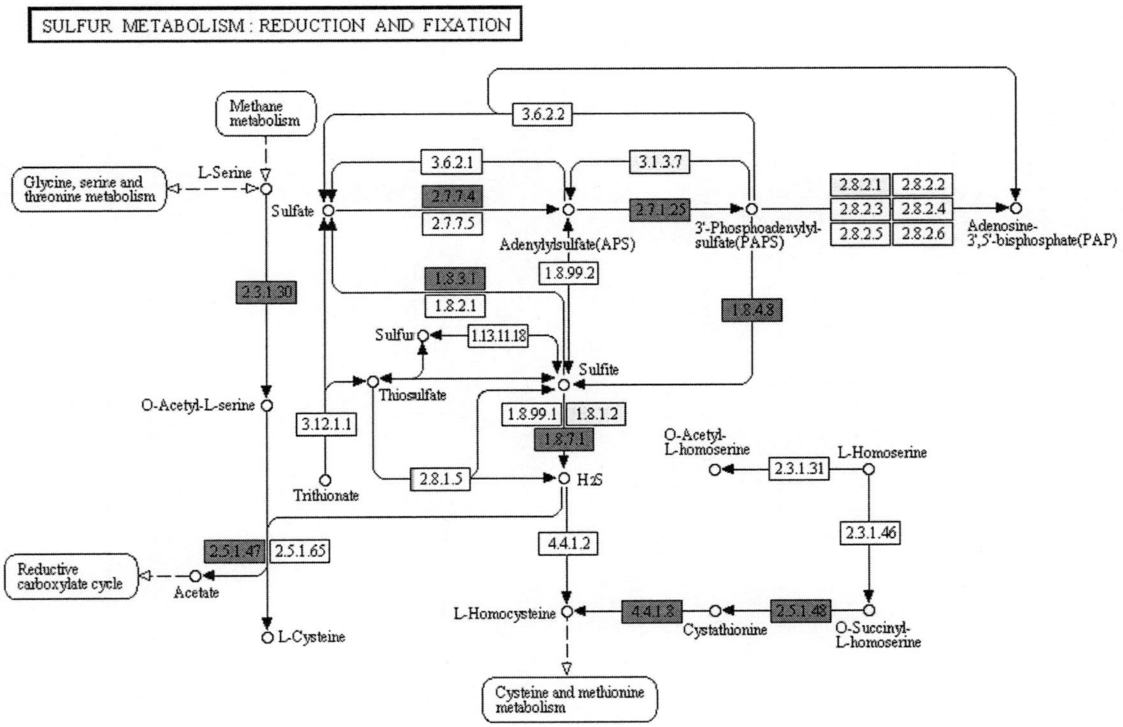

FIGURE 3: Dynamic Visualization of Gene Lists onto KEGG Pathway Maps. Dynamic KEGG pathway maps may be visualized to show the different proteins within a user-submitted gene list. Shown is the 'Sulfur Metabolism' dynamic pathway with the matching proteins submitted highlighted in red. In this example, the submitted gene list is drawn from literature characterizing *Chlamydomonas* under sulfur-deprived conditions [26].

across the conditions, and this colormap may be changed to display absolute expression, log expression, or log ratios of expression. The distance between any gene and the original query gene is displayed by hovering the mouse over the gene identifier of interest. Quantitative expression data (e.g. absolute counts) are provided for each experiment by hovering over the colormap. Whenever a description of a gene is available, this is displayed when hovering over the gene identifier as well. Links to external databases (e.g. JGI, KEGG) providing more information about the genes are provided with the results.

11.3.2 ABILITY TO RE-RUN ANALYSIS FOR SUBSETS OF GENES

Once a gene list is supplied and enrichment results have been returned, a subset of genes corresponding to those that contain a particular annotation may be isolated and re-run through the tool to be analyzed as a separate, smaller gene list. This allows users to select a particularly interesting group of functionally related genes and isolate them to see if they are also enriched for other functional terms. This also allows the user to prune large gene lists into more focused lists of functionally similar genes and removing some of the inherent noise associated with high-throughput experimental techniques and their resulting gene lists. This feature of the tool may be accessed by expanding the enrichment results of a particular annotation and selecting to re-run the analysis using only that subset of proteins. From this step, users may select which database types to query for enrichment (e.g. pathway, ontology, protein family).

11.3.3 EXPANDED ANNOTATION COVERAGE

The methods described to compensate for the incomplete annotation coverage of *Chlamydomonas* reinhardtii genes resulted in the addition of a vast number of unique annotations to the genome. While there is a strong overlap between pre-existing annotations and those assigned by inference, many new terms have also been added. The annotations derived by orthology,

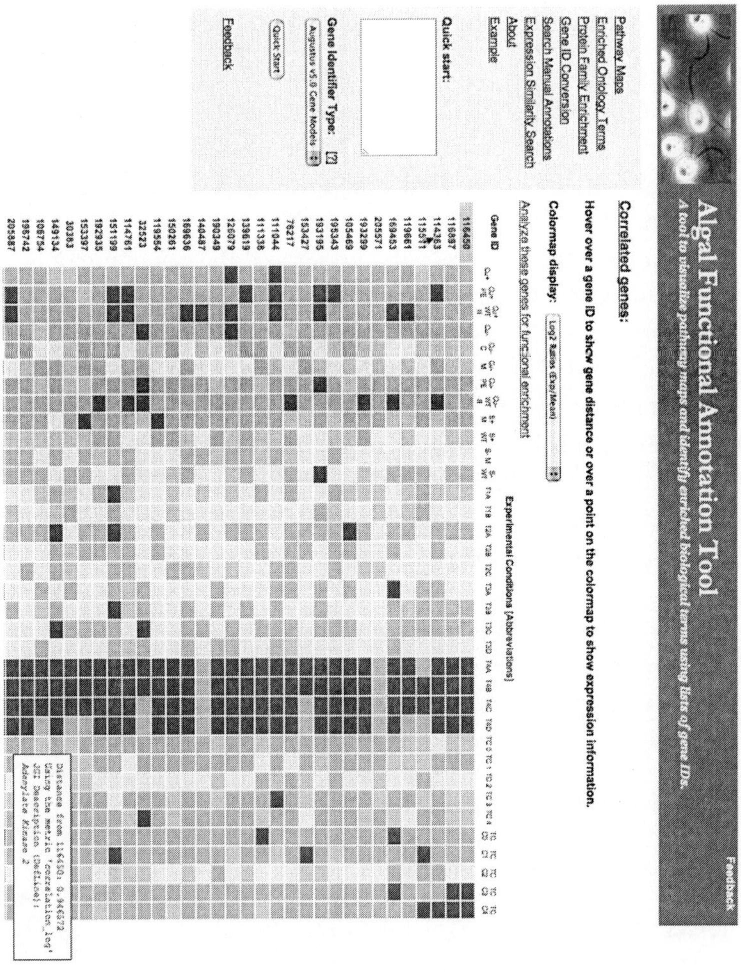

FIGURE 4: Expression Similarity Search Tool Results. An example of the results from the Gene Similarity Search Tool. Pairwise distances between resulting genes and the submitted gene are shown in the lower right corner when the mouse hovers over a gene of interest. Whenever applicable, a short description of the resulting gene is also shown when hovering over a gene. Expression data is shown when hovering over a point of the colormap.

however, are not mixed with the annotations attained directly to decrease the possibility of false positive associations of functional terms that may distort the analysis, and to permit a comparison with the functional terms derived directly from the *Chlamydomonas* annotation.

11.3.4 EXAMPLE: SULFUR-RELATED GENES

Using a filtered list of *C. reinhardtii* genes derived from transcriptome sequencing of the green alga under sulfur-depleted conditions [26], the Algal Functional Annotation Tool found enrichment for annotations related to sulfur metabolism, cysteine and methionine metabolism, and sulfur compound biosynthesis. For each annotation, the results may be expanded to reveal the genes containing that particular annotation. Furthermore, there is significant overlap between terms directly assigned to *C. reinhardtii* proteins and those inferred from A. thaliana orthology. Visualization of the sulfur metabolism KEGG pathway shows that a majority of the enzymes involved in this biological process is in the sample list, and the reactions they catalyze may be seen on the pathway map. The results for any enrichment analysis may be downloaded as a tab-delimited text file. Taking a gene found to be associated with the KEGG pathway 'Sulfur metabolism' by this enrichment analysis (JGI v. 3 ID 206154) as a starting input into the gene similarity search tool, the genes corresponding to sulfate transporter, methionine synthase reductase, and cysteine dioxygenase were found within the top 15 results using the correlation metric between log counts.

11.3.5 FUTURE DIRECTIONS

As with all tools that integrate data from multiple external sources, the power of analysis using the Algal Functional Annotation Tool is ultimately limited by the quality of the annotations within the primary databases. With the steady growth of knowledge in these annotation databases, the utility of the analyses provided is expected to increase in the future as more biological associations are assigned to genes. Additionally, as *Chlamydomonas* reinhardtii genes continue to be experimentally characterized, the assignment of manual annotations will also fill in the gaps left by automated annotation assignment and thus expand the annotation coverage throughout the genome, further improving the results generated by our portal. Lastly, the extensible nature of the Algal Functional Annotation Tool will allow us to add other algal organisms in the future using the same platform so that genomic data from other algal model organisms may be analyzed in a similar fashion as that currently available for *Chlamydomonas* reinhardtii.

11.4 CONCLUSIONS

The Algal Functional Annotation Tool is intended as a comprehensive analysis tool to elucidate biological meaning from gene lists derived from high-throughput experimental techniques. Annotation sets from a number of biological databases have been pre-processed and assigned to gene identifiers of the green alga *Chlamydomonas* reinhardtii, and this annotation data may be explored in multiple ways, including the use of enrichment tests designed for large gene lists. Furthermore, the site enables the visualization of proteins within pathway maps. Using several methods, such as inferring annotations from orthologous proteins of other organisms, the initially sparse annotation coverage of *C. reinhardtii* is alleviated, allowing for a more effective functional term enrichment analysis. Other functions of the tool include a batch gene identifier conversion tool and a manual annotation search tool. Lastly, similar genes based on expression across several conditions may be explored using the gene similarity search tool.

11.5 AVAILABILITY AND REQUIREMENTS

- Project name: Algal Functional Annotation Tool
- Public web service: http://pathways.mcdb.ucla.edu webcite; Free and no registration.
- Programming language: Perl/CGI
- Database: MySQL
- Software License: GNU General Public License

REFERENCES

1. Kanehisa M, Goto S, Furumichi M, Tanabe M, Hirakawa M: KEGG for representation and analysis of molecular networks involving diseases and drugs. Nucleic Acids Res 2010, (38 Database):D355-360.
2. Caspi R, Altman T, Dale JM, Dreher K, Fulcher CA, Gilham F, Kaipa P, Karthikeyan AS, Kothari A, Krummenacker M, et al.: The MetaCyc database of metabolic pathways and enzymes and the BioCyc collection of pathway/genome databases. Nucleic Acids Res 2010, (38 Database):D473-479.
3. Finn RD, Mistry J, Tate J, Coggill P, Heger A, Pollington JE, Gavin OL, Gunasekaran P, Ceric G, Forslund K, et al.: The Pfam protein families database. Nucleic Acids Res 2010, (38 Database):D211-222.
4. Huang da W, Sherman BT, Lempicki RA: Systematic and integrative analysis of large gene lists using DAVID bioinformatics resources. Nat Protoc 2009, 4(1):44-57.
5. Ingenuity Pathway Analysis (IPA), Ingenuity Systems [http://www.ingenuity.com]
6. Merchant SS, Prochnik SE, Vallon O, Harris EH, Karpowicz SJ, Witman GB, Terry A, Salamov A, Fritz-Laylin LK, Marechal-Drouard L, et al.: The *Chlamydomonas* genome reveals the evolution of key animal and plant functions. Science 2007, 318(5848):245-250.
7. Derelle E, Ferraz C, Rombauts S, Rouze P, Worden AZ, Robbens S, Partensky F, Degroeve S, Echeynie S, Cooke R, et al.: Genome analysis of the smallest free-living eukaryote Ostreococcus tauri unveils many unique features. Proc Natl Acad Sci USA 2006, 103(31):11647-11652.
8. Palenik B, Grimwood J, Aerts A, Rouze P, Salamov A, Putnam N, Dupont C, Jorgensen R, Derelle E, Rombauts S, et al.: The tiny eukaryote Ostreococcus provides genomic insights into the paradox of plankton speciation. Proc Natl Acad Sci USA 2007, 104(18):7705-7710.
9. Worden AZ, Lee JH, Mock T, Rouze P, Simmons MP, Aerts AL, Allen AE, Cuvelier ML, Derelle E, Everett MV, et al.: Green evolution and dynamic adaptations revealed by genomes of the marine picoeukaryotes Micromonas. Science 2009, 324(5924):268-272.
10. Blanc G, Duncan G, Agarkova I, Borodovsky M, Gurnon J, Kuo A, Lindquist E, Lucas S, Pangilinan J, Polle J, et al.: The Chlorella variabilis NC64A genome reveals adaptation to photosymbiosis, coevolution with viruses, and cryptic sex. Plant Cell 2010, 22(9):2943-2955.
11. *Chlamydomonas reinhardtii* v4.0, Joint Genome Institute [http://genome.jgi-psf.org/Chlre4/]
12. Rupprecht J: From systems biology to fuel--*Chlamydomonas reinhardtii* as a model for a systems biology approach to improve biohydrogen production. J Biotechnol 2009, 142(1):10-20.
13. Grossman AR, Croft M, Gladyshev VN, Merchant SS, Posewitz MC, Prochnik S, Spalding MH: Novel metabolism in *Chlamydomonas* through the lens of genomics. Curr Opin Plant Biol 2007, 10(2):190-198.
14. Beer LL, Boyd ES, Peters JW, Posewitz MC: Engineering algae for biohydrogen and biofuel production. Curr Opin Biotechnol 2009, 20(3):264-271.

15. Ghirardi ML, Dubini A, Yu J, Maness PC: Photobiological hydrogen-producing systems. Chem Soc Rev 2009, 38(1):52-61.
16. Hemschemeier A, Melis A, Happe T: Analytical approaches to photobiological hydrogen production in unicellular green algae. Photosynth Res 2009.
17. Finazzi G, Moreau H, Bowler C: Genomic insights into photosynthesis in eukaryotic phytoplankton. Trends Plant Sci 2010, 15(10):565-572.
18. Kruse O, Hankamer B: Microalgal hydrogen production. Curr Opin Biotechnol 2010, 21(3):238-243.
19. Scott SA, Davey MP, Dennis JS, Horst I, Howe CJ, Lea-Smith DJ, Smith AG: Biodiesel from algae: challenges and prospects. Curr Opin Biotechnol 2010, 21(3):277-286.
20. Radakovits R, Jinkerson RE, Darzins A, Posewitz MC: Genetic engineering of algae for enhanced biofuel production. Eukaryot Cell 2010, 9(4):486-501.
21. Eberhard S, Finazzi G, Wollman FA: The dynamics of photosynthesis. Annu Rev Genet 2008, 42:463-515.
22. Rochaix JD: Genetics of the biogenesis and dynamics of the photosynthetic machinery in eukaryotes. Plant Cell 2004, 16(7):1650-1660.
23. Harris EH: *Chlamydomonas* as a model organism. Annu Rev Plant Physiol Plant Mol Biol 2001, 52:363 406.
24. Marshall WF: Basal bodies platforms for building cilia. Curr Top Dev Biol 2008, 85:1-22.
25. Scholey JM, Anderson KV: Intraflagellar transport and cilium-based signaling. Cell 2006, 125(3):439-442.
26. Gonzalez-Ballester D, Casero D, Cokus S, Pellegrini M, Merchant SS, Grossman AR: RNA-seq analysis of sulfur-deprived *Chlamydomonas* cells reveals aspects of acclimation critical for cell survival. Plant Cell 2010, 22(6):2058-2084.
27. Miller R, Wu G, Deshpande RR, Vieler A, Gartner K, Li X, Moellering ER, Zauner S, Cornish AJ, Liu B, et al.: Changes in transcript abundance in *Chlamydomonas reinhardtii* following nitrogen deprivation predict diversion of metabolism. Plant Physiol 2010, 154(4):1737-1752.
28. Matthews L, Gopinath G, Gillespie M, Caudy M, Croft D, de Bono B, Garapati P, Hemish J, Hermjakob H, Jassal B, et al.: Reactome knowledgebase of human biological pathways and processes. Nucleic Acids Res 2009, (37 Database):D619-622.
29. *Chlamydomonas reinhardtii* v3.0, Joint Genome Institute [http://genome.jgi-psf.org/Chlre3/]
30. Thomas PD, Campbell MJ, Kejariwal A, Mi H, Karlak B, Daverman R, Diemer K, Muruganujan A, Narechania A: PANTHER: a library of protein families and subfamilies indexed by function. Genome Res 2003, 13(9):2129-2141.
31. Ashburner M, Ball CA, Blake JA, Botstein D, Butler H, Cherry JM, Davis AP, Dolinski K, Dwight SS, Eppig JT, et al.: Gene ontology: tool for the unification of biology. The Gene Ontology Consortium. Nat Genet 2000, 25(1):25-29.
32. Hunter S, Apweiler R, Attwood TK, Bairoch A, Bateman A, Binns D, Bork P, Das U, Daugherty L, Duquenne L, et al.: InterPro: the integrative protein signature database. Nucleic Acids Res 2009, (37 Database):D211-215.
33. Thimm O, Blasing O, Gibon Y, Nagel A, Meyer S, Kruger P, Selbig J, Muller LA, Rhee SY, Stitt M: MAPMAN: a user-driven tool to display genomics data sets

onto diagrams of metabolic pathways and other biological processes. Plant J 2004, 37(6):914-939.
34. Huang da W, Sherman BT, Lempicki RA: Bioinformatics enrichment tools: paths toward the comprehensive functional analysis of large gene lists. Nucleic Acids Res 2009, 37(1):1-13.
35. Tatusov RL, Fedorova ND, Jackson JD, Jacobs AR, Kiryutin B, Koonin EV, Krylov DM, Mazumder R, Mekhedov SL, Nikolskaya AN, et al.: The COG database: an updated version includes eukaryotes. BMC Bioinformatics 2003, 4:41.

This chapter was originally published under the Creative Commons Attribution License. Lopez, D., Casero, D., Cokus, S. J., Merchant, S. S., and Pellegrini, M. Algal Functional Annotation Tool: A Web-Based Analysis Suite to Functionally Interpret Large Gene Lists Using Integrated Annotation and Expression Data. BMC Bioinformatics 2011: 12(282).

CHAPTER 12

TRANSCRIPTOMIC ANALYSIS OF THE OLEAGINOUS MICROALGA *NEOCHLORIS OLEOABUNDANS* REVEALS METABOLIC INSIGHTS INTO TRIACYLGLYCERIDE ACCUMULATION

HAMID RISMANI-YAZDI, BERAT Z. HAZNEDAROGLU, CAROL HSIN, and JORDAN PECCIA

12.1 BACKGROUND

Important advantages of microalgae-based biofuels over first generation biofuels include algae's greater solar energy conversion efficiency compared to land plants [1], the ability of oleaginous microalgae to utilize non-arable land and saline or waste-water, and their high content of energy dense neutral lipids that can be readily transesterified to produce biodiesel [2,3]. Under stress conditions such as nutrient deprivation or high light intensity, several species of oleaginous microalgae can alter lipid biosynthetic pathways to produce intracellular total lipid contents between 30 to 60% of dry cell weight (DCW) [4]. Triacylglycerides (TAGs) are the dominant form of lipids produced under these conditions. The excess production of TAGs in microalgae is thought to play a role in carbon and energy storage and functions as part of the cell's stress response [5].

Due to the limited understanding of microalgae genetics and physiology, lipid metabolism from higher plants and bacteria have been the basis from which the accumulation of TAGs in microalgae has been modeled [5]. TAGs and polar membrane lipids are synthesized from fatty acids, that are primarily produced in the chloroplast [6]. The committed step in fatty acid biosynthesis starts with the conversion of acetyl CoA to malonyl CoA through the enzyme acetyl CoA carboxylase (ACCase). In some plants, there is evidence that both photosynthesis- and glycolysis-derived pyruvate could be endogenous sources of acetyl CoA pool for fatty acid biosynthesis [5]. Fatty acid production in *E. coli* is regulated through feedback-inhibition by long chain fatty acyl carrier proteins (ACP) [7,8], and a recent study in the microalgae *Phaeodactylum tricornutum* demonstrated that overexpression of genes that encode for the thioesterases that hydrolyze the thioester bond of long chain fatty acyl ACPs resulted in a significant increase in fatty acid production [9]. Recent nitrogen deprivation studies in the model, nonoleaginous microalga *Chlamydomonas reinhardtii* have also suggested an important role for lipases in restructuring the cell membrane under nitrogen limitation in order to supply fatty acids for TAG biosynthesis [10].

The stress-induced production of TAGs provides an opportunity to observe differential gene expression between high and low TAG accumulating phenotypes. Because multiple pathways are associated with the enhanced production of neutral lipids in microalgae, transcriptomic studies are an appropriate tool to provide an initial, broad view of carbon partitioning [11] and regulation of TAG biosynthesis during microalgae stress responses. However, the most promising strains thus far identified by growth experiments and lipid content screening [4,12] do not have sequenced, fully annotated genomes [13-15]. In microalgae, transcriptomic studies have instead focuses on model organisms that are not oleaginous but have sequenced genomes [10,16]. There is a growing number of oleaginous microalgae from which de novo transcriptomes have been assembled and annotated but comprehensive quantitative gene expression analysis in these microalgae has not yet been performed [14,17-19]. Recently, a de novo assembled-transcriptome was used as a search model to enable a proteomic analysis of the oleaginous microalga *Chlorella vulgaris* that demonstrated up-regulation of fatty acid and TAG biosynthetic pathways in response to nitrogen limitations [13].

In the present study, we quantitatively analyzed the transcriptome of the oleaginous microalga *Neochloris oleoabundans* to elucidate the metabolic pathway interactions and regulatory mechanisms involved in the accumulation of TAG. *N. oleoabundans* (a taxonomic synonym of *Ettlia oleoabundans*[20]) is a unicellular green microalga belonging to the *Chlorophyta* phylum (class *Chlorophyceae*). It is known to produce large quantities of lipids (35 to 55% dry cell weight total lipids and greater than 10% TAGs) [4,12,21] in response to physiological stresses caused by nitrogen deprivation. To produce differences in lipid enrichment, *N. oleoabundans* was cultured under nitrogen replete and nitrogen limited conditions and major biomolecules including total lipids, TAGs, starch, protein, and chlorophyll were measured. The transcriptome was sequenced and assembled de novo, gene expression was quantified, and comparative analysis of genes, pathways and broader gene ontology categories was conducted. The results provide new insight into the regulation of lipid metabolism in oleaginous microalgae at the transcriptomic level, and suggest several potential strategies to improve lipid production in microalgae based on a rational genetic engineering approach.

12.2 RESULTS

12.2.1 MAJOR BIOMOLECULE CONTENT AND COMPOSITION DIFFER BETWEEN THE NITROGEN REPLETE (+N) AND NITROGEN-LIMITED (−N) GROWTH ENVIRONMENTS

To track gene transcription in the oleaginous microalga *N. oleoabundans*, cells were first grown under +N and −N conditions as a method to produce differential cellular enrichments of TAGs. Cells were harvested after 11 days. This sampling time corresponded to below detection level concentrations for NO^{3-} and a reduction in growth rate in the –N reactors (Figure 1A, B). The maximum growth rate for the –N cultures was 113 ± 4 (std. err.) $mgl^{-1} day^{-1}$ and decreased to 34 ± 0.7 $mgl^{-1} day^{-1}$ once nitrogen became limited in the reactor. Total lipids extracted under the +N and −N scenarios revealed a statistically significant increase ($p < 0.05$) from 22% DCW in +N to 36% in the −N condition (Figure 1C). Extracted lipids were transesterified and fatty acid methyl esters (FAMEs) (FAMEs assumed to

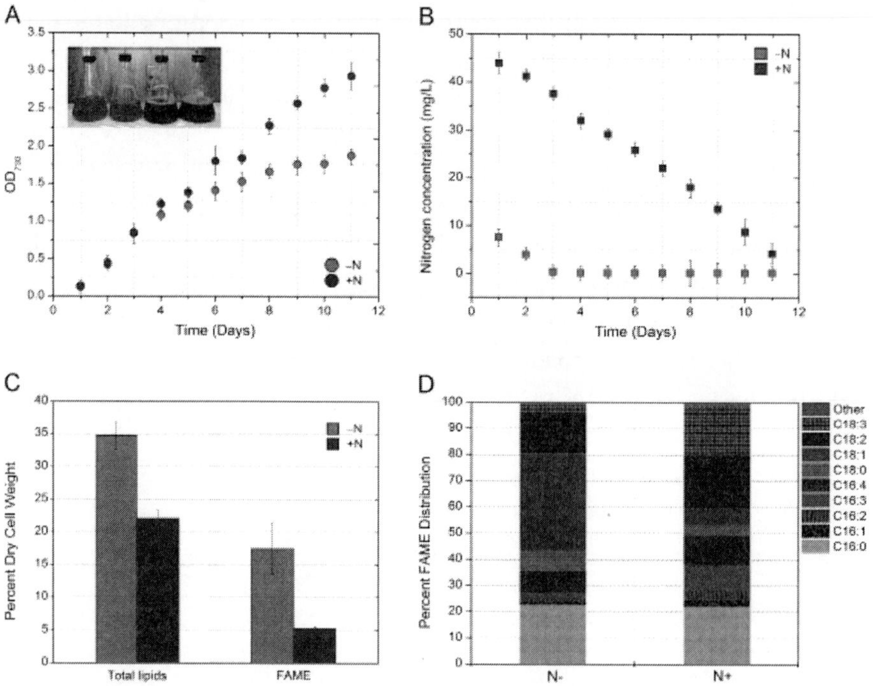

FIGURE 1: *N. oleoabundans* growth and lipid characteristics. (A) Growth curves under+N and−N conditions. Inset image represents the difference in culture appearance between the two growth condition; (B) Nitrate as N concentrations in the bioreactors during growth; (C) Cell weight enrichment of total lipids and fatty acid methyl esters (FAME, representative of TAGs) from cells harvested on day 11; and (D) Percentage distribution of FAME from cells harvested on day 11. All error bars represent one standard deviation.

be equivalent to TAGs content [22]), were quantified. Compared to the+N condition, the FAME or TAG content per cell mass increased by five times in the−N case ($p < 0.05$), demonstrating that the additional lipids produced during N limitations were mostly TAGs (Figure 1C). Estimates of total cell mass based on direct microscopic counts and DCW determinations revealed that the average mass of a cell in−N was 81% of that in+N, confirming that the change in TAG was independent of changes in DCW. FAME profiles are presented in Figure 1D, and show a 50% decrease in the proportion of unsaturated fatty acids (i.e. C16:2, C16:3, C16:4, C18:2, and C18:3) under nitrogen limitation. The most significant change was in the amount of oleic acid (C18:1), which increased over 5 times, while the quantity of α-linoleic acid (C18:3) decreased by 4.8-fold under –N conditions. This trend toward a greater proportion of C18:1 is consistent with prior investigations of the oleaginous microalgae *N. oleoabundans* and *Chlorella vulgaris* FAME contents under nitrogen limitations [13,22].

To aid in interpreting how photosynthetically fixed carbon was directed into major metabolic pathways, the chlorophyll, protein, and starch content of *N. oleoabundans* were also measured under the − N and + N scenarios (Table 1). Nitrogen deprivation lead to a reduction in nitrogen-containing chlorophyll content. This loss of chlorophyll is consistent with the light green color of chlorosis observed in the cultures under nitrogen limitation (Figure 1A inset). Also under nitrogen limitation, a decrease in cellular protein content and an increase in cellular starch content were observed. The observed changes in metabolite and biomolecule contents suggest the redirection of metabolism in *N. oleoabundans* during nitrogen limitation to reduce nitrogen-containing compounds (protein and chlorophyll) and favor the accumulation of nitrogen free storage molecules TAGs and starch.

TABLE 1: Culture density and cellular composition of major biomolecules of *N. oleoabundans* cells determined after 11 days of growth under nitrogen replete (+N) and nitrogen limited (–N) conditions

	+N	–N
Culture density (cells/mL)	$(6.1 \pm 0.2) \times 10^7$	$(3.8 \pm 0.2) \times 10^7$
Chlorophyll a (μg/mg)	$(119.3 \pm 12.6) \times 10^{-3}$	$(5.9 \pm 0.4) \times 10^{-3}$
Chlorophyll b(μg/mg)	$(42.6 \pm 5.5) \times 10^{-3}$	$(5.5 \pm 0.5) \times 10^{-3}$
Starch content (% DCW)	0.2 ± 0.1	4.0 ± 0.5
Protein content (% DCW)	37.9 ± 4.0	19.4 ± 17.1

12.2.2 DE NOVO TRANSCRIPTOME ASSEMBLY, ANNOTATION, AND EXPRESSION

In order to produce statistically reliable and comparable RNA-Seq data, cDNA library construction and sequencing was performed for each of the duplicate + N reactors, and each of the duplicate − N reactors. Over 88 million raw sequencing reads were generated and subjected to quality score and length based trimming; resulting in a high quality (HQ) read data set of 87.09 million sequences (average phred score of 35) with an average read length of 77 bp. By incorporating a multiple k-mer based de novo transcriptome assembly strategy (k-mers 23, 33, 63, and 83) [23], HQ reads were assembled into 56,550 transcripts with an average length of 1,459 bp and a read coverage of 1,444× (Figure 2C). Generated transcripts were subjected to searches against the National Center for Biotechnology Information's (NCBI) nonredundant and plant refseq databases [24], and the majority of transcripts showed significant mat-ches to other closely related green microalgae species (Figure 2A, B) including *C. variabilis* (~85% of all transcripts), *C. reinhardtii* (~2.6%), and *V. carteri* (~3.4%) (Figure 2A). With additional annotations by using KEGG services and Gene Ontology (GO), a total of 23,520 transcripts were associated with at least one GO term, and 4,667 transcripts were assigned with enzyme commission (EC) numbers. Overall, 14,957 transcripts had KO identifiers and were annotated as putative genes and protein families. This assembly provided a reliable, well-annotated transcriptome for downstream RNA-Seq data analysis.

Following the transcriptome assembly and annotation, HQ reads obtained from each experimental condition were individually mapped to the generated assembly in order to determine the transcript abundances as RPKM values. To determine fold change differences among +N and −N transcripts, non-normalized read counts were fed into the DESeq package (v1.5.1) and variance and mean dependencies were accounted for [25]. Based on the negative binomial distribution model used in DESeq package, 25,896 transcripts out of the total 56,550 non-redundant transcripts were up-regulated under the −N condition. Plotting transcript fold changes levels shows a high correlation among the biologically replicated sequencing runs as indicated by Euclidean distances (Figure 2D). Overall, 15,987 transcripts had significant differential regulation ($q < 0.05$) Figure 3A. A complete table of fold changes with significance level for all genes assessed is presented in Additional file 3.

We further investigated the alignment of HQ reads to the reference genomes of *C. reinhardtii* and *V. carteri* in order to improve and extend our transcriptomic analysis to the detection of splicing events and alternative isoform formation (Figure 3B, C). Although the majority of annotated or-

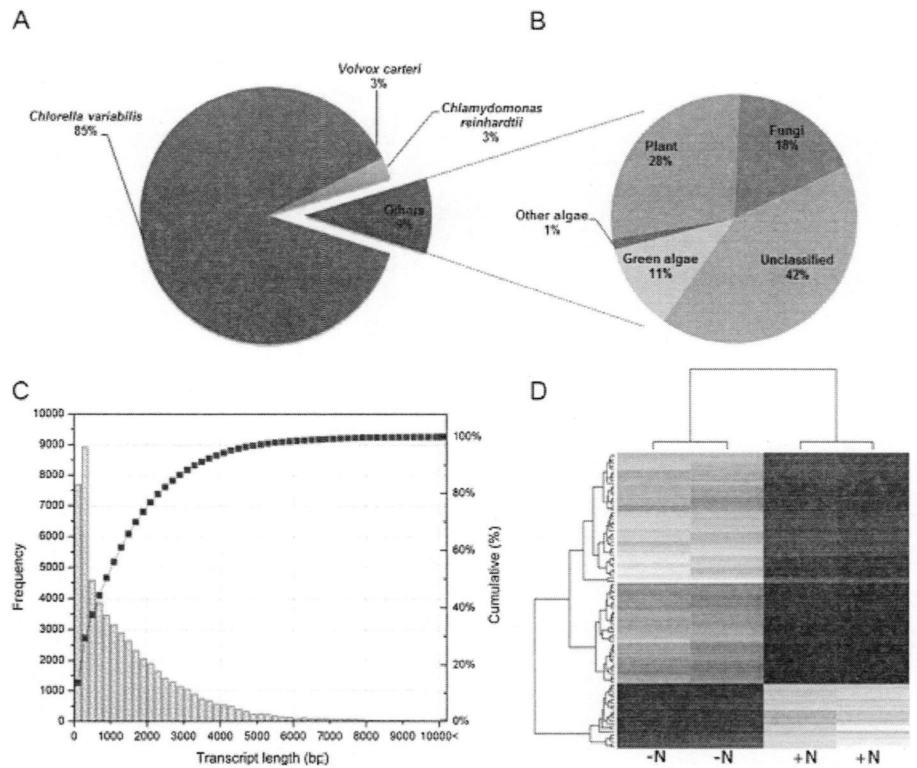

FIGURE 2: De novo assembly and mapping results. (A, B) Top-hit species distribution for BLASTX matches for the *N. oleoabundans* transcriptome; (C) Cumulative transcript length frequency distribution of the *N. oleoabundans* transcriptome assembly; (D) Heat map demonstrating the top 100 most differentially expressed genes in the biological replicates of +N and −N conditions.

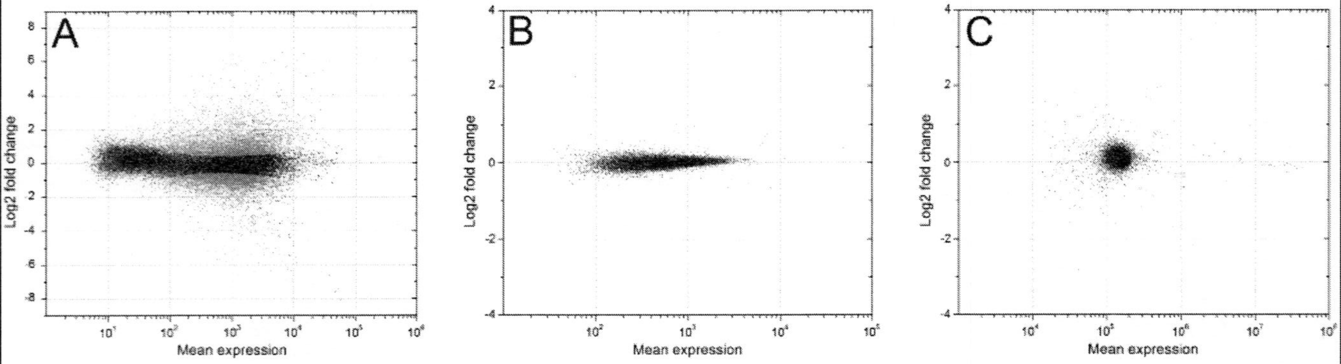

FIGURE 3: (A) MvA plot contrasting gene expression levels between the−N and+N scenarios based on reads mapped to the*N. oleoabundans*transcriptome. The x-axis represents the mean expression level at the gene scale, and the y-axis represents the log2 fold change from−N to+N. Negative fold changes indicate up-regulation of –N genes. Lighter gray dots are genes that are significant at a false discovery rate of 5%; (B) MvA plot for reads mapped to the *C. reinhardtii* genome; and (C) MvA plot for reads mapped to the *V. carteri* genome.

thologs were identified from these closely related microalgae species, very poor mappings (i.e. <5% of reads) were observed between the RNA-Seq data of *N. oleoabundans* and the genomes of *C. reinhardtii* and *Volvox carteri*. As a result, the number of transcripts annotated and evaluated for differential expression was suboptimal, and genomes from these most closely related organisms were not used for gene expression analysis.

12.2.3 CLUSTERING OF RELEVANT GO TERMS AND DIFFERENTIAL EXPRESSION

The transcripts annotated in+N and−N transcriptomes were first classified based on Gene Ontology (GO) terms. In the+N and –N datasets, respectively, 6,846 and 7,473 transcripts were classified into 306 and 218 broader GO term categories in accordance with the Gene Ontology Consortium [26]. An enrichment analysis of the broader GO terms was performed using the modified Fisher's Exact test in Blast2GO to quantitatively compare the distribution of differentially enriched GO terms between the+N case and the entire data set (Figure 4A), and between the –N case and the entire data set (Figure 4B). The functional categories enriched under+N were distinctly different from those enriched under the –N condition. In the+N case (Figure 4A), functional categories linked to carbon fixation, photosynthesis, protein machinery, and cellular growth were highly enriched compared to the –N condition; reflecting the higher growth rate, higher cell mass, and increased chlorophyll content observed in+N. Under –N conditions, genes associated with carboxylic acid and lipid biosynthetic process, NADPH regeneration, the pentose-phosphate pathway, phospholipid metabolic process, and lipid transport demonstrated a greater enrichment of transcripts than the overall dataset (Figure 4B). These enriched GO terms directly correlated with the observed increase of lipid accumulation in –N cells. Other major categories identified as significantly expressed under the –N condition included the synthesis of value added products such as terpenoids, pigments, and vitamins as well as cellular response to nitrogen starvation, nitrate metabolic process, and nitrate assimilation (Figure 4B). Genes involved in the latter three functional categories were exclusively expressed in the nitrogen-limited cells.

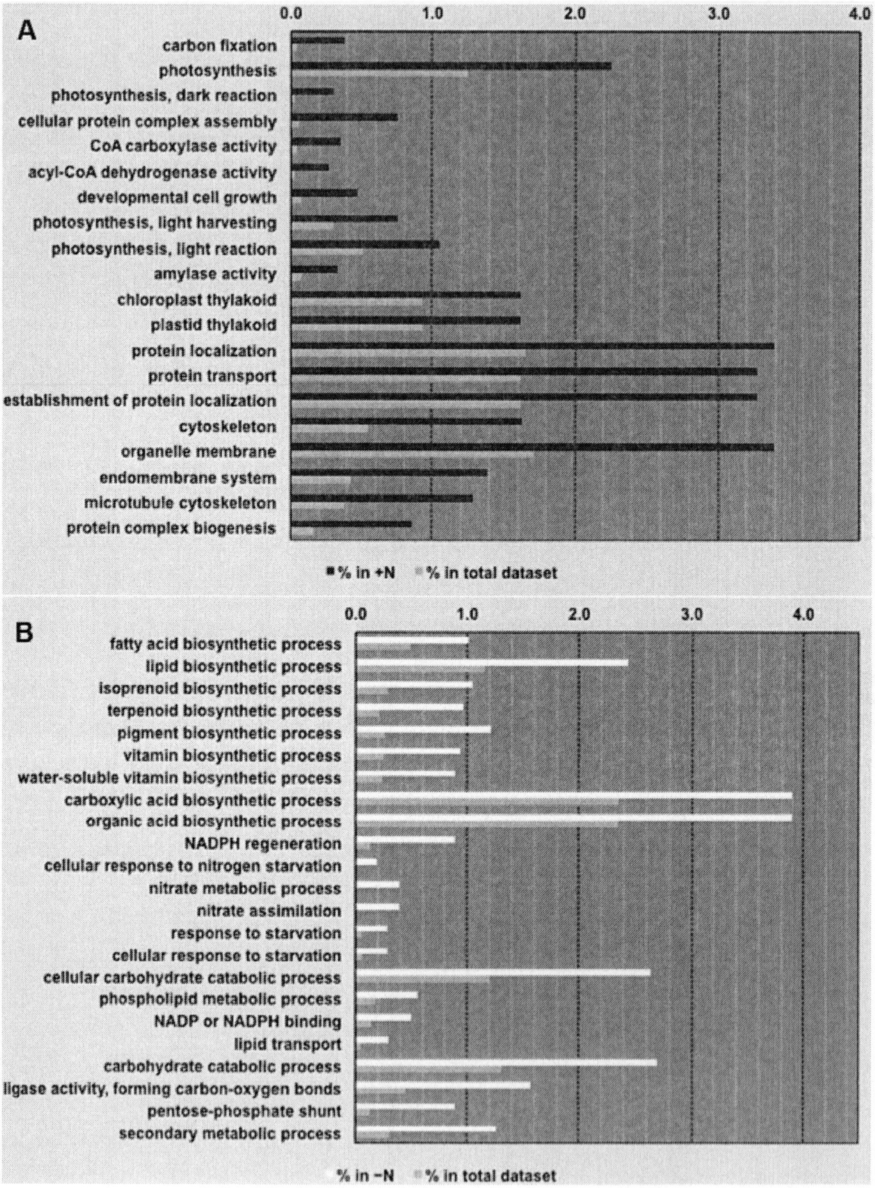

FIGURE 4: Over representation analysis of selected significant GO terms. (A) contains results for +N versus the full dataset and (B) contains results for −N versus the full dataset.

12.2.4 FATTY ACID BIOSYNTHESIS PATHWAY IS UP-REGULATED AND THE B-OXIDATION PATHWAY IS REPRESSED UNDER NITROGEN-LIMITING CONDITIONS

The majority of genes governing fatty acid biosynthesis were identified as being overexpressed in nitrogen limited cells as shown in the global metabolic pathway level and fatty acid biosynthesis module. The fold-change and abundances of identified transcripts for the components of fatty acid biosynthesis at the gene level are presented in Figure 5A. The first step in fatty acid biosynthesis is the transduction of acetyl-CoA into malonyl-CoA by addition of carbon dioxide. This reaction is the first committing step in the pathway and catalyzed by Acetyl-CoA Carboxylase (ACCase). While the gene encoding ACCase was repressed under the –N condition, the biotin-containing subunit of ACCase, biotin carboxylase (BC), was significantly up-regulated in response to nitrogen starvation. The BC catalyzes the ATP-dependent carboxylation of the biotin subunit and is part of the heteromeric ACCase that is present in the plastid—the site of de novo fatty acid biosynthesis [27]. To proceed with fatty acid biosynthesis, malonyl-CoA is transferred to an acyl-carrier protein (ACP), by the action of malonyl-CoA ACP transacylase (MAT). This step is followed by a round of condensation, reduction, dehydration, and again reduction reactions catalyzed by beta-ketoacyl-ACP synthase (KAS), beta-ketoacyl-ACP reductase (KAR), beta-hydroxyacyl-ACP dehydrase (HAD), and enoyl-ACP reductase (EAR), respectively. The expression of genes coding for MAT, KAS, HAD, and EAR were up-regulated, whereas the KAR encoding gene was repressed in –N cells. The synthesis ceases after six or seven cycles when the number of carbon atoms reaches sixteen (C16:0-[ACP]) or eighteen (C18:0-[ACP]). ACP residues are then cleaved off by thioesterases oleoyl-ACP hydrolase (OAH) and Acyl-ACP thioesterase A (FatA) generating the end products of fatty acid synthesis (i.e. palmitic (C16:0) and stearic (C18:0) acids). Genes coding for these thioesterases, i.e. FatA and OAH, were overexpressed in –N cells. The up-regulation of these thioesterase encoding genes, as previously reported in *E. coli* and the microalga *P. tricornutum*, is associated with reducing the feedback inhibition that partially controls the production rate of fatty acid biosynthesis

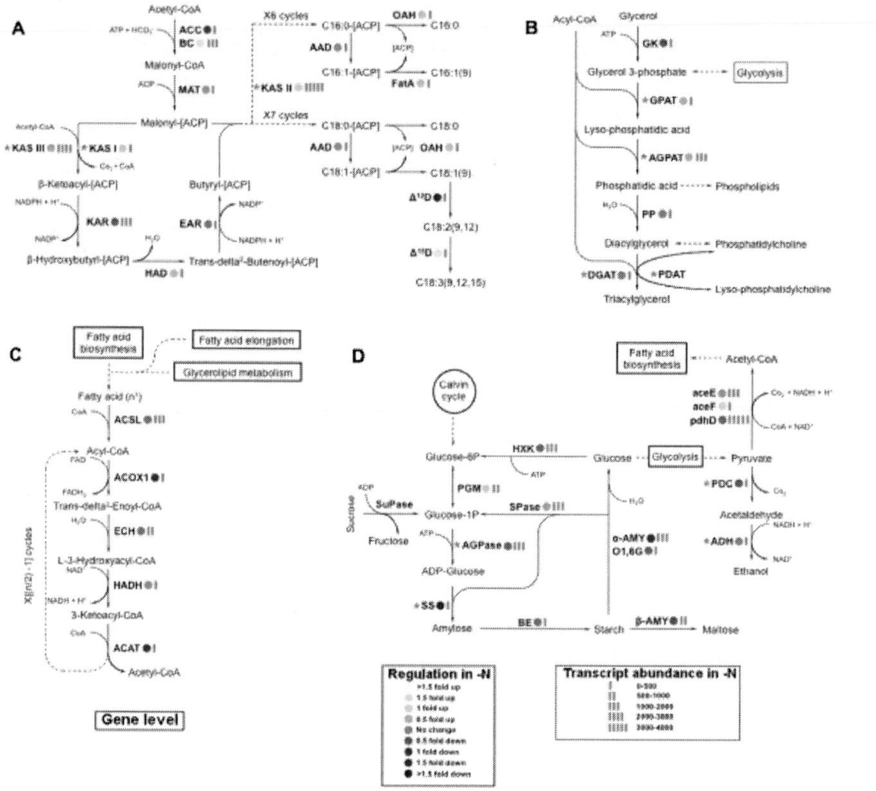

FIGURE 5: Differential expression of genes involved in (A) the fatty acid biosynthesis; (B) triacylglycerol biosynthesis; (C)β-oxidation; and (D) starch biosynthesis. Pathway were reconstructed based on the de novo assembly and quantitative annotation of the *N. oleoabundans* transcriptome. (A) Enzymes include: ACC, acetyl-CoA carboxylase (EC: 6.4.1.2); MAT, malonyl-CoA ACP transacylase (EC: 2.3.1.39); KAS, beta-ketoacyl-ACP synthase (KAS I, EC: 2.3.1.41; KASII, EC: 2.3.1.179; KAS III, EC: 2.3.1.180); KAR, beta-ketoacyl-ACP reductase (EC: 1.1.1.100); HAD, beta-hydroxyacyl-ACP dehydrase (EC: 4.2.1.-); EAR, enoyl-ACP reductase (EC: 1.3.1.9); AAD, acyl-ACP desaturase (EC: 1.14.19.2); OAH, oleoyl-ACP hydrolase (EC: 3.1.2.14); FatA, Acyl-ACP thioesterase A (EC: 3.1.2.-); Δ12D, Δ12(ω6)-desaturase (EC: 1.4.19.6); Δ15D, Δ15(ω3)-desaturase (EC: 1.4.19.-); (B) Enzymes include: GK, glycerol kinase (EC: 2.7.1.30); GPAT, glycerol-3-phosphate O-acyltransferase (EC: 2.3.1.15); AGPAT, 1-acyl-sn-glycerol-3-phosphate O-acyltransferase (EC:2.3.1.51); PP, phosphatidate phosphatase (EC: 3.1.3.4); DGAT, diacylglycerol O-acyltransferase (EC: 2.3.1.20); and PDAT, phopholipid:diacyglycerol acyltransferase (EC 2.3.1.158); (C) Enzymes include: ACS, acyl-CoA synthetase (EC: 6.2.1.3); ACOX1, acyl-CoA oxidase (EC: 1.3.3.6); ECH, enoly-CoA hydratase (EC: 4.2.1.17); HADH, 3-hydroxyacyl-CoA dehydrogenase (EC: 1.1.1.35); ACAT, acetyl-CoA C-acetyltransferase (EC: 2.3.1.16, 2.3.1.9); (D) Enzymes include: PGM, phosphoglucomutase (EC: 5.4.2.2); AGPase, ADP-glucose pyrophosphorylase (EC:

2.7.7.27); SS, starch synthase (EC: 2.4.1.21); BE, α-1,4-glucan branching enzyme (EC: 2.4.1.18); and HXK, hexokinase (2.7.1.1). Starch catabolism enzymes include: α-AMY, α-amylase (EC: 3.2.1.1); O1,6G, oligo-1,6-glucosidase (EC: 3.2.1.10); β-AMY, β-amylase (EC: 3.2.1.2); and SPase, starch phosphorylase (EC: 2.4.1.1). Ethanol fermentation via pyruvate enzymes include: PDC, pyruvate decarboxylase (EC: 4.1.1.1); and ADH, alcohol dehydrogenase (EC: 1.1.1.1). Enzymes aceE, pyruvate dehydrogenase E1 component (EC 1.2.4.1); aceF, pyruvate dehydrogenase E2 component (EC: 2.3.1.12); and pdhD, dihydrolipoamide dehydrogenase (EC 1.8.1.4), transforms pyruvate into acetyl-CoA. Key enzymes are shown with an asterisk (*) next to the enzyme abbreviations, and dashed arrows denote reaction(s) for which the enzymes are not shown. All presented fold changes are statistically significant, q value < 0.05.

[7,8], and results in the overproduction of fatty acids [9]. It has also been suggested that an increase in FatA gene expression and the associated acyl-ACP hydrolysis may aid in increased fatty acid transport from the chloroplast to the endoplasmic reticulum site where TAG assembly occurs [10,28]. Finally, for supplying reducing equivalents via NADPH to power fatty acid biosynthesis, genes encoding for the pentose phosphate pathway were strongly up-regulated in the –N condition (Table 2).

The altered expression of genes associated with the generation of double bonds in fatty acids reflects the observed increase in the proportion of unsaturated of fatty acids (Figure 1D), and the enrichment of C18:1 during nitrogen limitations. The acyl-ACP desaturase (AAD), which introduces a one double bond to C16:0/C18:0, and delta-15 desaturase, which converts C18:2 to C18:3, were significantly up-regulated in the −N case, whereas the delta-12 desaturase catalyzing the formation of C18:2 from C18:1was repressed during nitrogen limitation.

Under nitrogen limitations, 10 of the 13 genes associated with fatty acid degradation (α and β-oxidation pathways for saturated and unsaturated acids) were significantly repressed. Figure 5C demonstrates the typical β-oxidation pathway for saturated fatty acids, while Table 3 displays expression levels for additional peroxisomal genes associated with fatty acid oxidation, but not shown in Figure 5C. Before undergoing oxidative degradation, fatty acids are activated through esterification to Coenzyme A. The activation reaction, is catalyzed by acyl-CoA synthetase (ACSL), which was up-regulated in –N cells. The acyl-CoA enters the β-oxidation pathway and undergoes four enzymatic reactions in multiple rounds. The first three steps of the pathway; oxidation, hydration and again oxidation of acyl-CoA are catalyzed by acyl-CoA oxidase (ACOX1), enoly-CoA hydratase (ECH), and hydroxyacyl-CoA dehydrogenase (HADH), respectively. In the last step of the pathway, acetyl-CoA acetyltransferase (ACAT) catalyzes the cleavage of one acetyl-CoA, yielding a fatty acyl-CoA that is 2 carbons shorter than the original acyl-CoA. The cycle continues until all the carbons are released as acetyl-CoA. The expression level of ECH and HADH were unchanged and genes encoding for enzymes ACOX1 and ACAT catalyzing the first and last reactions in the cycle were identified as significantly repressed in −N cells.

Table 2. *N. oleoabundans* genes involved in the pentose phosphate pathway

Pentose phosphate pathway	**Log2FC**
Phosphogluconate dehydrogenase (decarboxylating) (PGD, EC: 1.1.1.44)	−1.13
Glucose-6-phosphate dehydrogenase (G6PD, EC: 1.1.1.49)	−1.41
Transketolase (tktA, EC: 2.2.1.1)	2.55
Transaldolase (talA, EC: 2.2.1.2)	−0.66
6-phosphofructokinase (PFK, EC: 2.7.1.11)	−0.45
Gluconokinase (gntK, EC: 2.7.1.12)	0.10
Ribokinase (rbsK, EC: 2.7.1.15)	0.11
Ribose-phosphate diphosphokinase (PRPS, EC: 2.7.6.1)	−0.10
Gluconolactonase (GNL, EC: 3.1.1.17)	−0.67
6-phosphogluconolactonase (PGLS, EC: 3.1.1.31)	0.07
Fructose-bisphosphatase (FBP, EC: 3.1.3.11)	−0.24
Fructose-bisphosphate aldolase (fbaB, EC: 4.1.2.13)	0.17
Ribulose-phosphate 3-epimerase (RPE, EC: 5.1.3.1)	−0.11
Ribose-5-phosphate isomerase (rpiA, EC: 5.3.1.6)	−0.34
Glucose-6-phosphate isomerase (GPI, EC: 5.3.1.9)	−1.21
Phosphoglucomutase (pgm, EC: 5.4.2.2)	−0.83

Negative Log2FC values represent up-regulation under nitrogen limitation. All presented fold changes are statistically significant, q value < 0.05.

TABLE 3: *N. oleoabundans* genes involved in catabolic pathways related to peroxisomal fatty acid oxidation, lysosomal lipases, and the regulation of autophagy

Enzyme encoding gene	Log2FC
Peroxisome	
α-oxidation	
2-hydroxyacyl-coa lyase 1 (HACL1, EC: 4.1.-.-)	0.35
Unsaturated fatty acid β-oxidation	
Peroxisomal 2,4-dienoyl-coa reductase (DECR2, EC: 1.3.1.34)	0.21
Δ(3,5)-Δ(2,4)-dienoyl-coa isomerase (ECH1, EC: 5.3.3.-)	−0.27
ATP-binding cassette, subfamily D (ALD), member 1 (ABCD1)	0.25
Long-chain acyl-coa synthetase (ACSL, EC: 6.2.1.3)	0.25
Other oxidation	
Peroxisomal 3,2-trans-enoyl-coa isomerase (PECI, EC: 5.3.3.8)	0.59
Carnitine O-acetyltransferase (CRAT, EC: 2.3.1.7)	0.30
NAD + diphosphatase (NUDT12, EC: 3.6.1.22)	0.47
Glycerolipid metabolism	
Triacylglycerol lipase (EC: 3.1.1.3)	0.33
Acylglycerol lipase (MGLL, EC: 3.1.1.23)	−0.13
Glycerophospholipid metabolism	
Phospholipase A1 (plda, EC: 3.1.1.32)	−1.26
Phospholipase A2 (PLA2G, EC: 3.1.1.4)	−0.31
Phospholipase C (plcc, EC: 3.1.4.3)	−0.10
Lysosome	
Lipases	
Lysosomal acid lipase (LIPA, EC: 3.1.1.13)	−0.48
Lysophospholipase III (LYPLA3, EC: 3.1.1.5)	0.20
Regulation of autophagy	
Unc51-like kinase (ATG1, EC: 2.7.11.1)	−0.53
5'-AMP-activated protein kinase, catalytic alpha subunit (snrk1, PRKAA)	−0.05
Vacuolar protein 8 (VAC8)	0.13
Beclin 1 (BECN1)	−0.59

TABLE 3: *cont.*

Enzyme encoding gene	Log2FC
Phosphatidylinositol 3-kinase (VPS34, EC: 2.7.1.137)	−1.26
Phosphoinositide-3-kinase, regulatory subunit 4, p150 (VPS15, EC: 2.7.11.1)	0.11
Autophagy-related protein 3 (ATG3)	0.11
Autophagy-related protein 4 (ATG4)	−0.16
Autophagy-related protein 5 (ATG5)	−0.27
Autophagy-related protein 7 (ATG7)	0.17
Autophagy-related protein 8 (ATG8)	−0.50
Autophagy-related protein 12 (ATG12)	−0.58

Negative log2 fold change (Log2FC) values represent up-regulation under nitrogen limitation. All presented fold changes are statistically significant, q value < 0.05.

12.2.5 NITROGEN LIMITATION AND THE REGULATION OF GENES ASSOCIATED WITH TAG BIOSYNTHESIS

TAG is the major storage lipid in oleaginous microalgae and in this study nitrogen limitations induced a five-fold increase in its intracellular content. Several genes involved in TAG biosynthesis displayed changes in their expression in response to nitrogen limitation. Biosynthesis of TAG in the chloroplast begins with two consecutive acyl transfers from acyl-CoA to positions 1 and 2 of glycerol-3-phosphate to form phosphatidic acid (PA), which is subsequently dephosphorylated to form 1,2-diacylglycerol (DAG) (Figure 5B). These reactions are catalyzed by enzymes glycerol-3-phosphate acyltransferase (GPAT), acyl-glycerol-3-phosphate acyltransferase (AGPAT), and phosphatidate phosphatase (PP), respectively. The last step in the pathway, catalyzed by diacylglycerol acyltransferase (DGAT), involves the transfer of third acyl group to the DAG 3 position. This final reaction is the only dedicated step in TAG synthesis since the preceding intermediates (i.e. PA and DAG), are also substrates for the synthesis of membrane lipids. Our results indicated that the expression of genes encoding GPAT and AGPAT was up-regulated in response to nitrogen starvation. However, the expression of gene encoding PP and DGAT remained relatively unchanged.

Though TAG biosynthesis in microalgae is believed to occur mainly through the glycerol pathway as described above, an alternative route known as the acyl CoA-independent mechanism has also been reported to take place in some plants and yeast [29]. In this mechanism, phospholipid is utilized as the acyl donor in the last step of TAG formation and the reaction is catalyzed by phospholipid:diacylglycerol acyltransferase (PDAT). We have recently found homologues of gene encoding for PDAT in the D. tertiolecta transcriptome, suggesting that the PDAT route could also play a role in microalgae TAG biosynthesis [14]. We did not however identify such homologues in the transcriptome of *N. oleoabundans*, making it unclear if PDAT contributes to TAG biosynthesis in this organism.

12.2.6 DURING NITROGEN LIMITATION GENES ASSOCIATED WITH LIPASES AND REGULATING AUTOPHAGY ARE UP-REGULATED

All three phospholipases encoding genes identified were overexpressed in –N, while only one of the two TAG lipase genes found, acylglycerol lipase, was overexpressed (Table 3). The overexpression of lipase genes during nitrogen deprivation in *C. reinhardtii* has been thought to be associated with the reconstruction of the cellular membrane for the purpose of channeling fatty acids to triacylglyceride production [10]. Triacylglyceride lipase, which is active in triacylglyceride hydrolysis was moderately repressed (log2 fold change = 0.33) under the –N scenario providing some support to the hypothesis that while membrane reconstruction was active, TAG degradation was reduced under nitrogen limitation (Table 3). Finally, genes associated with regulating autophagy and the 5' AMP-activated protein kinase gene (SnRK1 gene in plants) were overexpressed in the –N scenario (Table 3). SnRK1 is a global regulator of carbon metabolism in plants [30,31], and its up regulation—along with that of autophagy associated genes—further demonstrates the cells efforts to maintain homeostasis under –N conditions.

12.2.7 NITROGEN LIMITATION AFFECTS THE NITROGEN-ASSIMILATORY PATHWAY AT THE TRANSCRIPTOME LEVEL

We identified a number of genes that encode for components of the nitrogen assimilatory pathway (Table 4). Genes that encode for enzymes catalyzing the reduction of NO^{3-} to NH^{4+} and the biosynthesis of nitrogen-carrying amino acids were strongly expressed under nitrogen limitation [32,33]. Along with the pentose phosphate pathway, these genes were among the most up-regulated genes in –N cells of *N. oleoabundans*. The increased expression of these genes was consistent with their role in nitrogen uptake and assimilation, and the nitrogen limited growth environment from which cells were derived.

TABLE 4: *N. oleoabundans* genes involved in nitrogen assimilation

Nitrogen assimilation	Log2FC
High affinity nitrate transporters	−4.4
Ammonium transporters	−2.8
Nitrate reductase (NR, EC: 1.7.1.1)	−3.8
Ferredoxin-nitrite reductase (NiR, EC: 1.7.7.1)	−3.9
Glutamine synthetase (GS, EC: 6.3.1.2)	−2.3
Glutamate synthase (NADH) (GOGAT, EC: 1.4.1.13-14)	−1.4
Glutamate synthase (Ferredoxin) (EC: 1.4.7.1)	0.27
Glutamate dehydrogenase (GDH, EC: 1.4.1.3)	0.89
Aspartate aminotransferase (aspat, EC: 2.6.1.1)	−2.3
Asparagine synthetase (AS, EC: 6.3.5.4)	−1.5

Negative Log2FC values represent up-regulation under nitrogen limitation. All presented fold changes are statistically significant, q value < 0.05.

12.2.8 STARCH SYNTHESIS UNDER NITROGEN LIMITATIONS

While several genes associated with the preparatory steps in starch synthesis are up-regulated in the –N case, the genes encoding for key enzymes AGPase and starch synthase were repressed (Table 1). The degradative side of starch metabolism, specifically α-amylase which hydrolyzes starch to glucose, was also strongly repressed during nitrogen limitations. When coupled to the increased but still overall low starch contents in the –N case (Table 3), these findings suggest that the –N cells accumulated starch by repressing starch degradation. It is also notable that pyruvate kinase (log2FC=−0.21) and the three-enzyme pyruvate dehydrogenase complex for converting glucose to acetyl-CoA (to supply fatty acid synthesis) were up-regulated during nitrogen limitation (Figure 5D).

12.3 DISCUSSION

Oleaginous microalgae can accumulate large quantities of lipid under stress inducing growth conditions, making them a target organism for sustainable liquid biofuel production. In the present study, we induced TAG production and accumulation in *N. oleoabundans* through nitrogen deprivation, and investigated the expression of genes involved in TAG production at the transcriptome level. Mapping reads to the assembled and annotated transcriptome provided significantly more information than mapping reads to other microalgae for which the genome has been sequenced and annotated (Figure 3). While transcriptomic analysis is not substitute for detailed gene and pathway studies, it does provide a broad overview of the important metabolic processes from which to efficiently build hypotheses that can guide future detailed studies on improving lipid accumulation.

Our results suggest that under –N conditions, the altered expression of coordinated metabolic processes, many of which occur in the plastid, redirect the flow of fixed carbon toward biosynthesis and storage of lipids. These processes include up-regulation of de novo fatty acid and TAG synthesis, and concomitant repression of β-oxidation and TAG lipases. To supply precursors for lipid production, genes associated with the pyruvate dehydrogenase complex for converting pyruvate to acetyl CoA and lipases involved in the release free fatty acids from cell wall glycerophospholipids were overexpressed in the –N scenario. To power fatty acid production, strong overexpression under –N was observed in the pentose-phosphate pathway, which is primarily involved in supplying reducing equivalents for anabolic metabolism, including the production of fatty acids and assimilation of inorganic nitrogen [34].

12.3.1 TRANSCRIPTOME RESPONSE OF *N. OLEOABUNDANS* TO NITROGEN LIMITATION

A primary physiological response to nitrogen limitation is a decrease in cell growth, as observed with the three times reduction in *N. oleoabundans* growth rate. The transcript profile of nitrogen-starved *N. oleoabundans* clearly reflects the decrease in cell proliferation and stressed physiological status of the cells. Gene ontology terms related to cellular growth, photosynthesis, and protein machinery are significantly suppressed under –N conditions, and autophagy genes were up-regulated. The 5' AMP-activated protein kinase (SnRK1 gene in plants) was slightly overexpressed in the − N scenario. SnRK1 is activated under starvation conditions, including nitrogen depletion [31] and is a global regulator of starch and TAGs production in plants [30]. Overexpression of SnRK1 in the transgenic potato Solanum tubersum cv. Prairie [35] and Arabidopsis thaliana[36] has resulted in changes in starch and carbohydrate levels, thus confirming this gene's central role in carbon partitioning and suggesting that SnRK1 may be an important target for metabolic engineering efforts in oleaginous microalgae. We note also that genes encoding for the components of nitrogen assimilation are identified as the most significantly up-regulated genes in the transcriptome of nitrogen limited *N. oleoabundans*. Overexpression of nitrogen assimilation pathways under nitrogen limiting conditions has been previously reported in the transcriptome of other non-oleaginous microalgae species [10,33].

12.3.2 THE REGULATION OF FATTY ACID AND TAG BIOSYNTHESIS AND SUPPLY OF PRECURSORS

While under nitrogen deprivations, there has been considerable uncertainty expressed whether the increase in TAG content is due to a reduction in the mass of the cell, rather than increase in TAG production [2]. Both the measured increase in TAG content per cell dry weight reported here (which accounted for the loss of cell mass during nitrogen limitation), and the observed changes in the FAME profile unequivocally demonstrate the overproduction and accumulation of TAG in *N. oleoabundans* under nitrogen stress. Quantitative gene expression results also support these TAG production observations. In our study, most of the genes involved in the fatty acid biosynthetic pathway were up-regulated under –N conditions. The gene encoding for ACCase, the first enzyme in the pathway, was reported as down-regulated under –N. However, the biotin-containing subunit of ACCase, biotin carboxylase (BC), was significantly overexpressed. In photosynthetic organisms, two different forms of ACCase have been identified, one located in the plastid and the other located in the cytosol. The plastidal ACCase is a heteromeric multi-subunit enzyme that contains BC, whereas the cytosolic ACCase is a homomeric multifunctional protein that does not contain BC [27]. In our transcriptome analysis, we identified genes encoding for both forms of ACCase. In the plastid—the primary cite of lipid biosynthesis in microalgae—we have observed a significant increase in expression of the BC subunit of heteromeric isoform that catalyzes the very first step of carboxylation. On the other hand, the expression of homomeric ACCase, predominantly located in the cytosol where lipid biosynthesis does not typically occur, was repressed.

Although the overexpression of BC points to a key step in the pathway as a potential target to genetically engineer an improved oleaginous strain, mixed results for improving fatty acid synthesis in microalgae have been observed when ACCase is overexpressed [2]. Recent research has suggested that fatty acid synthesis may also be regulated by inhibition from the buildup of long chain fatty acyl ACPs [9]. Overexpressing genes that cleave ACP residues from the long chain fatty acyl ACPs is a condition observed in bacteria and recently in the microalga *P. tricornutum* to result in increased production of fatty acids [9]. In our study, genes encoding for these enzymes were highly overexpressed under the –N conditions. Therefore, a potential target for metabolic engineering in *N. oleoabundans* is the overexpression of thioesterases FatA and OAH that cleave off ACP residues.

Genes encoding enzymes involved in the steps downstream of fatty acid biosynthesis, including elongation and desaturation, have also displayed significant changes in transcription levels in response to nitrogen starvation. In particular, the genes encoding AAD and delta-15 desaturase, which catalyze the formation of double bond between the 9th, 10th, 14th, and 15th carbon, respectively, were up-regulated under –N conditions. A similar observation has been reported by Morin et al. [37], where the gene encoding delta-9 fatty acid desaturase is up-regulated in the oleaginous yeast Y. lipolytica cultured under nitrogen limitation. As observed here, and supported by gene expression levels, nitrogen limitation alter the lipid profile towards higher saturation (increase in C18:1, and decrease in C18:2 and C18:3). The increased proportion of saturated fatty acids in TAG has been demonstrated to improve cetane number and stability of resulting biodiesel [38].

Based on the lipid metabolism genes discovered from our transcriptome assembly, the acyl-CoA dependent mechanism is the major contributor to TAG biosynthesis in *N. oleoabundans*. In our study, two genes associated with biosynthesis of TAG show significant changes in their expression under –N condition: one encoding GPAT and the other one encoding AGPAT. These enzymes catalyze the acyl-CoA-dependent acylation of positions 1 and 2 of glycerol-3-phosphate, respectively. The acylation of glycerol-3-phosphate represents the first and committed step in glycerolipid biosynthesis, and likely the rate limiting step in the pathway as GPAT exhibits the lowest specific activity among all enzymes involved in the glycerol-3-phosphate pathway [39]. A recent proteomics study also reported significant up-regulation of TAG-related acyltransferases in parallel with accumulation of large quantities of lipid in C. vulgaris cultured under nitrogen limitation [13]. The overexpression of GPAT and AGPAT has been reported to increase seed oil accumulation in Arabidopsis and Brassica napus[40-42]. The up-regulation of these two genes also indicates an increase in the flow of acyl-CoA toward TAG biosynthesis. The final step of the TAG biosynthesis pathway is catalyzed by DGAT, the third acyltransferase. In our study, the gene encoding DGAT displays relatively no change in its expression under nitrogen limitation. This observation coupled with the significant increase in TAG production in the –N case, and previous proteomics studies that showed overexpression of DGAT in the C. vulgaris due to nitrogen limitation [13] provides evidence that DGAT expression in *N. oleoabundans* may be regulated post-transcriptionally. The post-transcriptional regulation of DGAT has previously been documented in the oilseed rape Brassica napus [43].

Finally, the enrichment of intracellular starch increased during the –N case. Although starch synthase and AGPase encoding genes were repressed in –N, the gene encoding for α-amylase, responsible for the hydrolysis of starch to glucose monomers, was also repressed. The concomitant accumulation of starch and lipids under nitrogen limitation has been reported in the nonoleaginous *C. reinhardtii*[44,45] and recently reported for *N. oleoabundans*[46]. This contrasts with recent reports in Micractinium pusillum where carbohydrate content was reduced and TAG production was increased under nitrogen limitation [19]. Genetic manipulations (sta6 mutant) that block starch synthesis in *C. reinhardtii* have resulted in a significant increase in TAG accumulation [47]. Under nitrogen limitation, the increased TAG content in *N. oleoabundans* and concomitant repression of starch synthase are analogous to the *C. reinhardtii* sta6 mutant. These results extend the idea of blocking starch synthesis for improvement of TAG production to the oleagenous microalga *N. oleoabundans*.

12.3.3 LIPID TURNOVER

In our study, several genes encoding enzymes involved in the intracellular breakdown of fatty acids and lipids are significantly repressed under –N (Table 3). Repressing β-oxidation is a clear strategy for maintaining a higher concentration of fatty acids within a cell. In contrast, most of the identified lipases (with the exception of triacylglycerol lipases) are overexpressed during nitrogen limitation. Upon closer examination, the up-regulated lipases are mostly phospholipases associated with hydrolyzing cell wall glycerophospholipids and phospholipids into free fatty acids, potentially for incorporation into TAGs. A known result of nitrogen limitation induced autophagy in *C. reinhardtii* is the degradation of the chloroplast phospholipid membrane [47,48]. Moreover, the overexpression of lipases during nitrogen limitation in *C. reinhardtii* has previously been hypothesized to be associated with the reconstruction of cell membranes [10]. In addition to phospholipases, we have identified an enriched number of transcripts for phospholipid metabolic processes and lipid transport in the –N case (Figure 4B). The up-regulation of genes encoding for enzymes that produce free fatty acids is also consistent with the fact that the PDAT enzyme associated with the acyl-CoA-independent mechanism of TAG synthesis (which utilizes phospholipids, rather than free fatty acids, as acyl donors) was not recovered in our assembled transcriptome.

12.4 CONCLUSIONS

Assembling the transcriptome and quantifying gene expression responses of Neochloris oleoabundans under nitrogen replete and nitrogen limited conditions enabled the exploration of a broad diversity of genes and pathways, many of which comprise the metabolic responses associated with lipid production and carbon partitioning. The high coverage of genes encoding for full central metabolic pathways demonstrates the completeness of the transcriptome assembly and the repeatability of gene expression data. Furthermore, the concordance of metabolite measurements and observed physiological responses with gene expression results lends strength to the quality of the assembly and our quantitative assessment. Our findings point to several molecular mechanisms that potentially drive the overproduction of TAG during nitrogen limitation. These include up-regulation of fatty acid and TAG biosynthesis associated genes, shuttling excess acetyl CoA to lipid production through the pyruvate dehydrogenase complex, the role of autophagy and lipases for supplying an additional pool of fatty acids for TAG synthesis, and up-regulation of the pentose phosphate pathway to produce NADPH to power lipid biosynthesis. These identified gene sequences and measured metabolic responses during excess TAG production can be leveraged in future metabolic engineering studies to improve TAG content and character in microalgae and ultimately contribute to the production of a sustainable liquid fuel.

12.5 METHODS

12.5.1 BIOREACTOR EXPERIMENTS

N. oleoabundans (UTEX # 1185) was obtained from the Culture Collection of Algae at the University of Texas (UTEX, Austin, TX, USA). Batch cultures were started by inoculation with 106 log growth phase cells into 1 liter glass flasks filled with 750 ml of Modified Bold-3 N medium [49] without soil extract. The concentration of nitrogen in the medium was adjusted to 50 mg as N l-1 (nitrogen replete; denoted as + N) and 10 mg as N l-1 (nitrogen limited; denoted as –N) using potassium nitrate (KNO3) as the sole source of nitrogen. These concentrations were chosen based on preliminary experiments that identified incubation times and nitrogen concentrations necessary to induce nitrogen depletion in the mid log-phase of the –N cultures and to ensure that the nitrogen-replete cultures never encountered nitrogen-limitation during the course of the experiment. For each nitrogen condition, cells were cultured in duplicate reactors. Reactors were operated at room temperature (25°C ± 2°C), and with a 14:10 h light:dark cycle of exposure to fluorescent light (32 Watt Ecolux, General Electric, Fairfield, CT, USA) at a photosynthetic photon flux density of 110 µmol-photon $m^{-2} s^{-1}$. Cultures were mixed by an orbital shaker at 200 rpm and continuously aerated with sterile, activated carbon filtered air at a flow rate of 200 ml min^{-1} using a mass flow controller (Cole-Parmer Instrument Company, IL, USA).

12.5.2 NITROGEN, BIOMASS AND BIOMOLECULE ANALYSIS

The nitrate concentration of culture media was determined daily by passage through a 0.2 µm pore-size filter and analysis on an ion chromatograph equipped with conductivity detection [50]. Microalgae growth was monitored daily by measuring the optical density of the cultures at 730 nm (OD730) using a spectrophotometer (HP 8453, Hewlett Packard, Palo Alto, CA, USA). Biomass samples for analysis of cellular constituents (starch, proteins, chlorophyll and lipids), and extraction of total RNA were harvested on day-11 by centrifugation at 10,000 g for 5 min at 4°C. Cell pellets were snap-frozen in liquid nitrogen and immediately transferred to −80°C until further analysis. The dry cell weight (DCW) of cultures was determined by filtering an aliquot of cultures on pre-weighed 0.45 µm pore size filters and drying the filters at 90°C until constant weight was reached. For analysis of starch content, 109 cells ml^{-1} were suspended in deionized water in 2 ml screw-cap tubes containing 0.3 g of 0.5 mm glass beads, and disrupted by two cycles of bead-beating at 4800 oscillations per minute for 2 min, followed by three freeze/thaw cycles. The suspension was then incubated in a boiling water bath for 3 min and autoclaved for 1 hour at 121°C to convert starch granules into a colloidal solution. After samples were cooled to 60°C, cell debris was removed by centrifugation at 4,000 g for 5 min. The concentration of starch in the supernatant was measured enzymatically using the Sigma Starch Assay Kit (amylase/amyloglucosidase method, Sigma-Aldrich, Saint Louis, MO, USA) according to the manufacturer's instruction. Chlorophyll a and b were measured by the N,N'-dimethylformamide method and calculated from spectrophotometric adsorption measurement at 603, 647, and 664 nm, as previously reported [51,52]. The total protein content of cells was determined with minor modifications to the original Bradford method [53] as described in [54]. Starch, chlorophyll, and protein measurements were performed in at least triplicates, and averages and standard deviations are reported as a percent of DCW.

The total lipid content of the cells was determined using a modified Bligh and Dyer method utilizing 2:1 chloroform:methanol [55]. To determine the profile of fatty acids, lipid samples were transesterified [56] and the resulting fatty acid methyl esters (FAME) were analyzed using a liquid chromatography-mass spectrometer (Varian 500-MS, 212-LC pumps, Agilent Technologies, Santa Clara, CA, USA) equipped with a Waters normal phase, Atlantis® HILIC silica column (2.1 × 150 mm, 3 μm pore size) (Waters, Milford, MA, USA), and atmospheric pressure chemical ionization [56]. Identification was based upon the retention time and the mass to charge ratio of standard FAME mixtures. The sum of FAME was used as a proxy for TAG content [22].

12.5.3 RNA EXTRACTION, CONSTRUCTION OF CDNA LIBRARIES AND DNA SEQUENCING

To control for cell synchronization, cells for the +N and –N conditions were harvested at the same time of day. Total RNA was extracted and purified separately from each of the two nitrogen replete and the two nitrogen limited cultures using the RNeasy Lipid Tissue Mini Kit (Qiagen, Valencia, CA,

USA). The quality of purified RNA was determined on an Agilent 2100 bioanalyzer (Agilent Technologies, Santa Clara, CA, USA). Isolation of mRNA from total RNA was carried out using two rounds of hybridization to Dynal oligo(dT) magnetic beads (Invitrogen, Carlsbad, CA, USA). Aliquots from mRNA samples were used for construction of the cDNA libraries using the mRNA-Seq Kit supplied by Illumina (Illumina, Inc., San Diego, CA, USA). Briefly, the mRNA was fragmented in the presence of divalent cations at 94°C, and subsequently converted into double stranded cDNA following the first- and second-strand cDNA synthesis using random hexamer primers. After polishing the ends of the cDNA using T4 DNA polymerase and Klenow DNA polymerase for 30 min at 20°C, a single adenine base was added to the 3' ends of cDNA molecules. Illumina mRNA-Seq Kit specific adaptors were then ligated to cDNA 3' ends. Next, the cDNA was PCR-amplified for 15 cycles, amplicons were purified (QIAquick PCR purification kit, Qiagen Inc., Valencia CA, USA), and the size and concentration of the cDNA libraries were determined on an Agilent 2100 bioanalyzer. Each of the four cDNA libraries (two nitrogen deplete and two nitrogen replete) was layered on a separate Illumina flow cell and sequenced at the Yale University Center for Genome Analysis using Illumina HiSeq 100 bp single-end sequencing. An additional lane was dedicated to sequencing PhiX control libraries to provide internal calibration and to optimize base calling. The sequence data produced in this study can be accessed at NCBI's Sequence Read Achieve with the accession number SRA048723.

12.5.4 RNA-SEQ DATA ANALYSES

For quality control, raw sequencing reads were analyzed by FastQC tool (v0.10.0) [57] and low quality reads with a Phred score of less than 13 were removed using the SolexaQA package (v1.1) [58]. De novo transcriptome assembly was conducted using Velvet (v1.2.03) [23] and Oases (v0.2.06) [59] assembly algorithms with a multi-k hash length (i.e. 23, 33, 63, and 83 bp) based strategy to capture the most diverse assembly with improved specificity and sensitivity [59,60]. Final clustering of transcripts were obtained using the CD-HIT-EST package (v4.0-210-04-20) [61] and a non-redundant contigs set was generated.

For transcriptome annotation, the final set of contigs was searched against the NCBI's non-redundant (nr) protein and plant refseq [24] databases using the BLASTX algorithm [62] with a cut off E-value≤10-6. Contigs with significant matches were annotated using the Blast2GO platform [63]. Additional annotations were obtained through the Kyoto Encyclopedia of Genes and Genomes (KEGG) gene and protein families database through the KEGG Automatic Annotation Server (KAAS) (v1.6a) [64]. Associated Gene Ontology (GO) terms as well as enzyme commission (EC) numbers were retrieved and KEGG metabolic pathways were assigned [65].

To determine transcript abundances and differential expression, high quality reads from each experimental condition were individually mapped to the assembled transcriptome using Bowtie software (v0.12.7) [66]. Reads mapping to each contig were counted using SAMtools (v0.1.16) [67] and transcript abundances were calculated as reads per kilobase of exon model per million mapped reads (RPKM) [68]. All differential expression analysis (fold changes) and related statistical computations were conducted by feeding non-normalized read counts into the DESeq package (v1.5.1) [25]. Separate sequence read datasets were used as inputs into the DESeq package where size factors for each dataset were calculated and overall means and variances were determined based on a negative binomial distribution model. Fold change differences were considered significant when a q-value<0.05 was achieved based on Benjamin and Hochberg's false discovery rate (FDR) procedure [69], and only statistically significant fold changes were used in the results analysis. In addition to individual enzyme encoding transcripts, contigs were pooled for each experimental condition and tested against the combined dataset to determine the enriched GO terms using the Gossip package [70] integrated in the Blast2GO platform. Significantly enriched GO terms (q-value<0.05) were determined for both +N and −N conditions.

Finally, reference guided mapping and differential expression was as also explored as a quantitation method. In this case, the Tophat package (v1.3.3) [71] was used to map high quality reads from each experimental condition against the genomes of closely related green algae species *Chlamydomonas reinhardtii* (version 169) and *Volvox carteri* (version 150) available through Phytozome (v7.0) [72]. Differential gene expression analysis was quantified using the Cufflinks package (v1.2.1) [73].

REFERENCES

1. Melis A: Solar energy conversion efficiencies in photosynthesis: Minimizing the chlorophyll antennae to maximize efficiency. Plant Sci 2009, 177:272-280.
2. National Renewable Energy Labs: A look back at the US Department of Engergy's aquatic species program: biodiesel from algae,report NREL/TP-580-24190. National Renewable Energy Labs; 1998.
3. Chisti Y: Biodiesel from microalgae. Biotechnol Adv 2007, 25:294-306.
4. Griffiths M, Harrison S: Lipid productivity as a key characteristic for choosing algal species for biodiesel production. J Appl Phycology 2009, 21:493-507.
5. Hu Q, Sommerfeld M, Jarvis E, Ghirardi M, Posewitz M, Seibert M, Darzins A: Microalgal triacylglycerols as feedstocks for biofuel production: perspectives and advances. Plant J 2008, 54:621-639.
6. Ohlrogge J, Browse J: Lipid biosynthesis. Plant Cell 1995, 7:957-970.
7. Davis MS, Solbiati J, Cronan JE: Overproduction of acetyl-CoA carboxylase activity increases the rate of fatty acid biosynthesis in Escherichia coli. J Biol Chem 2000, 275:28593-28598.
8. Lu X, Vora H, Khosla C: Overproduction of free fatty acids in *E. coli*: Implications for biodiesel production. Metab Eng 2008, 10:333-339.
9. Gong Y, Guo X, Wan X, Liang Z, Jiang M: Characterization of a novel thioesterase (PtTE) from Phaeodactylum tricornutum. J Basic Microbiol 2011, 51:666-672.
10. Miller R, Wu G, Deshpande R, Vieler A, Gartner K, Li X, Moellering E, Zauner S, Cornish A, Liu B: Changes in transcript abundance in Chlamydomonas reinhardtii following nitrogen deprivation predict diversion of metabolism. Plant Physiol 2010, 154:1737-1752.
11. Bourgis F, Kilaru A, Cao X, Ngando-Ebongue G-F, Drira N, Ohlrogge JB, Arondel V: Comparative transcriptome and metabolite analysis of oil palm and date palm mesocarp that differ dramatically in carbon partitioning. Proc Natl Acad Sci 2011, 108:12527-12532.
12. Li Y, Horsman M, Wang B, Wu N, Lan C: Effects of nitrogen sources on cell growth and lipid accumulation of green alga Neochloris oleoabundans. Appl Microbiol Biotechnol 2008, 81:629-636.
13. Guarnieri MT, Nag A, Smolinski SL, Darzins A, Seibert M, Pienkos PT: Examination of triacylglycerol biosynthetic pathways via de novo transcriptomic and proteomic analyses in an unsequenced microalga. PLoS One 2011, 6(10):e25851.
14. Rismani-Yazdi H, Haznedaroglu B, Bibby K, Peccia J: Transcriptome sequencing and annotation of the microalgae Dunaliella tertiolecta: Pathway description and gene discovery for production of next-generation biofuels. BMC Genomics 2011, 12(1):148.

15. Radakovits R, Jinkerson R, Darzins A, Posewitz M: Genetic engineering of algae for enhanced biofuel production. Eukaryot Cell 2010, 9:486-501.
16. Gonzalez-Ballester D, Casero D, Cokus S, Pellegrini M, Merchant SS, Grossman AR: RNA-Seq nalysis of sulfur-deprived Chlamydomonas cells reveals aspects of acclimation critical for cell survival. The Plant Cell Online 2010, 22:2058-2084.
17. Baba M, Ioki M, Nakajima N, Shiraiwa Y, Watanabe MM: Transcriptome analysis of an oil-rich race A strain of Botryococcus braunii (BOT-88-2) by de novo assembly of pyrosequencing cDNA reads. Bioresour Technol 2012, 109:282-286.
18. Wan L, Han J, Sang M, Li A, Wu H, Yin S, Zhang C: De novo transcriptomic analysis of an oleaginous microalga: Pathway description and gene discovery for production of next-generation biofuels. PLoS One 2012, 7(4):e35142.
19. Li Y, Fei X, Deng X: Novel molecular insights into nitrogen starvation-induced triacylglycerols accumulation revealed by differential gene expression analysis in green algae Micractinium pusillum. Biomass and Bioenergy 2012, 42:199-211.
20. Deason TR, Silva PC, Watanabe S, Floyd GL: Taxonomic status of the species of the green algal genus Neochoris. Plant Systematics and Evolution 1991, 177:213-219.
21. Pruvost J, Van Vooren G, Le Gouic B, Couzinet-Mossion A, Legrand J: Systematic investigation of biomass and lipid productivity by microalgae in photobioreactors for biodiesel application. Bioresour Technol 2011, 102:150-158.
22. Griffiths M, van Hille R, Harrison S: Selection of direct transesterification as the preferred method for assay of fatty acid content of microalgae. Lipids 2010, 45(11):1053-1060.
23. Zerbino DR, Birney E: Velvet: algorithms for de novo short read assembly using de Bruijn graphs. Genome Res 2008, 18:821-829.
24. Gianoulis TA, Raes J, Patel PV, Bjornson R, Korbel JO, Letunic I, Yamada T, Paccanaro A, Jensen LJ, Snyder M, et al.: Quantifying environmental adaptation of metabolic pathways in metagenomics. Proc Natl Acad Sci 2009, 106:1374-1379.
25. Anders S, Huber W: Differential expression analysis for sequence count data. Genome Biol 2010, 11:R106.
26. Ashburner M, Ball CA, Blake JA, Botstein D, Butler H, Cherry JM, Davis AP, Dolinski K, Dwight SS, Eppig JT, et al.: Gene Ontology: tool for the unification of biology. Nat Genet 2000, 25:25-29.
27. Sasaki Y, Nagano Y: Plant acetyl-CoA carboxylase: Structure, biosynthesis, regulation, and gene manipulation for plant breeding. Biosci Biotechnol Biochem 2004, 68:1175-1184.
28. Pollard M, Ohlrogge J: Testing models of fatty acid transfer and lipid synthesis in spinach leaf using in vivo oxygen-18 labeling. Plant Physiol 1999, 121:1217-1226.
29. Dahlqvist A, Stahl U, Lenman M, Banas A, Lee M, Sandager L, Ronne H, Stymne H: Phospholipid: diacylglycerol acyltransferase: an enzyme that catalyzes the acyl-CoA-independent formation of triacylglycerol in yeast and plants. Proc Natl Acad Sci USA 2000, 97:6487-6492.
30. Baena-Gonzalez E, Rolland F, Thevelein JM, Sheen J: A central integrator of transcription networks in plant stress and energy signalling. Nature 2007, 448:938-942.
31. Ghillebert R, Swinnen E, Wen J, Vandesteene L, Ramon M, Norga K, Rolland F, Winderickx J: The AMPK/SNF1/SnRK1 fuel gauge and energy regulator: structure, function and regulation. FEBS J 2011, 278:3978-3990.

32. Kang L-K, Hwang S-PL, Gong G-C, Lin H-J, Chen P-C, Chang J: Influences of nitrogen deficiency on the transcript levels of ammonium transporter, nitrate transporter and glutamine synthetase genes in Isochrysis galbana (Isochrysidales, Haptophyta). Phycologia 2007, 46:521-533.
33. Morey J, Monroe E, Kinney A, Beal M, Johnson J, Hitchcock G, Van Dolah F: Transcriptomic response of the red tide dinoflagellate, Karenia brevis, to nitrogen and phosphorus depletion and addition. BMC Genomics 2011, 12(1):346.
34. Neuhaus HE, Emes MJ: Nonphotosynthetic metabolism in plastids. Annu Rev Plant Physiol Plant Mol Biol 2000, 51:111-140.
35. McKibbin RS, Muttucumaru N, Paul MJ, Powers SJ, Burrell MM, Coates S, Purcell PC, Tiessen A, Geigenberger P, Halford NG: Production of high-starch, low-glucose potatoes through over-expression of the metabolic regulator SnRK1. Plant Biotechnol J 2006, 4:409-418.
36. Jossier M, Bouly J-P, Meimoun P, Arjmand A, Lessard P, Hawley S, Grahame Hardie D, Thomas M: SnRK1 (SNF1-related kinase 1) has a central role in sugar and ABA signalling in Arabidopsis thaliana. Plant J 2009, 59:316-328.
37. Morin N, Cescut J, Beopoulos A, Lelandais G, Le Berre V, Uribelarrea J-L, Molina-Jouve C, Nicaud J-M: Transcriptomic analyses during the transition from biomass production to lipid accumulation in the oleaginous yeast Yarrowia lipolytica. PLoS One 2011, 6(11):e27966.
38. Bamgboye AI, Hansen AC: Prediction of cetane number of biodiesel fuel from the fatty acid methyl ester (FAME) composition. Int Agrophysics 2008, 22:21-29.
39. Coleman RA, Lee DP: Enzymes of triacylglycerol synthesis and their regulation. Prog Lipid Res 2004, 43:134-176.
40. Zou JT, Katavic V, Giblin EM, Barton DL, MacKenzie SL, Keller WA, Hu X, Taylor DC: Modification of seed oil content and acyl composition in the brassicaceae by expression of a yeast sn-2 acyltransferase gene. Plant Cell 1997, 9:909-923.
41. Taylor DC, Katavic V, Zou JT, MacKenzie SL, Keller WA, An J, Friesen W, Barton DL, Pedersen KK, Giblin EM, et al.: Field testing of transgenic rapeseed cv. Hero transformed with a yeast sn-2 acyltransferase results in increased oil content, erucic acid content and seed yield. Molecular Breeding 2002, 8:317-322.
42. Jain RK, Coffey M, Lai K, Kumar A, MacKenzie SL: Enhancement of seed oil content by expression of glycerol-3-phosphate acyltransferase genes. Biochem Soc Trans 2000, 28:958-961.
43. Nykiforuk CL, Furukawa-Stoffer TL, Huff PW, Sarna M, Laroche A, Moloney MM, Weselake RJ: Characterization of cDNAs encoding diacylglycerol acyltransferase from cultures of Brassica napus and sucrose-mediated induction of enzyme biosynthesis. Biochimica Et Biophysica Acta-Mol Cell Biol Lipids 2002, 1580:95-109.
44. Ball SG, Dirick L, Decq A, Martiat J-C, Matagne RF: Physiology of starch storage in the monocellular alga Chlamydomonas reinhardtii. Plant Sci 1990, 66(1):1-9.
45. Wattebled F, Ral JP, Dauvillee D, Myers AM, James MG, Schlichting R, Giersch C, Ball SG, D'Hulst C: STA11, a Chlamydomonas reinhardtii locus required for normal starch granule biogenesis, encodes disproportionating enzyme. Further evidence for a function of alpha-1,4 glucanotransferases during starch granule biosynthesis in green algae. Plant Physiol 2003, 132:137-145.

46. Giovanardi M, Ferroni L, Baldisserotto C, Tedeschi P, Maietti A, Pantaleoni L, Pancaldi S: Morphophysiological analyses of Neochloris oleoabundans; (Chlorophyta) grown mixotrophically in a carbon-rich waste product. Protoplasma
47. Wang ZT, Ullrich N, Joo S, Waffenschmidt S, Goodenough U: Algal lipid bodies: Stress induction, purification, and biochemical characterization in wild-type and starch-less Chlamydomonas reinhardtii. Eukaryot Cell 2009, 8:1856-1868.
48. Martin NC, Goodenough UW: Gametic differentiation in Chlamydomonas reinhardi. I. Production of gametes and their fine structure. J Cell Biol 1975, 67:587-605.
49. Karampudi S, Chowdhury K: Effect of media on algae growth for bio-fuel production. Notulae Scientia Biologicae 2011, 3:33-41.
50. APHA, AWWA, WEF: Standard methods for the examination of water and wastewater. 18th edition. Washington, D.C: APHA, AWWA, WEF; 2005.
51. Arnon DI: Copper enxymes in isolated chloroplasts. Polyphenoloxidae in Beta vulgaris. Plant Physiol 1949, 24:1-15.
52. Mochizuki N, Brusslan JA, Larkin R, Nagatani A, Chory J: Arabidopsis genomes uncoupled 5 (GUN5) mutant reveals the involvement of Mg-chelatase H subunit in plastid-to-nucleus signal transduction. Proc Natl Acad Sci 2001, 98:2053-2058.
53. Bradford MM: A rapid and sensitive method for the quantitation of microgram quantities of protein utilizing the principle of protein-dye binding. Anal Biochem 1976, 72:248-254.
54. Kruger NJ: The Bradford method for protein quantitation. In The Protein Protocols Handbook. Edited by Walker JM. New Jersey: Humana Press; 2002::15-21.
55. Bligh EG, Dyer WJ: A rapid method of total lipid extraction and purification. Can J Biochem Physiol 1959, 37:911-917.
56. Soh L, Zimmerman J: Biodiesel production: the potential of algal lipids extracted with supercritical carbon dioxide. Green Chem 2011, 13:1422-1429.
57. Andrews S: FastQC. Babraham: Bioinformatics; 2011.
58. Cox M, Peterson D, Biggs P: SolexaQA: At-a-glance quality assessment of Illumina second-generation sequencing data. BMC Bioinformatics 2010, 11(1):485.
59. Schulz MH, Zerbino DR, Vingron M, Birney E: Oases: Robust de novo RNA-seq assembly across the dynamic range of expression levels. Bioinformatics
60. Surget-Groba Y, Montoya-Burgos JI: Optimization of de novo transcriptome assembly from next-generation sequencing data. Genome Res 2010, 20:1432-1440.
61. Li W, Godzik A: Cd-hit: a fast program for clustering and comparing large sets of protein or nucleotide sequences. Bioinformatics 2006, 22:1658-1659.
62. Altschul SF, Madden TL, Schäffer AA, Zhang J, Zhang Z, Miller W, Lipman DJ: Gapped BLAST and PSI-BLAST: a new generation of protein database search programs. Nucleic Acids Res 1997, 25:3389-3402.
63. Conesa A, Götz S, Garcia-Gomez JM, Terol J, Talon M, Robles M: Blast2GO: a universal tool for annotation, visualization and analysis in functional genomics research. Bioinformatics 2005, 21:3674-3676.
64. Moriya Y, Itoh M, Okuda S, Yoshizawa AC, Kanehisa M: KAAS: an automatic genome annotation and pathway reconstruction server. Nucleic Acids Res 2007, 35(suppl 2):W182-W185.
65. Ogata H, Goto S, Sato K, Fujibuchi W, Bono H, Kanehisa M: KEGG: Kyoto Encyclopedia of Genes and Genomes. Nucleic Acids Res 1999, 27:29-34.

66. Langmead B, Trapnell C, Pop M, Salzberg S: Ultrafast and memory-efficient alignment of short DNA sequences to the human genome. Genome Biol 2009, 10(3):R25.
67. Li H, Handsaker B, Wysoker A, Fennell T, Ruan J, Homer N, Marth G, Abecasis G, Durbin R, Subgroup GPDP: The Sequence Alignment/Map format and SAMtools. Bioinformatics 2009, 25:2078-2079.
68. Mortazavi A, Williams BA, McCue K, Schaeffer L, Wold B: Mapping and quantifying mammalian transcriptomes by RNA-Seq. Nat Meth 2008, 5:621-628.
69. Benjamini Y, Hochberg Y: Controlling the false discovery rate: a practical and powerful approach to multiple testing. J Royal Stat Soc Ser B (Methodological) 1995, 57:289-300.
70. Bluthgen N, Kielbasa SM, Herzel H: Inferring combinatorial regulation of transcription in silico. Nucleic Acids Res 2005, 33:272-279.
71. Trapnell C, Pachter L, Salzberg SL: TopHat: discovering splice junctions with RNA-Seq. Bioinformatics 2009, 25:1105-1111.
72. Goodstein DM, Shu S, Howson R, Neupane R, Hayes RD, Fazo J, Mitros T, Dirks W, Hellsten U, Putnam NÄ, et al.: Phytozome: a comparative platform for green plant genomics. Nucleic Acids Res 2012, 40:D1178-D1186.
73. Roberts A, Pimentel H, Trapnell C, Pachter L: Identification of novel transcripts in annotated genomes using RNA-Seq. Bioinformatics 2011, 17:2325-2329.
74. Yamada T, Letunic I, Okuda S, Kanehisa M, Bork P: iPath2.0: interactive pathway explorer. Nucleic Acids Res 2011, 39(suppl 2):W412-W415.

This chapter was originally published under the Creative Commons Attribution License. Rismani-Yazdi, H., Haznedaroglu, B. Z., Hsin, C., and Peccia, J. Algal Transcriptomic Analysis of the Oleaginous Microalga Neochloris Oleoabundans Reveals Metabolic Insights into Triacylglyceride Accumulation. Biotechnology for Biofuels 2012: 5(74).

AUTHOR NOTES

CHAPTER 1

Acknowledgments

The authors are grateful for financial support of the Natural Science Foundation of China (30960304), of the Key Programme for Bioenergy Industrialization of Jiangxi Provincial Department of Science and Technology (2007BN12100), of the International Science and Technology Cooperation Programme of China (2010DFB63750), of the International Science and Technology Cooperation Programme of Jiangxi Provin-cial Department of Science and Technology (2010EHB03200), of the research programe of State Key Laboratory of Food Science and Technology, Nanchang university (SKLF-TS-201111 and SKLF-TS-200814), of 948 Programe of State Forestry Bureau (2010-4-09), and National High-tech R&D Program of China (2012AA101800-03, 2012AA021704, and 2012AA021205).

CHAPTER 2

Acknowledgments

Financial support provided by the University of Edinburgh through the Innovation and Knowledge Transfer Award is greatly appreciated. The authors would like to thank the Royal Society of Edinburgh for providing a grant to enable collaborations with other universities within this field of research. Comments of two anonymous reviewers are greatly appreciated.

CHAPTER 3

Acknowledgements

The authors are indebted to the following persons for their assistance, input and advice (alphabetical order): Bart Dehue, Dr. Carlo Hamelinck, Willem Hettinga, Dr. Monique Hoogwijk, Arno van den Bos.

CHAPTER 5

Acknowledgments
We would like to thank the entire algal biofuels research team at the University of Texas at Austin for collaboration on this research and OpenAlgae LLC for financial support.

CHAPTER 6

Acknowledgements
We gratefully acknowledge the funding support provided by the University of Georgia's Biorefining and Carbon Cycling Program for this work.

CHAPTER 7

Acknowledgments
Rafael Luque gratefully acknowledges Ministerio de Ciencia e Innovacion, Gobierno de España for the concession of a Ramon y Cajal contract (ref. RYC-2009-04199) and Consejeria de Ciencia e Innovacion, Junta de Andalucia for funding under project P10-FQM-6711. Funding from projects CTQ-2010-18126 andCTQ2011-2894-C02_02 (MICINN) is also gratefully acknowledged.

CHAPTER 8

Acknowledgments
This work was supported in part by OpenAlgae.

CHAPTER 9

Acknowledgements
We wish to thank members of the Algae Biotechnology Laboratory at The University of Queensland for their valuable comments during writing of this article, and the Australian Research Council and the Australian Endeavour Award Program for financial support.

CHAPTER 10

Authors' contributions

MP conceived the analysis and main features of the tool. DL wrote and tested the code, constructed the annotation database, designed the user interface, and wrote the initial draft of the manuscript. SC provided the implementation of the hypergeometric distribution function. DC provided Pfam data and compiled the expression data. SM provided access to the expression data and tested the tool. All authors read, edited and approved the final manuscript.

Acknowledgements and Funding

We acknowledge funding of this work by the US Department of Energy under contract DE-EE0003046 awarded to the National Alliance for Advanced Biofuels and Bioproducts.

CHAPTER 11

Acknowledgments

The authors are grateful to Andrzej Walichnowski for help with paper editing, Joanne Schiavoni for formatting, and Michael Shillinglaw for figure preparation. This chapter was written within the scope of the Genome Canada TUFGEN project, and support from all funding partners is gratefully acknowledged.

CHAPTER 12

Competing interests

The authors declare that they have no competing interests.

Authors' contributions

HR-Y carried out the growth experiments, conducted the transcriptome sequencing, and participated in the study design and in the preparation of the manuscript. BH assisted with the growth experiments and biomolecule measurements, performed the bioinformatics analysis, and assisted in the preparation of the manuscript. CH participated in the algal growth and biomolecule measurement. JP conceived the study, participated in the study design, and oversaw manuscript drafting. All authors read and approved the final manuscript.

Authors' information

Hamid Rismani-Yazdi and Berat Z. Haznedaroglu denote equal authorship.

Acknowledgments

This research was supported by the Connecticut Center for Advanced Technologies under a Fuel Diversification Grant, by the National Science Foundation Grant #0854322, and by the Yale Climate and Energy Institute and Yale Institute for Biospheric Studies. We acknowledge the Yale University Biomedical High Performance Computing Center and the NIH Grant# RR19895, for providing access to computational facilities.

INDEX

D

E

F

G

H

I

K

L

M

R

S

T

U

W

Y